Personhood, Ethics, and Animal Cognition

Peter –
Thanks very much for your support of this project. I greatly appreciate it.

[illegible]
8/1/12

Personhood, Ethics, and Animal Cognition

SITUATING ANIMALS IN HARE'S TWO-LEVEL UTILITARIANISM

Gary E. Varner

OXFORD
UNIVERSITY PRESS

OXFORD
UNIVERSITY PRESS

Oxford University Press, Inc., publishes works that further
Oxford University's objective of excellence
in research, scholarship, and education.

Oxford New York
Auckland Cape Town Dar es Salaam Hong Kong Karachi
Kuala Lumpur Madrid Melbourne Mexico City Nairobi
New Delhi Shanghai Taipei Toronto

With offices in
Argentina Austria Brazil Chile Czech Republic France Greece
Guatemala Hungary Italy Japan Poland Portugal Singapore
South Korea Switzerland Thailand Turkey Ukraine Vietnam

Published by Oxford University Press, Inc.
198 Madison Avenue, New York, New York 10016

www.oup.com

Library of Congress Cataloging-in-Publication Data
Varner, Gary E. (Gary Edward), 1957-
Personhood, ethics, and animal cognition : situating animals in Hare's two level utilitarianism / Gary E. Varner.
p. cm.
Includes bibliographical references (p.).
ISBN 978-0-19-975878-4 (alk. paper)
1.Animal rights. 2.Cognition in animals. 3.Ethics. 4.Utilitarianism. 5.Hare, R. M. (Richard Mervyn)I. Title.
HV4708.V37 2012
179'.3—dc23 2011036088

1 3 5 7 9 8 6 4 2

Printed in the United States of America
on acid-free paper

On another occasion it happened that a certain heathen came before Shammai and said to him, "Make me a proselyte, on condition that you teach me the whole Torah while I stand on one foot." Thereupon he [Shammai] repulsed him with the builder's square which was in his hand. When he went before Hillel, he [Hillel] said to him, "What is hateful to you, do not do to your neighbor: that is the whole Torah, while the rest is commentary thereon; go and learn it."

—TALMUD, *SHABBAT 31A (TRANSLATION FROM WATTLES 1996, P. 48)*

{ CONTENTS }

{ LIST OF TABLES AND FIGURES }

{ PREFACE }

In the fall of 2001 I taught a graduate course on "The Work of Peter Singer." We began by reading R. M. Hare's *Moral Thinking*, and I made the overarching theme of the seminar to ask how far Hare's theory supported Singer's conclusions on a range of issues in animal ethics. The rationale was that Singer was Hare's graduate student at Oxford and Singer now claims to accept a version of Hare's two-level utilitarianism. Ever since, I have been working on how Hare's theory applies to the range of ways in which humans and animals interact. I originally presented the result to Oxford University Press as a single draft manuscript under the working title *Harey Animals: Situating Animals in the Two-Level Utilitarianism of R. M. Hare*. Although the title was rejected, the manuscript was accepted, but it seemed there were two projects, worth separating out into two volumes.

Hare's theory is complicated in a number of ways, but the complications yield a more nuanced treatment of issues in animal ethics than competing theories. This first volume is a detailed treatment of the theory and those complications. Although I have tried to make it accessible to non-philosophers, this volume is intended primarily for philosophers, and my hope is that both ethical theorists and those focusing on animal ethics will read it. The ethical theorists will find a fresh defense of Hare's theory and extensions of it that Hare himself did not foresee. The applied ethicists will find the kind of systematic application of his theory to questions about personhood and the moral status of animals that Hare himself never undertook.

The second volume, titled *Sustaining Animals: Envisioning Humane, Sustainable Communities* (Varner forthcoming), will focus on applying the theory to a broad range of issues in animal ethics: animal agriculture, which is treated more briefly in part III of the present volume, but also pet ownership and working animals, scientific research on animals, and wildlife/ecosystem management. My hope is that the sequel will be read by a range of people interested in animal ethics, including applied ethicists, but also animal activists, veterinarians, animal scientists, psychologists, and others interested in animal ethics. The emphasis in that volume will be less on defending Hare's theory than on emphasizing the ways in which it sheds light on issues in animal ethics, and readers wishing a more detailed philosophical defense of the theory will be directed back to this volume.

{ ACKNOWLEDGMENTS }

A first draft of this book was completed during a sabbatical in academic 2006–2007. While half of the funding for that teaching release came from my home institution (Texas A&M), the other half was funded by the National Science Foundation (NSF grant #0620808). Of course, the opinions expressed herein are mine and do not necessarily reflect the views of Texas A&M or the NSF. A travel grant from A&M's Glasscock Center for Humanities Research allowed me to consult with poultry welfare experts at Michigan State University in October 2008.

Having worked on this project for a decade, it is impossible for me to remember the names of everyone who has helped out with it in some way, so I will not attempt to list everyone. That said, I must mention: the BLAB (Brains, Learning, and Animal Behavior) interdisciplinary reading group at Texas A&M, which slogged through my first draft of the manuscript during 2006–2007; philosophy graduate students Adam Shriver and Steve Campbell; animal scientist Ted Friend and psychologist Jim Grau; and my erstwhile philosophy colleague Colin Allen.

Text for portions of various chapters has been taken from the following previous publications.

"A Harean Perspective on Humane Sustainability" appeared in *Ethics and Environment* 15 (2010), pp. 31–49. Some text from this essay is included in chapter 10 of this book with permission of Indiana University Press.

"Personhood, Memory, and Elephant Management" appeared in Christen Wemmer and Catherine Christen, eds., *Elephants and Ethics: The Morality of Coexistence* (Johns Hopkins University Press, 2008), pp. 41–68. Some text from this essay is included in chapter 8 of this book with permission of The Johns Hopkins University Press.

"Utilitarianism and the Evolution of Professional Ethics" appeared in *Science and Engineering Ethics*, 14 (2008), pp. 551–573. Some text from this essay is included in chapters 1 and 4 of this book with permission of Springer Netherlands.

My review of Jean Kazez, *Animalkind: What We Owe to Animals* (Wiley-Blackwell, 2010), originally appeared in *Notre Dame Philosophical Reviews*: http://ndpr.nd.edu/review.cfm?id=21569. Some text from that review is included in chapter 9 of this book with permission of *Notre Dame Philosophical Reviews*.

My review of Victoria Braithwaite, *Do Fish Feel Pain?* (Oxford University Press, 2010) originally appeared in *Environmental Ethics*, 33 (2011), pp. 219–222.

Some text from this review is included in chapters 5 and 10 of this book with permission of Environmental Philosophy, Inc.

Finally, while I never met R. M. Hare, working on his theory and its application to animal issues for a decade has made me feel like I know one facet of him very well. I hope that my enthusiasm for that facet shines through in my writing.

{ ABBREVIATIONS }

I have adopted two of Hare's conventions. One is referring to locations in this book using the section/subsection numbers. The other is that of referring to Hare's major works with the following acronyms:

LM = *The Language of Morals* (1952)
FR = *Freedom and Reason* (1963)
MT = *Moral Thinking* (1981)

I will also use the following acronyms to refer to a couple of collections of his essays, which include almost everything else by Hare to which I will refer:

EB = *Essays on Bioethics* (1993)
OP = *Objective Prescriptions* (1999)

For works by Hare not included in any of the above books, I will use the standard (author year, page number) citation method.

{ PART I }

Hare's Two-Level Utilitarianism

{1}

Introduction

§1.1. Shammai and Hillel

In the famous story told in the epigraph for this book, Rabbi Hillel tells an impatient potential convert that a version of what is now popularly known as the golden rule is "the whole Torah" while "the rest is commentary thereon," but, Hillel immediately adds, referring to the commentary: "go and learn it." The story is one in a series illustrating Hillel's renowned patience in contrast to his contemporary Rabbi Shammai. In that context, Hillel seems to be both having fun with the impatient convert and displaying his own renowned patience. You cannot convert to Judaism by accepting a single command, so Hillel is playing a trick on the impatient convert when, rather than running him off with a stick as Shammai did, he states "the whole Torah" in a single sentence. At the same time, however, in telling him that he must also "go and learn it"—where "it" refers to a lengthy "commentary" on that single command—Hillel's patience is manifest. For if the convert is intrigued enough by the golden rule, then he will study the entire Torah, and in due time his conversion may become real.

I am a convert to utilitarianism, and when someone asks me about my views in ethics, I sometimes feel like Shammai, other times like Hillel. If they insist that I put it in a single sentence, and they are unwilling to listen to a substantial commentary on that sentence, then I might as well run them off with a stick. For "the right thing to do is whatever will maximize aggregate happiness" is no more the whole of utilitarianism than the golden rule is the whole of the Torah. On the other hand, if they are sincerely interested in understanding, I begin by saying that "Utilitarianism is a family of theories with one thing in common: they all hold that, *at least ultimately*, the right thing to do is whatever will maximize aggregate happiness."[1] For while "do what will

[1] An exception would be G. E. Moore (1903), who believed that the existence of beauty—even unperceived beauty—adds intrinsic value to the world, and still called himself a utilitarian. That Moore's utilitarianism was non-standard is underscored by the fact that he felt compelled to give it a special name: he called his view "*ideal* utilitarianism." In my first book, I stated that I was "a utilitarian

maximize aggregate happiness" expresses the principle behind utilitarian thinking in ethics, the qualification—"at least ultimately"—points to the complexities involved in specifying what that principle implies in practice, what it would mean to live a life inspired by the principle.

As Jeffrey Wattles points out, in the story of Hillel and the impatient convert, the word translated as "commentary," *perusha*, is also used to mean "specification" (1996, p. 49). So while Hillel believed that the golden rule somehow expresses the whole of Judaism, he was telling the impatient convert that to understand what the golden rule implies in practice is a complicated business, requiring a treatment as long as the Torah. The remaining chapters of this book run longer than the Torah, and in those chapters I provide a lengthy, new commentary on or specification of the principle of utility, one inspired by the work of R. M. Hare (1919–2002). What converted me to utilitarianism was reading and re-reading Hare's main work on utilitarianism, *Moral Thinking: Its Levels, Method and Point* (1981), over half a dozen times over two decades. So why am I offering a new commentary on or specification of the principle of utility?

First, as the title conveys, this book focuses on two things that Hare had little to say about: personhood and animals. Hare avoided talking in terms of "personhood," and he never published a systematic treatment of how his theory applies to issues in animal ethics. In contrast, in chapters 6 and 7 I will defend the moral legitimacy of distinguishing among "persons," "near-persons," and "the merely sentient" within Harean two-level utilitarianism, and in chapters 9 and 10 I will provide a detailed illustration of the theory's application to some questions in animal ethics. This new perspective on Harean utilitarianism is timely, since Hare's most famous graduate student, Peter Singer, is known as the philosophical father of the modern animal rights movement. Singer now says that he endorses Hare's version of utilitarianism, he defends utilitarianism in ways that parallel Hare's defense, and he has invoked Hare's distinction between "critical-level" thinking and thinking in terms of "intuitive-level rules" in response to certain objections to his conclusions on several issues. Throughout my discussion of personhood and animals, I will ask to what extent Harean utilitarianism supports the more abolitionist conclusions associated with Singer's writings.

This book also draws on empirical research in ways that Hare did not. Obviously, contemporary research on consciousness and cognition in non-human animals is relevant to the questions of which animals are sentient (chapter 5) and which, if any, might be "persons" or at least "near-persons" (chapter 8). As

at heart" (Varner 1998, p. 88), but there I defended a version of biocentric individualism, arguing, in chapter 3, that non-conscious organisms have interests, the satisfaction of which creates intrinsic value. I have subsequently repudiated the core argument of that chapter, however (see Varner 2003, pp. 415–416), and so I am now a standard, sentientist utilitarian.

detailed in chapter 4, however, contemporary research on human moral reasoning also provides new support for Hare's two-level utilitarianism.

Apart from bringing empirical research to bear in ways that Hare did not, I intend to improve on Hare's theory in three ways. Chapter 2 provides a fresh defense of two-level utilitarianism, including a formal presentation of how I understand Hare's core argument in *Moral Thinking*. Chapter 3 significantly expands Hare's account of how intuitive-level rules function in moral thinking, and chapter 4 provides a systematic overview of objections to the theory, building on that expanded account.

The remainder of this introductory chapter discusses the concept of personhood, and provides a general overview of Hare's theory. In that overview, I emphasize the place of animals and how, in particular, Harean utilitarianism can incorporate aspects of both animal welfare and animal rights views. I hope you will find this introductory chapter sufficiently intriguing to justify reading the rest of this book. I'm not predicting that you'll be a convert by the time you reach the final section, but I believe that this book sheds new light on the complexities inherent in the principle of utility, and on knotted questions about personhood and the moral status of animals.

§1.2. A Properly Philosophical Conception of Personhood

In day-to-day speech, "person" is often used synonymously with "human being." The law, political philosophy, ethical theory, and common sense, however, all recognize other senses of the term, in which being biologically human is neither necessary nor sufficient for personhood. In this section, I nail down what I take to be the proper form of a philosophical conception of personhood.

In table 1.1, I list some of the ways that I have seen the term "person" defined, grouping them into four families. Dictionaries always give, as one definition of "person," something like "human being," but they also always give variations on what I characterize as family two definitions. These define personhood descriptively, but in terms of various cognitive capacities rather than species membership. Common sense employs a family two definition when it counts both gods and intelligent space aliens as persons, despite the fact that neither are human; and when we "personify" a puppet, cartoon character, or animal, we represent it as having the suite of characteristics typical of normal, adult humans, without thinking of it as a member of our species.

Dictionaries also usually include at least one version of a family three definition. These define "person" normatively rather than descriptively; they use the term to denote any entity that deserves a certain kind of treatment or respect. Thus, in addition to natural persons (individual human beings), governments and corporations are persons in a legal sense and have certain legal rights and powers. Although dictionaries seldom include other examples of family three definitions, readers familiar with the philosophical literatures on animal ethics and abortion will be familiar with the other examples given in table 1.1. These philosophical definitions of "person" are, like the legal definition, normative rather than descriptive.

Obviously, my list is not exhaustive. Some theists would want to add "has a soul" to the family two list, Kantians would want to add "should be treated as 'an end-in-itself' rather than merely as a means to others' ends" to the family three list, and so on. I have limited my family three list to three items that I find useful in formulating what Hare calls "intuitive-level rules" for humans and animals, and I have limited my family two list to "a biographical sense of self" and three other important cognitive capacities that, I believe, come with it (see §7.7).

What are we to make of this welter of definitions? One reaction would be to consign "person" to the scrapheap of overly ambiguous expressions that impede rather than further rational discussion of an issue. Compare "natural," which has many inconsistent meanings. As Holmes Rolston (1979) notes, these range from a sense in which humans could not possibly act unnaturally (where "natural" means "in accordance with the laws of physical nature") to one in which they could not possibly intend to do what is natural (where "natural" means "not the result of human intention"). In still another, evolutionary sense, a behavior is "natural" if it expresses "any adaptive trait that human beings displayed during their evolutionary emergence as a species" (Varner 1998, p. 119). On this definition, undesirable or immoral things can turn out to be "natural." Thus it seems that appealing to "what is natural" can only muddle the issue unless we can agree on how to fill in the ellipsis in "natural in the sense of . . ."

The abortion debate might be taken to show that "person" is a similarly unhelpful term. It seems that appealing to the definition of "person" won't help at all if under one definition a fetus obviously *is* a person (because it is a member

TABLE 1.1. Four Families of Definitions of the Term "Person"

1. "Person" =df a member of the species *Homo sapiens*.

2. "Person" =df an entity with certain cognitive characteristics, e.g., it is

(A) "rational and self-conscious,"

(B) a moral agent (capable of thinking about right and wrong and adjusting its behavior accordingly),

(C) autonomous (in the sense of having second-order desires), or

(D) an individual with "a biographical sense of self."

3. "Person" =df an entity that deserves a special kind of treatment or respect, e.g., it

(A) "has legal rights and/or duties,"

(B) "has moral rights," or

(C) is "not replaceable."

4. [*Schema*] "Person" =df an individual who deserves special treatment or respect of a type or types listed in 3(A–C) *because* he, she, or it has some capacity or capacities listed in 2(A–D).

of our species), while under another definition it obviously is *not* (yet) a person (because it lacks the sophisticated cognitive abilities suggested by the characteristics in the family two definitions). For such reasons, David DeGrazia suggests that, "at least in any context where someone's personhood is reasonably debatable . . . there is nothing you can do with the concept of personhood that you can't do as well or better with whatever more specific concepts are immediately relevant" (1997, p. 311—italics removed). I agree with DeGrazia that, in commonsense usage, "personhood is a vague concept that is not analyzable into necessary and sufficient conditions" (p. 301). And for that reason, I agree with him that "where someone's personhood is reasonably debatable" we can only make headway by deciding what specific capacities or properties are relevant to the question of how we should treat the individual, and then asking whether or not the individual has those capacities or properties.

As I mentioned in §1.1, Hare avoided talking in terms of "personhood." He warned that "most of the disputes about this allegedly crucial question of whether the foetus is a person are going to be a waste of time and can never get anywhere" (*EB*, p. 169). As Hare himself also noted, however, rather than serving as a premise in an argument about a moral issue like abortion, a philosopher's stipulative definition of "person" can serve to *summarize an argument* on the issue (*EB*, p. 171). In this way, a philosophically useful definition of "person" fits the schema that I provide in table 1.1 for a fourth family of definitions. Such definitions do not substitute for arguments, but they focus our attention on the inference in need of justification. They provide a thumbnail sketch of a philosopher's answer to the question of why some special kind of moral significance described in a family three definition would be justified by possession of some kind of cognitive capacity described in a family two definition.

This is what I mean by a "properly philosophical" conception of personhood.[2] In chapters 6 and 7 I will defend the following family four definition:

> *"Persons" =df individuals who deserve special respect of the types listed in 3(A–C) because they have capacity 2(D)—that is, persons are individuals who deserve legal and moral rights, and should be treated as "not replaceable," because they have a biographical sense of self.*

[2] Locke is often credited with first articulating the modern philosophical conception of personhood when he defined a person as "a thinking intelligent being, that has reason and reflection, and can consider itself as itself, the same thinking thing, in different times and places" ([1690a] no date, book II, chapter 27, paragraph 9). Note, however, that this is a family two (descriptive) definition of "person," and Locke nowhere gives a family three (normative) definition. This famous definition was given in his *An Essay Concerning the Human Understanding*, a book that has nothing to do with ethics and political philosophy. In his *Second Treatise of Government* ([1690b] 1952), which was published the same year but gives Locke's central account of "the social contract," he never gives what I call a family three definition of "person." There he only uses the term "person" to refer to human beings, to their bodily "persons," and to authorities created by the state.

Such a definition is clearly stipulative; in giving it, I am not attempting to capture the commonsense meaning of "person." Instead, such a definition explicitly relates the normative and descriptive dimensions of personhood, calling attention to the inference in need of defense. It is the fully fleshed out argument that, if sound, would justify the special kind of treatment or respect for individuals with the specified cognitive capacities.

In the context of Hare's theory, specifically, there are other reasons for retaining the term "person" and putting it to such a specialized use (see §1.9). These can only be explained after giving an overview of Hare's theory, however, which in turn requires, first, some words about the nature of academic philosophy and how philosophers go about defending an ethical theory.

§1.3. The Nature of Academic Philosophy

A fairly standard taxonomy of philosophy divides the field into three areas: (1) metaphysics, which studies the essential or ultimate nature of things; (2) epistemology, or the theory of how we know things (which includes logic, the study of inferences); and (3) value theory, which includes aesthetics, political philosophy, and ethics. In each of these areas, there is some apparent overlap between philosophy and science. After all, physics—the "hardest" of the sciences—studies "the essential or ultimate nature" of matter, and psychologists, anthropologists, and political scientists study both "how we know things" and how we make aesthetic, political, and ethical judgments.

The overlap is mirrored in the etymology of the word "philosophy." In all major European languages, it is derived from the combination of two Greek words: *philo*, meaning "love of," and *sophia*. The latter is usually translated as "wisdom," but in ancient Greek it referred to all knowledge that was systematic and acquired through the exercise of human reason. This ruled out knowledge derived from divine revelation, guesswork or intuition, and also practical "know-how" (*techné*), but it included everything that we now think of as science. Thus prominent scientists like Galileo and Newton continued to refer to themselves as "natural philosophers" through the eighteenth century, and today the "hardest" of scientists still receive a Ph.D. (Doctor of Philosophy) degree.

Following the codification of the experimental method during the scientific revolution (which could arbitrarily be dated to 1642, the year that Galileo died and Newton was born), one scientific discipline after another separated from philosophy by defining its subject matter in a way that made it subject to observation and, therefore, experimentation. For example, by making atomic and sub-atomic particles visible, physics wrested some metaphysical turf away from philosophy (although philosophers continue to dominate discussions of the "essential nature" of other things, such as agents and works of art), and the ascendence of behaviorism in mid-twentieth-century psychology can be interpreted as an attempt to make the soul (*psyche*) subject to observation and experimentation. Psychologists have also seized some of philosophy's traditional turf insofar as they study how people make logical inferences, and for their part anthropology and sociology have seized more turf by applying the scientific method to the study of how people make moral and aesthetic judgments.

From this historical perspective, philosophy looks like the land of the leftovers. As explained in the etymology above, it is committed to systematic, rational inquiry. But it has been robbed of everything that the sciences have appropriated, and philosophy is left with whatever aspects of metaphysics, epistemology, and value theory can be studied rationally, but without relying on observation and experimentation. This has led to the following standard conception of how the philosophical study of ethics differs from the ways in which anthropologists and sociologists study ethics: the scientists are said to be doing *descriptive* ethics, whereas the philosophers are doing *normative* ethics; the scientists are said to be studying how people *in fact* believe and act, what principles they *in fact* endorse, whereas the philosophers are studying what people *ought* to believe, what principles are *best justified*, and whether or not people in fact endorse them or act accordingly.

The idea that it is a fallacy to infer what ought to be from what is—which has been variously described as committing "the naturalistic fallacy" (Moore 1903) or ignoring the "is/ought gap" (Hume [1739–1740] 1978)—makes this seem like a hospitable division of labor. From the philosophers' perspective, scientific studies of what principles people in fact believe and how they in fact act do not intrude on the philosophers' turf, and most scientists seem happy to leave normative ethics to the philosophers. So as long as it is possible to rationally inquire about *what ought to be* in a way that transcends the empirical study of what *is*, philosophy seems sure to maintain a monopoly on normative ethics.[3] Unfortunately, the picture gets muddier when we look at how philosophers go about justifying ethical theories and principles.

[3] I should note that within academic philosophy, the logical positivists of the early twentieth century are usually seen as capitulating in the turf war. They are said to have followed Hume ([1739–1740] 1978) in dividing all meaningful statements into two kinds: (1) "relations of ideas," which can be shown to be true or false in virtue of the meanings of the terms used and the logical relations established among them in the statement, and (2) "matters of fact," which are the province of science. Then, because all of the non-trivial questions in traditional philosophy (about ethics, the nature and existence of God, aesthetics, etc.) fit into neither category, they are said to have limited philosophy to the analysis and clarification of language. On this view, as Wittgenstein ([1921] 1961) famously put it, academic philosophers' job is to cure people of an illness by getting them to stop asking questions that cannot be given meaningful answers.

For a more nuanced perspective on the logical positivists, one that has them adopting a pragmatist approach to questions of value, see the appendix to Bryan Norton's *Sustainability: A Philosophy of Adaptive Ecosystem Management* (2005).

§1.4. Methods of Justification in Philosophical Ethics

Generally, philosophers have adopted one of four approaches. They either:

1. Appeal to God's will,
2. Claim that certain moral principles are self-evident,
3. Use what is called "the method of reflective equilibrium," or
4. Appeal to the logic of moral discourse.[4]

The first two approaches are generally out of favor today, but it's worth saying something about why.

When used to justify ethical principles, appeals to God's will can take two importantly different forms. One involves endorsing the Divine Command Theory. This is the view that what *makes* something right is just God's willing that it be done. This approach makes sense in a religious tradition that conceives of God as inherently good, but it has the uncomfortable implication that *anything* would be right if God happened to will it. It also seems to trivialize God's goodness in the following way. On this view, saying that someone is good just means that they do what God wills; but then saying that *God* is good just means that "God does what God wills," which sounds like a backhanded compliment. The other way that appeals to God's will can be used to justify an ethical theory or principle does not involve holding that God's willing something is what *makes* it right, but only that God's will is an infallible guide to what is right. This works if it is true both that God is all-knowing and that He would not mislead us about what is right, but this method effectively punts on the question of what *makes* something right, on what ultimately justifies an ethical principle or theory.

The second method, claiming that certain principles are self-evident, does answer that question, but in a way that is ultimately unsatisfying. For one thing, precious few substantive moral principles are candidates for being self-evident. A good candidate would be the principle of universalizability, or the golden rule, both of which are discussed below. But these principles are formal rather than substantive. That is, they express logical requirements on all moral judgments, rather than declaring any particular class of things to be right or wrong, and so these formal principles instead get invoked by philosophers seeking to justify their theories using the fourth method. Intuitionist philosophers like W. D. Ross (1930) have proposed short lists of substantive moral principles, at least some of which are plausible candidates for what are called "hypernorms" in the professional ethics literature. Hypernorms express

[4] Some would list contract theory and discourse ethics as distinct methods of justifying ethical principles, but I think that these are best categorized either as a variation on the method of reflective equilibrium (in the case of John Rawls, for instance) or as a pragmatist-inspired appeal to the logic of moral discourse (in the case of Jürgen Habermas, for instance).

values that seem to be endorsed by all known human cultures, such as non-maleficence, truthfulness, and promise keeping. Intuitionism is ultimately unsatisfying, however, because it is inevitable that such general principles will conflict, for instance when lying or breaking a promise is necessary to prevent harm to others, and intuitionism offers no principled way of deciding such hard cases.

For the foregoing reasons, few contemporary ethical theorists are intuitionists, and few use appeals to God's will to justify moral principles. By far the most widely used approach has come to be known as "the method of reflective equilibrium." John Rawls gave it this label in his widely influential book, *A Theory of Justice* (1971, pp. 20–21). Philosophers use this method when they argue that, for some range of cases, their theory or principle conforms to our intuitive (or "pre-theoretical") moral judgments better than competing theories. In this way, the method of reflective equilibrium is analogous to theory justification in science. In science, a theory is defended by showing that it fits various publicly observable data, on which all participants to the debate agree, better than competing theories. In the method of reflective equilibrium, shared moral intuitions are, in effect, the data. (Norman Daniels later [1979] described "*wide* reflective equilibrium" as being achieved when a theory not only "fits" the pre-theoretical moral intuitions in question, but is also consistent with the best available science.)

An illustration of the method in practice comes from a standard tactic employed by the critics of utilitarianism. A utilitarian, it is claimed, cannot hold that such things as slavery and knowingly punishing the innocent are inherently wrong, because under some circumstances, at least, they would maximize aggregate happiness. Philosophers critical of utilitarianism have dreamed up all kinds of "test cases" in which it is very clear that breaking the promise you made to a dying friend or hanging an innocent man will maximize aggregate utility. These philosophers appeal to their readers' strongly contrary intuitions to turn them against the theory.

In a scathing (1973) review of Rawls's book, and throughout his subsequent published works, Hare argued that the method of reflective equilibrium is deeply flawed, because our intuitive judgments are, at least in the beginning, a product of the moral education we receive as children. Unless we have independent reasons for believing that the intuitions inculcated in us during that education were sound, or that some of the intuitions that we have subsequently acquired are themselves sound, the method of reflective equilibrium provides no rational support for a principle or theory. Rawls notes that the attempt to systematize our intuitions using the method of reflective equilibrium itself alters our intuitions: over time, what seemed intuitive comes to seem counter-intuitive, and vice versa. In this way, applying the method of reflective equilibrium can lead us away from the set of intuitions that were inculcated in us as children. That such shifts sometimes result in a set of intuitions that are more internally consistent does not show that the resulting set is more rationally justified, however, unless we can give some independent reason for thinking that at least some of the intuitions involved were sound to begin with.

Hare sometimes called his theory "Kantian" utilitarianism (*EB*, pp. 209, 216), which sounds like an oxymoron. Kant was adamantly opposed to utilitarianism in ethics, but not because it runs counter to strongly held moral intuitions, for Kant's own views were strongly counterintuitive. Of lying, for instance, he held that it was wrong, no matter what the consequences. Even if a would-be murderer asked you where his intended victim was hiding, Kant held that it would be wrong to lie to him ([1797] 1909). (In fairness to Kant, he did not say that you should tell the would-be murderer where his intended victim was, he just said that you shouldn't lie. So it would be OK, on Kant's view, to keep quiet.) Of punishment, he held that releasing a criminal from punishment or even reducing a sentence was always wrong, no matter what "advantage it promises"; and even if society were disbanding and no good would come of it, Kant held that "the last murderer remaining in prison" should be executed ([1797] 1996, pp. 105, 106). These views on lying and punishment are both highly counterintuitive. Almost everyone thinks that lying would be justified if the consequences of not lying would be horrendous, and almost everyone thinks that prosecutors are sometimes justified in granting immunity for a crime in order to prevent a greater crime going unpunished. What makes Hare's theory Kantian is that he, like Kant, abandons the method of reflective equilibrium in favor of appealing to the logic of moral discourse. Both philosophers defended their theories without relying on the moral intuitions of their readers, and both believed that where their theories diverged from our moral intuitions, it was the intuitions and not the theory that needed work.

On one (admittedly oversimplified) reading, Kant claimed that all substantive moral truths can be derived from the logical requirements of *universalizability*, the requirement to judge similar cases similarly, and *overridingness*, the notion that in making a moral judgment one is claiming that this judgment should override other types of norms (of prudence, law, and etiquette).[5] Hare claimed that moral judgments actually have an additional logical property: they are, in a special sense, *prescriptive*. Moral judgments are commonly called

[5] For more on these logical requirements, see §2.4. Kant's emphasis on "pure" moral philosophy in his most widely read book on moral philosophy, the *Groundwork of the Metaphysics of Morals* ([1785] 1948), suggests that he thought that all substantive moral truths could be derived from these two logical requirements of moral discourse alone. Even there, however, Kant says that moral philosophy "requires anthropology for its *application* to man" (p. 79, italics in original). That is, while the categorical imperative can be derived from the logic of moral discourse alone, its application to human beings requires empirical knowledge "of human nature (and indeed of many other things)" (Paton 1948, p. 14). Hare is much more explicit about this than Kant is in the *Groundwork*. Hare held that the logic of moral discourse forces us to think like utilitarians, but that what utilitarianism requires depends on facts about human nature and the human condition.

prescriptive to contrast them with descriptive statements; moral judgments describe what ought to happen, rather than what *in fact* happens. But when Hare claimed that moral judgments are prescriptive, he meant something much stronger and more controversial: he meant that to sincerely assent to a moral judgment, one must act accordingly and welcome others acting accordingly. Hare claimed that Kant's famous categorical imperative, "Act only on that maxim through which you can at the same time will that it should become a universal law" ([1785] 1948, p. 88), actually expressed the combination of these three logical properties of moral judgments, and he claimed that, *pace* Kant, these logical properties of moral judgments force us to think—at least "ultimately"—like utilitarians.

§1.5. Hare's Two-Level Utilitarianism

A thorough exposition of Hare's arguments in support of his claims that moral judgments are prescriptive in his special sense and that the three logical properties of moral judgments together force us to think like utilitarians will involve complications that are best reserved for the next chapter. For present purposes, however, a brief and very rough sketch of his argument can be given in terms of the golden rule. Hare states that the theory he articulates in *Moral Thinking* (1981) "provides the logical basis of the Golden Rule" (*EB*, p. 166) and that, properly understood, the golden rule expresses the three logical properties of moral judgments (*EB*, chapter 10, passim). This makes the story of Hillel in the epigraph all the more appropriate, for Hare would say that while the golden rule somehow expresses the whole of ethical thinking, specifying what the golden rule implies in practice is a complicated business.

"Do unto others as you would have them do unto you" is an easy way for parents to teach children that if they believe their way of treating others is morally OK, then they are logically committed to believing that it would be OK if roles were reversed. This is the requirement of universalizablity, but according to Hare, the golden rule, properly understood, implies the following:

> *To judge that what you are doing in a given situation is morally right requires that you sincerely prefer to live through the experiences of everyone involved in that scenario rather than living through all of their experiences under the alternative of not doing what you judge is right.*

Hare argues that this forces you to think like a utilitarian, for if you had to live through everyone's experiences in turn, rather than just your own, you would choose to maximize aggregate happiness, since you would have to enjoy or suffer all of the benefits and costs in some order or other. One of the main objections to Hare concerns this claim. People often say that in this choice situation, they would not choose to maximize the ratio of benefits to costs and would instead operate according to some other strategy, such as Rawls's "maximin" rule (which says, in effect, "maximize the minimum possible payoff").

This objection is important, and will be discussed in detail in §2.3. For the purpose of getting a general overview of Hare's theory in this introductory chapter, however, all we need to understand is that Hare's argument is Kantian, because it relies on an analysis of the logic of moral discourse rather than the method of reflective equilibrium. In particular, he argues that moral judgments have three logical features: they are universalizable, overriding, and (in his special sense) prescriptive. According to Hare, these three features force us, when taking the moral point of view, to adopt an attitude of generalized or disinterested prudence. So on Hare's theory, just as accounts of prudence stress not discounting the future, morality involves not discounting others' experiences—morality, on Hare's view, is "universalized prudence" (*MT*, p. 100; q.v. pp. 222–223). And when we take this disinterested perspective on interest satisfaction, according to Hare, we would make the same choices that a utilitarian would make.

That is how, according to Hare, the logic of moral discourse forces us to think like utilitarians. For several reasons, however, Hare argues that real-world human beings need non-utilitarian principles for day-to-day decision making, and this is what puts the "at least ultimately" into Hare's version of utilitarianism. The first reason is that extremely detailed information would be needed to apply the theory correctly, but human beings are usually lacking some or much of the relevant information. An extreme example would be that if the child drowning in the pond was the young Adolf Hitler, then apparently you did the wrong thing by saving him, even though you could not have known the consequences of what you did. In much less dramatic ways, real world humans are constantly forced to act in ignorance of many relevant details that only an "archangel" (Hare's term for the perfect utilitarian thinker) could know. Humans also have limited data-processing abilities and we make mistakes, so even if all of the relevant information could somehow be made available to us, we would still come to the wrong conclusions in a range of cases. Finally, Hare stresses the extent to which we are inclined to "cook the data" in favor of self-interest. So even when the relevant information is both available and clearly indicates a course of action, we are prone to mistakenly believe that doing what is in our own best interest is also doing what is best on the whole, what maximizes aggregate happiness; and we are prone to do this in mundane, commonly encountered situations.

This is why Hare's theory has two levels. The logic of moral discourse forces us to think like utilitarians at what he calls the "critical level" of moral thinking, but good utilitarian reasons can be given for training ourselves to think most of the time in terms of what he calls "intuitive-level" rules. For purposes of moral education, these rules need to be fairly simple, like "Don't lie." At the same time, it is often difficult (if not impossible) to express the rule that one is trying to convey in a specific formulation in words. And given humans' tendency to self-deceit, the rules also need to be, so to speak, deontological in "flavor" (my expression); that is, they need to be internalized in a way that results in us being diffident about violating them, even when, after careful critical thinking (which is explicitly utilitarian), we believe that we are justified in doing so. For this reason, Hare even finds rights claims—which have been characterized as "trump cards against utilitarian arguments"—useful in formulating some intuitive-level principles (*MT*, pp. 147–156; *EB*, pp. 24–30).

We nevertheless *must* use explicitly utilitarian thinking in several kinds of circumstances, including:

1. In novel cases (which intuitive-level rules are not designed to handle, and on which they give little or no guidance), and
2. When intuitive-level rules conflict.

In these two kinds of circumstances, we would be paralyzed if we relied exclusively on intuitive-level thinking, and so we must rely on critical thinking, however imperfect we are at doing it. We must also must rely on critical thinking:

3. To select and amend intuitive-level rules over time and in light of new information and experiences.

Such improvements in intuitive-level rules could never occur if all we had to rely on were our existing intuitions.

Enough has now been said to make clear why I said that utilitarians "ultimately" hold that the right thing to do is whatever will maximize aggregate happiness. For in the case of Hare's two-level utilitarianism, when we do intuitive-level thinking, we do not apply the principle of utility, and we may even think in terms of respecting rights, with all of the "deontological flavor" that way of thinking has. But the justification for having and using such non-utilitarian rules at the intuitive level is given through explicitly utilitarian thinking at the critical level, and the specific content of intuitive-level rules is determined through explicitly utilitarian thinking. Hare uses the images of "the archangel" and "the prole" to help us understand why humans need both kinds of thinking (*MT*, chapter 3). "The archangel" would have super-human knowledge, super-human powers of critical thinking, and no human weaknesses, so the archangel would have no need for intuitive-level thinking. "The prole" would be dreadfully ignorant and uninformed, incapable of critical thinking, and would have human weaknesses in an extreme degree. So the prole would be incapable of critical thinking. Normal human beings are neither archangels nor proles: we may act on moral "auto pilot" most of the time, but we are capable of using critical thinking in unusual cases and to determine when an intuitive-level rule should be abandoned or amended.

There is one more kind of case in which we *should*, according to Hare, rely on critical thinking, even though intuitive-level thinking would not be paralyzed in this kind of situation:

4. When something is ruled out by the intuitive-level rules that one has internalized, but both (a) it seems clear that the violation will maximize aggregate happiness and (b) one can trust one's judgment that this is so.

This fourth kind of case has been another main focus of Hare's critics, and their objections will be discussed in chapter 4 (§§4.7–4.11). I will conclude this section by very briefly considering what Hare would say about a pair of such cases.

Remember that "a standard tactic employed by the critics of utilitarianism" employs the method of reflective equilibrium by describing a case in which your intuitive response will be the opposite of what the theory implies. Here is one kind of example that has commonly been used this way:

> Suppose that you are a famous transplant surgeon, and that your transplants always work. You have five patients, each of whom needs a transplant. One needs a heart, one a brain, two need one lung apiece, and one needs a liver. One of your patients, Bloggs, has come in today to find out the results from some lab work. You know from the results of the lab work that Bloggs would be a perfect donor for each of your five other patients, and you know that there are no other available donors. So you ask Bloggs if he would be willing to be cut up and have his organs distributed. He declines your kind offer, but you realize that you could easily overpower Bloggs and cut him up without his consent, and without your crime ever being discovered.[6]

The critic expects that you will have a strong intuition that it would be wrong to cut up Bloggs, but it also seems clear that, if the facts are exactly as stipulated in the case, doing so would maximize aggregate utility. Critics of utilitarianism argue that other theories better match our intuitive judgments about many such cases.

How would Hare respond? The first thing to notice about such cases is that they are completely unrealistic—the specified circumstances would never arise in the real world. More fundamentally, however, the fact that critical thinking would lead to a counterintuitive conclusion in such a case is exactly what Hare's theory predicts. For in his two-level version of utilitarianism, one of the functions of critical thinking is to *override* intuitive-level thinking in such unusual cases.

When we turn to real-world cases that seem to fall into the fourth category above, it becomes all the more clear how this is so. In *Moral Thinking*, Hare himself describes one such realistic type of case:

---

[6] I borrowed this particular formulation of the case and the name "Bloggs" from Rob Streiffer, who uses it (along with some other cases) in his teaching. In British English, "Bloggs" is often used as a generic name (like "Smith" or "Jones"), and I've been told that some materials used by American students to prepare for standardized tests also use the name.

> A person falls overboard from a ship in a wartime convoy; if the master of the ship leaves his place in the convoy to pick him up, he puts the ship and all on board at risk from submarine attack; if he does not, the person will drown. In the film *Cruel Sea* (adapted from Monsarrat, 1951) a somewhat similar case occurs; the commander of a corvette [cruiser] is faced with a situation in which if he does not drop depth-charges the enemy submarine will get away to sink more ships and kill more people; but if he does drop them he will kill the survivors in the water. In fact he drops them, and is depicted in the film as suffering anguish of mind. And we should think worse of him if he did not. (*MT*, p. 29)

In terms of how certain the outcomes are described as being, this type of case is also unrealistic: for instance, the master of the convoy does not know with certainty that the person who has fallen overboard will drown (she or he might be picked up by the enemy and survive the war in a prison camp). Otherwise this type of case, in which a person in a position of authority must decide to override a deeply ingrained rule that they would normally follow automatically, is all too realistic.

Hare's theory fits such real-world cases right down to the anguish that he expects the commander who ordered the depth charges to feel. For although most people never find themselves in command of ships, good intuitive-level rules for ship captains would require them to go to great lengths to save people who fall overboard. Then, since intuitive-level rules are internalized in a way that makes us diffident about violating them, even when clear critical thinking indicates that we are justified in doing so, we would expect the commander of the corvette to "suffer anguish of mind" over his decision. That is, Hare's theory implies both that ultimately the right thing to do in such cases is to override a good intuitive-level rule, and that a person who does the right thing will continue to have the intuition that what they did was wrong. This is because a person who could calmly do what duty ultimately demands in such difficult circumstances would not have internalized the relevant intuitive-level rules properly; they would not, for him, have a "deontological flavor." As Hare puts it on the same page, concerning a case in which one is justified in breaking a promise to one's children, "Would not the man who could break his promise to his children without a twinge of compunction be, not a better, because more rational, person, but a morally worse one than most of us who are afflicted in this way?"

§1.6. Hare's Theory and the Naturalistic Fallacy

At the end of §1.3, I said that "a hospitable division of labor" was achieved by reserving the domains of descriptive and normative ethics for science and philosophy, respectively. I can now explain why Hare's approach muddies this tidy division of labor in two related ways.

First, Hare appears to be doing descriptive ethics, at least in addition to normative ethics, when he claims it as a virtue of his theory that it explains a lot about what humans think and say about morality. This was implicit in a number of things that I said in the preceding section. For instance, I said that Hare's theory "predicts" that people will find counterintuitive the conclusion that the physician ought to cut Bloggs up against his will in order to save five terminal patients, and I said that "Hare's theory fits such real-world cases right down to the anguish that he expects the commander who ordered the depth charges to feel." Hare often makes similar claims in support of his theory. Indeed, he devotes all of chapter 4 in *Moral Thinking* to arguing that if he is right about the logic of moral discourse and how humans need to use both the intuitive and the critical levels of moral thinking, then he can explain why other philosophers have proposed various other theories, including Kant's theory, rights views, emotivism, and relativism—he doesn't mention virtue theory, but it explains that, too, as noted below in §4.4 and §12.4(A). Hare describes this as an application of the "hypothetico-deductive" method (*MT*, p. 13), which was the label that Karl Popper gave to the process of comparing how well competing scientific theories are corroborated by available observations.

When he makes such claims, Hare is defending his theory the way a scientist would defend a theory of descriptive ethics: he is claiming that it is a good theory of how people in fact reason about ethics. This opens him up to a kind of objection that, according to the traditional division of labor, a philosopher should not have to face. For if Hare were offering a purely normative theory of ethics, pointing out that "people don't actually think that way" would cut no ice; he could simply reply that his is a theory about how people *ought* to think about ethics, not a theory about how they *in fact* think about ethics. So a thorough defense of Hare's theory needs to address the following two obvious facts. Although people seem to operate on intuition most of the time, it is not obvious that the system they are using fits Hare's description of intuitive-level thinking. And when people decide unusual cases, in which their intuitions conflict or offer little guidance, they do not always report that they employ the kind of explicitly utilitarian reasoning that constitutes Harean critical thinking. In response to this objection, I will discuss (in chapters 3 and 4) a range of recent empirical research on moral reasoning. Although some of it was conducted before he died in 2002, I assume that Hare was unaware of this research. The research does not uniquely support Hare's two-level utilitarianism, but I will argue that it significantly strengthens it as a theory of descriptive ethics.

Aside from raising the question of its descriptive adequacy, Hare's approach appears to commit a fallacy that is famous among philosophers, and which they are quick to accuse social scientists of committing. This is the fallacy of inferring what ought to be from what is. As noted at the end of §1.3, the idea that this is a fallacy is precisely what makes plausible the traditional division of labor between philosophy and science. Developmental psychologist Lawrence Kohlberg himself warned against it, describing the mistake as:

> a confusion between the idea that "everyone has their own values," and the idea that "everyone ought to have their own values." In other words, the value-relativity position often rests on logical confusion between matters of fact (there are no standards accepted by all men), and matters of value (there are no standards which all men ought to accept), that is, it represents the "naturalistic fallacy." . . . Philosophers who are aware of these logical confusions do not generally accept ethical relativity and assume that there is a rational enterprise termed "normative ethics." But ethical and cultural relativism has a very powerful hold on social scientists which is not explicable [sic] in terms of the facts of cultural relativity. (Kohlberg 1971, pp. 156–157)

Hare is certainly not an ethical relativist, since according to his theory all genuinely moral thinking is (at least ultimately) utilitarian. But his Kantian appeal to the logic of moral discourse appears, at least at first blush, to commit the naturalistic fallacy. For when Hare claims that moral judgments have three logical features, he claims to be accurately *describing* what it means to treat a judgment as a moral judgment, yet he infers from this that we *ought* to judge accordingly.

Hare was certainly aware of the so-called "is/ought gap"—he knew that it is a fallacy to infer normative claims from purely descriptive claims. But when Hare claims that moral judgments have three logical features, he is not making a normatively neutral claim. He claims to be accurately *describing* what it means to treat a judgment as a moral judgment. Since it is a claim about what moral ought-claims *mean*, however, this claim is also *normative*. For if it is true, this description of the logical requirements to which all genuinely moral judgments must conform constrains what we can consistently say when making moral judgments. In this way, Hare moves from a descriptive account of the logic of moral discourse to claims about what we ought to say—indeed, what we *must* say—if we are to be rational and consistent in our moral thinking. But he does it without committing the naturalistic fallacy or ignoring the is/ought gap.

Hare's theory is highly controversial, for the foregoing reasons and others as well. However, I know of no other ethical theory that does such a good job of explaining why its own critics' tactics work so well, while shedding significant light on the history of philosophical thought about ethics. I also find that Hare's theory sheds light on various real-world controversies. For instance, as detailed in §4.6, I show how it explains why certain controversies, for example over the use of torture in "the war on terror," are both inevitable and unresolvable. And all of part III of this book will be devoted to the animal welfare/animal rights debate, which is, I think, far more tractable, in part because Hare's theory can incorporate elements of both.

§1.7. Incorporating Animal Welfare and Animal Rights into Hare's Theory

Popular and philosophical accounts of this distinction cut the pie very differently, and to fully appreciate the way in which Hare's theory can incorporate elements of both animal welfare and animal rights positions, it is necessary to have both conceptions of the distinction before us.

In non-philosophical contexts, "animal rightists" are typically distinguished from "animal welfarists" in terms of what political ends and means the two endorse, rather than in terms of their underlying philosophical commitments. Accordingly, animal rightists are typically portrayed as seeking the abolition of various practices, whereas animal welfarists are said to seek reform of problematic practices rather than elimination of animal use. Animal rightists are portrayed as willing to employ illegal and even violent tactics, whereas animal welfarists are portrayed as "working within the system." It is also important to acknowledge that, perhaps because the term was coined by people in fields targeted by the animal rights movement, in non-philosophical contexts animal welfarists are typically portrayed as reasonable, well-informed people, while animal rightists are portrayed as irrational and poorly informed people acting in the grip of their emotions. Self-professed animal welfarists typically seek to defend some uses of animals while emphasizing that they take seriously a moral imperative to take animals' interests into account—that they are not neo-Cartesians who deny all significance to animals' lives and suffering. So it is not surprising that they wish to self-consciously distance themselves—in both their colleagues' and in the public's eyes—from self-professed animal rights activists, whose actions have sometimes richly earned them the above stereotype (as detailed in Jasper and Nelkin 1992).

As philosophers understand the distinction, however, it is not about political means and ends, but about basic commitments in ethical theory. Taking an animal welfare stance means thinking about our treatment of animals in purely utilitarian terms (and, usually, adopting a hedonistic, rather than preference-based conception of "happiness," at least when it comes to animals). Taking an animal rights stance, by contrast, involves attributing moral rights to animals, where those rights are conceived of as "trump cards" against utilitarian arguments. The "trump card" metaphor is due to philosopher Ronald Dworkin (1977), but in day-to-day talk about ethics, appeals to moral rights commonly function this way. For instance, to say that an aggressive preacher "has a right to free speech" means, at least in part, that we would not be justified in infringing that right just because the aggregate suffering of listeners and passers-by outweighs the enjoyment that the preacher and his followers get from his invective. Philosophically speaking, adopting a true animal rights stance involves extending such "trump cards" to animals. Tom Regan's *The Case for Animal Rights* (1983) is the paradigm statement of such a position.

Regan argues that if individuals "have moral rights," then we cannot justify harming them on the sole basis that doing so maximizes aggregate happiness; respect for animals' rights requires non-utilitarian reasons for involuntarily imposing harm, or significant risk of harm. (For a more detailed overview of the animal welfare/animal rights distinction and of Regan's position, see Varner 1998, chapter 5, or Varner 2002a.)

According to Hare, all moral thinking is (at least ultimately) utilitarian, and in this way his theory shares the basic philosophical commitment of animal welfare views as philosophers conceive of them. Given the way that they function in the moral thinking of real-world human beings, however, Harean intuitive-level rules must have a deontological "flavor," and Hare finds rights claims useful in formulating some intuitive-level principles designed to govern our treatment of human beings (*MT*, pp. 147–156; *EB*, pp. 24–30). This makes it possible to incorporate an animal rights component into Hare's theory, if it can be shown that sound critical thinking would support extending similar intuitive level protections to at least some animals. I will argue that this is the case, at least with regard to what I call "near-persons," and at least in modern, affluent societies.

By "near-persons," I mean individuals who lack the full-blown biographical sense of self required by my stipulative definition of "person" (§1.2), but who nevertheless have a robust, conscious sense of their own past, present, and future.[7] Chapter 8 will survey several lines of empirical research that, taken together, suggest that a range of non-human animals may be near-persons. In this way, the position adopted in this book resembles "an animal rights position," according to the popular and political characterization of the animal welfare/animal rights distinction. For I will argue in part III that sound critical thinking supports distinguishing between these animals and the merely sentient in the intuitive-level system of rules that we adopt, and, in some cases, it will be appropriate to express the special respect that is called for in terms of rights.

At the same time, I doubt that any non-human animal will ever qualify as a person under my stipulative definition. This is because a biography is a story, and storytelling is more likely to be a uniquely human characteristic than the more familiar candidates of tool making and language use. As detailed in chapter 6, our identities are tied to stories that we can tell about ourselves which begin with where we came from, explain where we are now, and include both aspirations for the future and an understanding of our own mortality. Communities, too, have biographies, and part of our identities as humans is a function of being able to tell and understand these stories. For this reason, we are unlikely ever to find a full-blown biographical sense of self in any non-human animal. Telling stories about ourselves and other beings requires far more than

---

[7] Readers familiar with the animal rights literature will recognize that the so-called "problem of marginal cases" arises with regard to human beings who do not qualify as persons on my view. This problem is discussed in §§7.6 and 9.7.

the simple syntax that has been taught to some great apes, cetaceans, and parrots. Without something very much like a human language, it may be impossible to represent one's life as a story, even to oneself.

In this way my position differs from what are sometimes characterized as "extreme" animal rights positions that attribute the very same moral status to both humans and animals. I also believe that sound critical thinking supports adopting substantially different intuitive-level rules governing our treatment of animals that are "merely sentient," and this drives my view closer to the animal welfare camp, at least where merely sentient animals are concerned. By the "merely sentient" I mean individuals that "live entirely in the present." This may turn out to be a hypothetical construct. For perhaps no sentient individual literally lives entirely in the present. Some conscious sense of at least the immediate future seems necessary in order to have conscious desires or yearnings at all, since to desire something is to want a change. So just as the difference between persons and near-persons is a matter of degree, the difference between individuals that I consider near-persons and those that I use as examples of "the merely sentient" may turn out to be a matter of degree. Accordingly, it is impossible to avoid characterizing near-persons in somewhat vague and imprecise terms: for example as "having *a robust* conscious sense of their future" or "having *a relatively robust* conscious sense of their own future." (See also the footnote on Tom Regan's notion of "a subject of a life" in §7.1.)

By adopting the position that the lives of animals that are near-persons have greater moral significance than those of animals that are merely sentient, a Harean perspective on animal ethics can simultaneously endorse what the popular and political conception of the distinction would see as an animal rights stance with regard to near-persons, and what it would see as an animal welfare position with regard to the merely sentient. In particular, I will argue (in §9.5) that Hare was correct (in "Why I am only a Demi-Vegetarian," *EB* chapter 15) to endorse humane, slaughter-based agriculture, but that it would be appropriate to limit this to animals that are merely sentient. I will also emphasize (in chapter 10, passim) that a Harean analysis of proposed agricultural reforms must be sensitive to background ecological, technological, economic, and cultural conditions in the target societies.

§1.8. "Moral Standing" or "Moral Considerability" versus "Moral Significance"

When I say that "the lives of animals that are near-persons have greater moral significance than those of animals that are merely sentient," I intend the term "moral significance" in a way that has become standard since Kenneth Goodpaster's essay "On Being Morally Considerable." There Goodpaster emphasized that there is a distinction

> between what might be called a *criterion of moral considerability* and a *criterion of moral significance.* The . . . latter, which might easily get confused with the former, aims at governing *comparative* judgments of moral "weight" in cases of conflict. Whether a tree, say, deserves any moral consideration is a question that must be kept separate from the question of whether trees deserve more or less consideration than dogs, or dogs than human persons. We should not expect that the criterion for having "moral standing" at all will be the same as the criterion for adjudicating competing claims to priority among beings that merit that standing. In fact, it may well be an insufficient appreciation of this distinction which leads some to a preoccupation with rights in dealing with morality. I suspect that the real force of attributions of "rights" derives from comparative contexts, contexts in which moral considerability is presupposed and the issue of strength is crucial. (1978, pp. 311–312)

In the literatures on environmental ethics and animal ethics, it has become standard to distinguish between "moral standing" and "moral significance" in Goodpaster's terms. To assert that an individual (or other entity) "has moral standing" or "is morally considerable" is to assert that it has a claim to be *considered* in the deliberations of moral agents. This is analogous to the legal concept of "standing." To say that someone (or something, like a corporation or a government) "has standing" in the legal sense just means that they can initiate a lawsuit on their own behalf or at least have one initiated in their name. "Moral significance," by contrast, is a matter of degree. In part II of this book, I will be arguing that the lives of persons have greater moral significance than those of near-persons, whose lives in turn have greater moral significance than those of the merely sentient. Obviously, to say this is not to deny that non-persons or the merely sentient have moral standing. I will argue that a Harean utilitarian must recognize that all sentient animals have moral standing, whether or not they have the special cognitive abilities characteristic of persons or near-persons, and for the reasons given in chapter 5, probably all vertebrates (and maybe some invertebrates) are sentient because they can consciously experience pain.

Perhaps less obviously, in saying that persons' lives have special moral significance vis-à-vis those of near-persons and the merely sentient, I do not mean that the interests of persons ought always to prevail over those of non-persons. For a utilitarian, at least when thinking at what Hare calls the critical level, this cannot be ensured (§9.5). Nevertheless, good reasons can be given for embodying special forms of respect for persons and near-persons in the intuitive level rules that a Harean utilitarian adopts for day-to-day use and in our legal system, which I understand as a system of intuitive-level rules (see §§9.7 & 7.6).

Also, I describe persons' and near-persons' lives as "having special moral significance," rather than as "being more valuable" in order to guard against a fundamental misunderstanding. Saying that one life "is more valuable" than another can easily be read as saying that the life in question is preferable or better from a moral point of view. In utilitarian terms, this would be to say that the life in question adds more positive value to the world than is added by those to which it is being compared. But when I say that persons' lives "have greater moral significance" than those of non-persons, my point is not that a person's life is better or preferable to that of a non-person. Rather, I mean that persons' special cognitive abilities make it possible to both benefit and harm them in ways that non-persons cannot be benefited and harmed. This can make persons' lives preferable to those of non-persons, but it can also make their lives go so much the worse for them. Perhaps Mill was right that "it is better to be a human being dissatisfied than a pig satisfied" ([1861] 1957, p. 14), but surely a satisfied pig's life would be preferable to certain wretched states of human affairs. When I say that persons' and near-persons' lives "have special moral significance," I mean to say that their special cognitive abilities make their lives more (so to speak) *morally charged* than those of the merely sentient. This means that we should take special care in our dealings with them, and this is why good ILS (intuitive-level system) rules will specify different kinds of respectful treatment for persons, near-persons, and the merely sentient.

§1.9. Harean Reasons for Retaining the Term "Person"

It should now be clear why I said (at the end of §1.2) that in Hare's theory there are good reasons for retaining the term "person" rather than consigning it to the scrapheap of ambiguous words that impede clear communication.

First, for educational purposes, intuitive-level rules must be stated succinctly, and "person" is a helpful shorthand, just like "bachelor." It's true that, in any context where someone's bachelorhood is "reasonably debatable," there is nothing we can do with it that we couldn't do as well or better with "unmarried male." "Bachelor" still achieves some economy of expression, however, as does "person" vis-à-vis "an individual with a biographical sense of self" (let alone "individuals who deserve legal and moral rights, and should be treated as 'not replaceable,' because they have a biographical sense of self").

Second, as DeGrazia notes (1997, p. 314), the commonsense notion of personhood seems to have normative connotations built into it alongside its descriptive meaning. For precisely this reason, "person" will be useful in formulating intuitive-level rules in a way that "individual with a biographical sense of self" will not. An equally short neologism would not carry the normative connotation that "person" has in day-to-day discourse.

Finally, as I will argue in chapter 8, in day-to-day life we encounter no non-humans whose status as individuals with a biographical sense of self is "reasonably debatable." So it is no problem that commonsense tends to equate "persons" with "human beings." Indeed, in some contexts this connotation will be helpful in formulating useful ILS rules, e.g. when educating young children.

§1.10. Coda

This introductory chapter has provided a general overview of Hare's complicated version of utilitarianism. I hope you find this bird's-eye view intriguing enough to warrant reading the rest of this book. It was also necessary to provide such a general overview before discussing in detail, in the next two chapters, Hare's Kantian defense of the principle of utility for what he calls critical thinking, and his notion of intuitive-level rules. The final chapter of part I provides an overall assessment of Harean utilitarianism as a theory of descriptive and normative ethics. Part II discusses in detail the place of personhood and animals in Harean utilitarianism, and part III discusses the theory's application to questions about animal ethics.

At this point, a host of objections and questions may come to mind. Hopefully, they will all be addressed somewhere in the remaining chapters. Just like the golden rule, the principle of utility can be stated very succinctly, but specifying what it implies in practice, what it means to live a life inspired by the principle, involves a discussion more lengthy than the Torah. To fairly assess utilitarianism as an ethical theory, one must study such a lengthy commentary or specification.

{2}

Hare on the Logic of Moral Discourse

§2.1. Universal Prescriptivism

In two scenes from the 1982 movie *Star Trek: The Wrath of Khan*, Mr. Spock, the eminently logical alien from the planet Vulcan, expresses the thought that utilitarian thinking in ethics is somehow required by logic alone. Early in the film, Spock declares that "logic clearly dictates that the needs of the many outweigh the needs of the few," and Kirk seems to agree, adding: "Or the one." And at the end of the film, after Spock knowingly exposes himself to a lethal dose of radiation while preventing the ship's nuclear reactor from exploding and killing the whole crew, Spock says, "Don't grieve, Admiral. It is logical. The needs of the many outweigh . . ." When Spock's speech trails off, Kirk completes his thought: ". . . the needs of the few," and Spock adds, "Or the one." Years ago, when teaching introductory symbolic logic, I used these scenes to illustrate what I then took to be a mistaken understanding of what logic can do. Utilitarianism is a substantive ethical theory, I would tell my students, and it cannot be established through logic alone. Hare's work in ethical theory finally convinced me that Spock was substantially correct, however, and in this chapter I set out why.

Hare calls his theory that all genuine moral judgments have three logical properties "universal prescriptivism." I say all "genuine" moral judgments, because the words that we use to express moral judgments, including "ought," "right" and "wrong," "good" and "bad," "desirable," and so forth, all are used in non-moral contexts as well, and Hare's theory only applies to their use in a specifically moral sense. According to Hare, the English "must" comes closest to an "idealized artificial 'ought'" that would always have these three logical properties (*MT*, p. 24). In this chapter, I take a closer look at the three properties: universalizability, overridingness, and (in Hare's special sense) prescriptivity, and at Hare's argument that these three properties together force us to think like utilitarians. The next chapter will discuss intuitive-level rules and will extend Hare's treatment of them in some significant ways, and chapter 4 will provide an overall assessment of how plausible

Hare's theory is. The focus in this chapter is on Hare's argument that the logic of moral discourse forces us, at least when doing critical thinking, to think like utilitarians.

As we saw in §§1.5 and 1.6, Hare's theory implies that humans should (and it predicts that they will) rely on "intuitive-level moral thinking" most of the time. Nevertheless, we must rely on what he calls "critical thinking" in the four types of cases referred to in §1.5. Of particular interest in this book is the third kind of case:

3. To select and amend intuitive level rules over time and in light of new information and experiences.

In part III of this book (and in Varner forthcoming) I will consider in detail what kinds of intuitive-level rules should be chosen to govern our treatment of non-human animals, but the present chapter will have almost nothing to say about animals. The next chapter will focus on intuitive-level rules and will begin to show how Hare's account of them sheds light in various ways on questions about our treatment of animals. Hare's argument that logic forces us to think like utilitarians applies only to the four types of cases listed in §1.5, and (as was also explained in §1.5), the intuitive-level rules that we rely on day to day may be entirely non-utilitarian in flavor. Nevertheless, Harean moral thinking is always—at least ultimately—utilitarian, because these intuitive-level rules are themselves selected and amended in accordance with what Hare calls "critical thinking," and this chapter focuses on Hare's argument that the logic of moral discourse forces us to think like utilitarians when doing critical thinking.

In the final section of this chapter, I will present a formal reconstruction of Hare's argument. As the intervening sections will emphasize, the premises of that argument are controversial, but that formal proof will show that Hare's claim is a valid inference from his claims about the logical requirements on moral judgments. It will be helpful, however, to first revisit the informal, golden rule version of Hare's argument that was discussed briefly in §1.5. Discussing various aspects of that argument in greater detail and a Harean response to one objection to it will help set the stage for the formal version of Hare's argument that I will present in the final section.

§2.2. The Golden Rule Argument for Utilitarianism

According to Hare, the golden rule, properly understood, requires that if you judge that what you are doing in a given situation is morally right, then you must sincerely prefer to live through the experiences of everyone affected by your action rather than live through all of the experiences that they would have had under any alternative. This, according to Hare, provides a commonsense expression of universal prescriptivism, and Hare argues that this forces you to think like a utilitarian.

His argument is that if you had to live through the experiences of everyone affected, then you would choose to maximize the ratio of benefits to harms, since you would have to enjoy or suffer all of them in some order or other. To illustrate how this reasoning is supposed to work, he offers the following example:

> I have to give a bun to either Jones or to Smith, and Jones wants it more than Smith. Let us call Jones-with-bun J_1, Smith-without-bun S_1, Jones-without-bun J_2, and Smith-with-bun S_2. Jones prefers J_1 to J_2 more than Smith prefers S_2 to S_1. Suppose then that I put myself in their two positions in the two outcomes, four positions in all. My choice is going to be between J_1 plus S_1 (Jones-with-bun plus Smith-without-bun) and J_2 plus S_2 (Jones-without-bun plus Smith-with-bun). It is not necessary to imagine myself occupying the four positions simultaneously; that would be asking too much. We might follow a suggestion of C. I. Lewis and suppose that I have a choice between occupying J_1 and S_1 in random order and occupying J_2 and S_2 in random order. Clearly, given the assumed strengths of Smith's and Jones' preferences, my own preference will be for the first of these alternatives . . . (*MT*, pp. 128–129)

In this way, according to Hare, the logical requirements on moral judgment make it equivalent to "universalized prudence" (*MT*, p. 100): just as accounts of prudence stress not discounting the future, morality involves not discounting others' experiences. The parallels between prudence and morality make the golden rule a useful tool for helping children learn to see their actions from a moral point of view.

Discussing several things about the above passage will help to clarify how Hare's argument is supposed to work. First, the passage illustrates a crucial point about applying the golden rule. When we ask children to "Do unto others as you would have them do unto you," we aren't really asking them to think about what *they* would want if *they* were in other persons' positions, *with their own preferences intact*. Getting the child to imagine that is the first step toward understanding the golden rule, but ultimately we want them to think about what *those other persons' preferences* are. I have often used the following rather crude example to illustrate this point. Suppose that I'm having you over for dinner and I'm trying to decide what to serve for dessert. Suppose also that I could serve either ice cream or cake, and that while I always prefer ice cream to cake, I know that you are lactose intolerant. I would be misunderstanding the golden rule if I declared that it required me to serve ice cream in that situation. Similarly, in Hare's example above, my being indifferent to buns wouldn't suffice to show that it didn't matter who got the bun.

Also notice that the above passage is misleading in the following way. Hare's claim is that the golden rule, properly understood, requires me to judge as if I had to live through the experiences of everyone affected by our actions.[1] That is, "We are to do to the others affected, taken together, what we wish were done to us if we had to be all of them by turns in random order" (1993, p. 158). In the above passage, Hare asks me to imagine having to experience either "Jones-with-bun plus Smith-without-bun" or "Jones-without-bun plus Smith-with-bun," but since I am one of the people affected by my action, the choice is really between having to experience either "me-giving-the-bun-to-Jones plus Jones-with-bun plus Smith-without-bun" or "me-giving-the-bun-to-Smith plus Jones-without-bun plus Smith-with-bun." The parallel point about utilitarianism is that everyone's interests count equally, and that means that the agent's own interests count, although for no more than any other affected individual's interests.

Of the bun case, Hare concludes that "Clearly, given the assumed strengths of Smith's and Jones' preferences" one would choose to have the series of experiences in which Jones gets the bun, and Hare claims that this illustrates how the logic of moral discourse requires us to think like utilitarians. In the bun case, it seems clear enough that I would choose to give it to the one who will enjoy it more, but (as I once heard Paul Thompson say) "Anytime you hear a philosopher begin a statement with 'Clearly . . .,' grab your wallet." For as illustrated in the next section, it is not always so clear that the universalized prudence embodied in Hare's reading of the golden rule would lead me to the same conclusions as a utilitarian.

§2.3. Prudence and Moral Judgment

Assuming that one is indifferent about who gets the bun and that neither Smith nor Jones will be affected by one's giving it to the other, the choice one faces in the bun case might be represented as in the top half of table 2.1. In that situation, applying the golden rule, which expresses the perspective of universalized prudence, does seem to lead us to the utilitarian conclusion. But suppose that, instead of giving away a bun, we are considering instituting human slavery. If we had to live through the experiences of everyone affected by our decision, it seems very unlikely that we would choose to institute human slavery. Perhaps

---

[1] We might as well say that it requires us to choose as if we had to experience *the entire lives* of others, but this further complication is unnecessary. Saying that we must choose as if we were going to experience that part of their lives that is affected by our decisions suffices, because this makes us "accountable" (so to speak) for every bit of their lives that our decisions affect. By hypothesis, then, the remainder of their lives will be exactly the same, no matter what we do, and therefore requiring us to experience their entire lives would not affect our decisions.

TABLE 2.1. The Bun and Slavery Cases

THE BUN CASE	EFFECTS ON			
Options	Chooser	Smith	Jones	Aggregate happiness
I give the bun to Smith:	0	+1	0	+1
I give the bun to Jones:	0	0	+2	+2
THE SLAVERY CASE	**EFFECTS ON**			
Options	Chooser	Smith	Jones	Aggregate happiness
The status quo:	0	0	0	0
Enslaving Jones:	+4	+4	-6	+2

Source: The bun case is described by Hare in *MT*, pp. 128–129. The slavery case is original to this book.

if the status quo was very bad and the contemplated form of slavery was benign, we would choose the peculiar institution, but we can imagine cases in which the effects on the slaves would be very bad and yet the institution would still maximize aggregate happiness vis-à-vis the status quo. Suppose, for instance, that the situation were accurately represented as in the lower half of table 2.1. As long as slavery maximizes aggregate happiness vis-à-vis the available alternatives, the principle of utility favors it, and yet it seems that people who had to experience all of the effects of their decision would *not* choose to institute slavery in all such cases. This suggests that in applying the golden rule as Hare understands it, we would *not* think like utilitarians.

Whenever I present Hare's theory to an audience, someone suggests that if we had to live through the experiences of everyone affected by our decisions, we would instead apply what Rawls calls a "maximin" strategy. This is the strategy of choosing so as to maximize the minimum payoff that one could receive. In the above example, that would explain why one would not choose the policy that would enslave Jones. But for the reasons given in the remainder of this section, this is not the rational strategy to apply when doing Harean critical thinking, and a closer look at Rawls's own treatment of the maximin strategy and what he calls "the original position" suggests that even he should agree. (Which means that, if Harean critical thinking expresses a correct understanding of the golden rule, the maximin strategy is not the rational one to use when applying the golden rule either.)

§2.3(A). RAWLS'S MAXIMIN PRINCIPLE

Rawls's political theory holds that institutions are just only if they would be chosen by rational, self-interested individuals writing a social contract from behind a "veil of ignorance." The veil excludes all knowledge of one's identity, including one's personal preferences, what socioeconomic advantages and talents one enjoys as accidents of birth, and so on. Rawls held that it would be rational under such conditions to agree to arrangements that would result in inequities in things like money and political power, so long as these arrangements improve the position of the least well-off in society.

Rawls characterized this strategy of maximizing one's minimum possible payoff as an "unusual rule," however, one that "does not . . . generally apply." In his typically nuanced and guarded way, Rawls did not claim that the maximin rule would be the uniquely rational strategy under certain circumstances; he claimed only that "there appear to be three chief features of situations that give [it] plausibility," and that "the paradigm situation for following the maximin rule is when all three features are realized to the highest degree. This rule does not, then, generally apply, nor of course is it self-evident. Rather, it is a maxim, a rule of thumb, that comes into its own in special circumstances." Two of the circumstances are that:

C1. "the situation is one in which a knowledge of likelihoods is impossible, or at best extremely insecure," and
C2. "grave risks are involved." (1971, pp. 154–155)

But in a significant number of cases in which Harean critical thinking is called for, one or both of these conditions will not be met.

Remember that on Hare's theory we should only engage in critical thinking in four kinds of cases (§1.5). More will be said later about cases of the first three types, but since the slavery case is presented by critics of utilitarianism as an instance of the fourth type, I will focus here on this type of case. Cases of this type arise when something is ruled out by the intuitive-level rules that one has internalized, yet both of the following are true:

(a) it seems clear that violating the intuitive-level rules will maximize aggregate happiness, and
(b) one can trust one's judgment that this is so.

Some real-world cases are of this type. Consider, for instance, lying to your mother on her death bed. Every day, in hospitals around the world, this kind of scenario yields concrete cases in which it is true both that violating a good intuitive-level rule (against lying) would clearly maximize aggregate happiness and that one can trust one's judgment that this is so.[2] But this would not be the

[2] A set of intuitive-level rules is good to the extent that it approximates the set that an archangel would prescribe for the target population (see §3.2). For a relevant qualification (that such exceptions are actually built into the intuitive-level rule against lying), see §3.4. This is one of the reasons that the first two types of cases in which critical thinking must be relied upon are probably more rare than we would at first expect.

kind of case that meets Rawls's criteria for applying the maximin rule: for (if her doctors are good) your knowledge of the probabilities involved is quite reliable, and (depending on the lie you're telling her) there are no "grave risks" involved if the physicians are wrong and mom lives to discover your lie. In such cases, people commonly do lie, and they commonly give utilitarian reasons for doing so. Some people choose not to lie, but Hare's theory is consistent with this. For if good intuitive-level rules will prohibit lying, then, according to his theory, we should be diffident about lying even when clear critical thinking calls for it.

Other real-world cases of this type come closer to fitting Rawls's criteria for applying the maximin rule. An example would be the wartime man overboard case from §1.5. Here "grave risks" are involved if the captain decides to leave the convoy for a rescue attempt, and it is unclear how likely it is that this will result in the ship being sunk. So the above two conditions might be met in such a case, but here again people commonly give utilitarian reasons for violating the good intuitive-level rules that prescribe saving people who fall overboard. Other people would praise a captain who took the risk and succeeded in saving the man without getting the ship sunk, but Hare's theory is consistent with this too. For if good intuitive-level rules require saving people who fall overboard, then people are going to have the intuition that saving him was a good thing, and people generally believe that taking risks to save others' lives is heroic (at least when it succeeds).

Perhaps in part for the foregoing reasons, the focus of those who deny that we would reason like utilitarians when doing Harean critical thinking tends to be on hypothetical cases in which, as in the slavery case in the bottom half of table 2.1, it seems clear both that a certain option would maximize aggregate utility and that almost no one would choose that option. But such hypothetical cases never meet Rawls's first condition for applying the maximin rule. For, as Hare repeatedly emphasizes, in hypothetical cases "we can assume as much knowledge as we need" (*MT*, p. 122; see also *MT*, pp. 62–65, 114). Accordingly, I define a "strictly hypothetical" case as follows:

> *To treat a case as "strictly hypothetical" is to assume that, unless uncertainty about some things is stipulated in the description of it, everything there is to know about the case is included in the statement of facts, and we know both that this is the case and all of the stated facts with absolute certainty.*

That explains why, when the facts mentioned in a hypothetical case are redescribed and one changes one's mind about it, one tends to say that the redescription *changed the case* into a new one, about which one's opinion is different, rather than that one was wrong about the case *(the very same case)* initially. In real-world cases, on the other hand, we tend to say that we were wrong about the case *(the very same case)* when we judged it in ignorance of facts that later came to light and changed our judgment about it. In real-world cases, the facts are what they are, whether or not we know them all. In hypothetical cases, by contrast, there are no "facts" apart from those stipulated in the description we happen to be working under at any moment. So when it comes to hypothetical cases, unless uncertainty over certain things is stipulated in the description of the case, the case never meets Rawls's criteria for applying the maximin strategy.

This helps to explain why it is consistent with Hare's theory that people do not tend to think like utilitarians when faced with the slavery case as described in table 2.1. For treating the case as *strictly hypothetical* involves assuming that, unless uncertainty about some things is stipulated in the description of it, we know everything there is to know about it, and with absolute certainty. But then the schematic description of the slavery case in table 2.1 is so unrealistic that it is difficult to take it seriously at all, and it would be perfectly reasonable to refuse to engage in critical thinking about it. In any remotely realistic case, there would be plenty of additional details that one would need to know about the case before one could trust oneself to engage in critical thinking about it. So if one refuses to treat the case as strictly hypothetical and instead approaches it the way one would a real-world case, then according to Hare's theory one *should* refuse to engage in critical thinking about it. For in any real-world case, one could not be sure, without knowing much, much more about it, what would maximize aggregate happiness.

§2.3(B). A CRUCIAL DIFFERENCE BETWEEN HAREAN CRITICAL THINKING AND RAWLS'S "ORIGINAL POSITION"

So in a significant range of cases in which Hare's theory calls on us to do critical thinking, the above two conditions do not obtain. But there is a third condition that Rawls places on the use of the maximin strategy, and this one could *never* obtain in the context of Harean critical thinking. Rawls states the condition thus:

> C3. "the person choosing has a conception of the good such that he cares very little, if anything for what he might gain above the minimum . . . that he can, in fact, be sure of by following the maximin rule." (1971, p. 154)

In the original position, this is ensured because people necessarily have what Rawls calls a "thin" conception of the good. This is because they do not know what position in society they will occupy when the "veil of ignorance" is lifted, and therefore Rawlsian contractors are forced to reason about what it would be prudent for them to agree to, based solely on the distribution of what he calls "primary goods." These are "things which it is supposed a rational man wants whatever else he wants," which Rawls assumes fall into four categories: "rights and liberties, opportunities and powers, income and wealth . . . [, and] a sense of one's own worth" (1971, p. 92). Critics have alleged that this list shows that his conception of primary goods is not as "thin" as he makes it sound, because only in modern, Western cultures do people place a premium on "rights and liberties," and only in capitalist economies does income count as a primary good. In response, Rawls has effectively admitted that his theory of justice is only justified by the method of reflective equilibrium with respect to pre-theoretic intuitions that are widely shared in modern democracies and liberal socialist regimes (Rawls 1990, pp. xv–xvi).

For present purposes, the important thing to notice is that Harean critical thinking necessarily employs a "thick" conception of the good. This is because when doing critical thinking, one judges as if one had to experience the effects of the decision in question from the perspective of every individual who is actually affected by it, and doing this requires one to adopt, sequentially, every affected person's particular conception of the good, in all of its details. Without doing so, one could not experience events in the way that particular individuals do. So the third condition that Rawls places on use of the maximin rule can never be met when one is doing critical thinking.

Relatedly, notice that while contractors in Rawls's "original position" do not know whose life they will live out after the veil of ignorance is lifted, they know that they will live out the life of one and only one member of their society. This is crucially different from the situation of someone engaged in Harean critical thinking. For according to Hare's interpretation of the golden rule, to sincerely believe that what you are doing in a given situation is morally right, you must sincerely prefer to live through the experiences of everyone affected by your decision rather than live through all of the experiences they would have under any alternative. This difference between the two choice situations changes things in a dramatic way, and this is why it is incorrect for Hare to say, near the end of his description of the bun case, that "an alternative would be to suppose myself to have a choice between having an equal chance of being in J_1 and S_1, and having an equal chance of being in J_2 and S_2" (*MT*, p. 129). In the slavery case, that would amount to the following. If I choose the status quo (line one), then I am giving myself an equal chance of living one of three possible lives as a free man. If I choose the institution of slavery (line two), then I am giving myself an equal chance of living one of three lives, either the free life of "the chooser," the free life of Smith, or the enslaved life of Jones. Either way, however, I will live out just one life, and *having an equal chance of living just one of several different lives* is a very different prospect from *having a 100% chance of living all three lives in succession.*

When the question is not which *individual's life* I will have to live after the Rawlsian veil of ignorance is lifted, but what *series of lives* I would prefer to live, adopting a maximin strategy would be imprudent. It would be just like holding that, within an individual life, the satisfaction of an academic career is not worth the sacrifice involved in getting through graduate school; it is because I am going to live through every phase of my life that the enjoyment of some phases can make up for the suffering endured in others. Similarly, if I think of myself as experiencing all of the effects of my decision, then the enjoyment of some individuals can make up for the suffering endured by others.

Relatedly, in a famous criticism of utilitarian thinking in ethics, Rawls argues that "utilitarianism fails to take seriously the distinction between persons" (1971, p. 187). It is true that in reality, no individual experiences the aggregated happiness, but Hare's view does not require that anyone actually experience all of the lives. His claim is only that the logic of moral discourse requires us to choose as if we were going to experience them all. And, as explained in the next subsection, there is no conceptual impossibility in that.

§2.3(C). THE CONCEIVABILITY OF "PERSPECTIVE HOPPING"

It would be like remembering what past lives were like with the proviso that in your former lives you were never aware that you had been and would again be reincarnated. It would be what Steve Campbell calls "perspective-hopping": "The process of perspective-hopping would be akin to having a dream in which one experiences the dream as reality as long as it lasts but, upon waking, vividly recalls the dream as well as her past experience (of previous dreams and the waking moments between them) and is able to distinguish the dream from her waking experience" (Campbell 2006, p. 41). All of this sounds totally fantastic, of course, but if Hare is correct that the golden rule requires us to choose as if we had to experience, in random order, all of the effects that our actions will have on others, then this is exactly what we are being asked to imagine.

The fact that real-world human beings are unable to do this on anything approximating the grand scale that perfect utilitarian thinking would require just underlines the fact that we are not archangels. It is easy to be skeptical about our ability to draw any conclusions about what it would be "prudent" to choose under such fantastic conditions, but note how the fantastic case compares to the following real-world choice situation that is fairly commonly encountered.

In contemporary medicine, a number of procedures are performed on patients under the influence of amnesic drugs. When I turned 50, for instance, I was treated to a colonoscopy, during which I was given both a painkiller and an amnesic. When I asked what the rationale for using an amnesic was, she said, "It's a painful procedure, and if you don't remember it, you won't be so anxious the next time you need to undergo it." The fact that I would not remember it afterward did not lead me to conclude that it didn't matter what I would experience during the procedure, however. And even being completely confident about the efficacy of the amnesic, I would not agree to forgo the anesthetic. That is, I would choose to avoid conscious suffering, even if I knew for certain that I would not remember it later.

Now imagine a slight variation on that real-world case. Suppose that the amnesic drug also caused temporary retrograde amnesia; that is, suppose that while under its influence I could recall nothing about my past. The experiences that I would have under that amnesic drug would hardly count as "mine," because I would not know who I was while having them and I would not remember them afterward. Offering me this special kind of amnesic drug would make me no more likely to forgo the anesthetic, however. For many males there is a real-world parallel to this imagined case. In U.S. hospitals, circumcisions are still commonly performed on male infants without any kind of anesthesia. By all appearances, the procedure is very painful. Thanks to the phenomenon of infantile amnesia, no one who was circumcised without anesthesia as an infant can remember it, but even if it had no long-term negative effects on us, we wouldn't conclude that it didn't matter whether or not an anesthetic was used.

The foregoing examples illustrate how we sometimes make value judgments about phases of our lives that are disconnected from the rest. From here it is a small step to imagining Campbell-style perspective hopping. So I conclude that, however idealized the choice situation, it makes sense to ask what strategies or principles one would use when doing Harean critical thinking and that, for the reasons given in the preceding subsection, it would not be Rawls's maximin principle. This does not prove that people would rely on utilitarian thinking, but the discussion in this section at least blunts the force of the objections discussed here. The overall plausibility of Hare's theory will be assessed in chapter 4.

§2.4. Universalizability and Overridingness as Logical Properties of Moral Judgments

Thus far in this chapter, I have approached Hare's argument that the logic of moral discourse forces us to think like utilitarians informally, via his interpretation of the golden rule and the role that prudence plays in its application. In the remaining sections, I will adopt a more formal approach. In this and the following section, I will discuss in greater detail the three logical properties that all genuine moral judgments have, according to Hare. This will set the stage for a formal reconstruction of Hare's argument in §2.6.

It is uncontroversial to say that universalizability is a logical property of moral judgments. To see what is meant by claiming that this is a *logical* property, compare the words "all" and "some." There is no controversy about what logical properties these words have. Everyone will agree that a person who one moment says that "All gavagais are black" and a moment later says that "Some gavagais are not black" is guilty of a logical inconsistency; and everyone understands this, even if they have no idea what "gavagais" means (the word "gavagai" was coined by Willard Van Orman Quine to illustrate problems about indeterminacy of translation). Hare says that the moral words are similar to "all" and "some" insofar as their "logical properties exhaust their meaning" (*MT*, p. 3). This is why we can sometimes spot logical inconsistencies in moral judgments even when we do not understand the substantive claims being made. For an example involving moral judgment, imagine "gavagai" being used as a transitive verb. If Mervyn says both that "It was wrong for Jack to gavagai Bill" and "It was not wrong for Jack to gavagai Jill," then the universalizability of moral judgments means that Mervyn's two judgments are logically inconsistent unless he can cite what he claims to be a morally relevant difference between the two cases. We do not immediately charge people with inconsistencies if we expect that they could do this, but suppose that we asked Mervyn and he replied, "No, I don't see any morally relevant difference between the two cases, I just judge that it was wrong for Jack to gavagai Bill but not for him to gavagai Jill." *Then*, I trust, we would all agree that Mervyn's judgments were logically inconsistent. Graduates of the free-love-and-nickel-beer school of pedagogy might caution against taking him to task for it, but when Mervyn refuses to even try to distinguish the two cases, he is at least choosing to wallow in the appearance of contradiction rather than attempting to drag himself out of the logical muck. This is what Hare means when he claims that moral judgments have the logical property of universalizability.

Universalizability is also presumed in the legal doctrine of *stare decisis*, under which judges departing from precedent are required to "distinguish the instant case," and Hare claims that all "value judgments" also have the logical property of being universalizable (*MT*, p. 22). This claim is more controversial, but if we look at how we respond to various real-world examples, it sounds plausible. Suppose that I loan you my mountain bike and when you return it, you tell me that it's not a good mountain bike. Then you borrow Lynn's mountain bike and you tell her that it's a good one because it has both front and rear suspension. Assuming that mine does, too, when I ask for an explanation, I am challenging the logical consistency of your value judgment. In such a case, it might be easy for you to meet the challenge (maybe the suspension on my bike is too stiff while hers is just right), but in doing so you are responding to a charge of inconsistency, one that is justified by the fact that such judgments are, like moral judgments, universalizable. The same could be said about aesthetic judgments. If Mervyn says that he likes country music but I know that he hates Ray Wylie Hubbard, I can challenge him with: "Ray Wylie's music is classified as country, so why don't you like him?" Again, it might be easy for him to meet the challenge (he might reply, "I like country music that celebrates patriotism and Christianity and Ray Wylie's music does neither"), but the challenge is a logical one, nevertheless.

This is where the property of overridingness comes in: it is what distinguishes moral judgments from other kinds of value judgments. Both Kant and

Hare claimed that to treat a norm as a *moral* norm is to think that it ought to trump other types of norms, including norms of prudence, law, etiquette, and aesthetics. So if you hold that it would be morally wrong to do what the law requires in a given situation, then you could not consistently say that what you "really ought to do," or what you ought to do "all things considered," is follow the law. Again, this could be expressed as the principle: "Where moral and non-moral norms conflict, what you ought to do, all things considered, is what is morally right." This principle is also self-evident in the sense that one who fails to acknowledge it does not understand what it means to make a distinctively moral judgment. Other judgments formulated using the word "ought" may not be subject to it, but this logical requirement is part of what it means to make a distinctively *moral* judgment.

Almost everyone agrees that both universalizability and overridingness are logical requirements on all moral judgments. Kant's innovative move was to argue that a specific moral principle could be generated from these two logical requirements. Most philosophers think that these requirements are purely formal and cannot by themselves be action-guiding. But Kant argued that these two requirements entail what he called "the categorical imperative," which he characterized as strongly anti-utilitarian. Hare claimed that Kant misunderstood the logical requirements of moral judgments and that universalizability and overridingness, along with a third requirement, which he called *prescriptivity*, actually entail utilitarianism.

§2.5. Prescriptivism and Morality as a System of Imperatives

It is important to see that Hare means something specific, special, and controversial when he describes moral judgments as "prescriptive." The word is commonly used to describe sentences that express moral judgments, by way of distinguishing them from "descriptive" sentences that make factual claims. Descriptive statements tell us what is the case, rather than telling people what they ought to do. Moral judgments are commonly said to be "prescriptive" insofar as they tell people what they ought or ought not to do, regardless of what they in fact do. That moral judgments are prescriptive in this sense is uncontroversial, but in this sense "prescriptive" just means "normative." So I use "normative" as my contrast term to "descriptive," reserving the word "prescriptive" for the special, controversial use that Hare gives it.

Given how important this special sense of "prescriptive" is to Hare's defense of utilitarianism, it is astonishing how little space he devotes to a basic characterization of it in his capstone work, *Moral Thinking*. There he simply says:

> The prescriptivity of moral judgements can be explained formally as the property of entailing at least one imperative. But since this presupposes an

> understanding of "imperative," which perhaps cannot be achieved without first explaining "prescriptive," a fuller but more informal explanation may be more helpful. We say something prescriptive if and only if, for some act *A*, some situation *S* and some person *P*, if *P* were to assent (orally) to what we say, and not, in *S*, do *A*, he logically must be assenting insincerely. (*MT*, p. 21)

He then gives an example from which it is clear that he means by this roughly the following:

> *One sincerely assents to a moral judgment only if one would act consistently with it wherever it applies, at least other things being equal (i.e. if there are no other relevant moral judgments that one sincerely assents to).*

To claim that moral judgments are prescriptive in this sense is highly controversial and counterintuitive. For if I cannot sincerely assent to a moral judgment without acting consistently with it, how is weakness of the will possible? That is, if moral judgments are prescriptive in Hare's specific sense, then it should be impossible for one to fail to do what one sincerely believes that one ought to do, and yet such "moral backsliding" is a pervasive feature of human life.

In fact, however, the problem is even more pervasive, for moral judgments are just one of many kinds of statements that Hare takes to be prescriptive. According to Hare, all value judgments are universal prescriptions, including aesthetic ones (*MT*, p. 22) and ones regarding "the best" hotels (*MT*, pp. 21–22). Moreover, Hare maintains that every desire of an individual can be represented in a prescription, albeit a non-universalizable one (*MT*, p. 56; 1968, p. 51; 1972, pp. 98–99). (The problem is not as serious for non-moral value judgments, however, because these other types of value judgment are not overriding, as moral judgments are.)

Debate over the connection between believing a value judgment and acting accordingly is as old as Western philosophy, dating back at least to the "Socratic paradox," and Hare discusses weakness of the will at some point in each of his main works (*LM*, p. 169; *FR* chapter 5; *MT*, pp. 57–60). These constitute one of his responses to the problem under discussion, as he explains how some cases of what appear to be weakness of the will really are not. For instance, if one says, "Of course it's wrong to steal" while stealing, one could be reporting a common moral judgment without endorsing it, or if you say, "Of course I believe that it's wrong to steal," you could be reporting that you have internalized an intuitive-level rule against stealing (*MT*, pp. 58–59), which you are nevertheless going ahead and violating after doing some careful critical thinking. Such cases provide no counterexamples to prescriptivism, but it is still true that one cannot sincerely assent to what Hare calls "critical-level" moral principles without acting accordingly, which sounds counterintuitive. (As described in the footnote to §9.2, in order to treat intuitive-level rules as expressing "genuine" moral judgments, Hare has to qualify the property of "overridingness.")

Hare has available another, more direct response, however, rooted in some of his earliest work, in "Imperative Sentences" (1949) and *The Language of Morals* (1952), where he offered a detailed analysis of the logical properties of imperative sentences. In *Moral Thinking*, Hare avoids the complications introduced by these earlier works, but the counterintuitiveness of his prescriptivism is attenuated by attention to some of those details, as described in the remainder of this section.

§2.5(A). ASSENTING TO AN IMPERATIVE

English grammar texts standardly distinguish three "moods": the indicative, imperative, and interrogative. Indicative sentences are used to say that something is, was, will, could, or would (not) be the case. An imperative sentence, by contrast, "tells us to make something the case" (Hare 1949, p. 27). This kind of speech act consists in "*telling someone to* do something, in contrast to telling him (or her) *that* something is the case" (*OP*, p. 43, italics in original). Hare argues that these two are the fundamental moods, because all interrogatives can be rephrased as imperatives. For instance, "'Who is at the door?' can be translated 'Name the person who is at the door'" (*LM*, pp. 4–5). Thus, according to Hare, language is used in two basic ways:

1. To tell people what is true, and
2. To tell people to make something true.

The latter is the unique function of imperatives.[3]

With regard to imperatives, Hare's characterization of prescriptivity sounds entirely plausible for the following reasons. It seems uncontroversial to say that to agree with or to affirm an indicative is to *believe it*, but then, since imperatives are neither true nor false (they have no truth value at all), what could it mean to agree with or to affirm an imperative? If Hare is right that the function of imperatives is to tell people to make things true, then to agree with an imperative would seem to involve *doing something*. Hence the propriety of Aristotle's doctrine that the conclusion of a practical syllogism is an action (*Nicomachean Ethics* 1147b30).

Since imperatives can be directed to either ourselves or others, however, it would be misleading to say only that to agree with an imperative is to *act* accordingly. Instead, we should say that to agree with an imperative is to *will* that what it commands be done. In the case of an imperative directed to oneself, this

---

[3] Hare emphasizes that *telling* someone what to do and *getting* them to do it are two different things. The unique function of imperatives is the former, not the latter: "It is true that commands [imperatives] are designed to produce an action, or a will to action, in the hearer; but even this does not necessarily make them more evocative than other sentences. If you want a man to take off his trousers, you will more readily succeed by saying 'A scorpion has just crawled up your trouser-leg' than by saying 'Take off your trousers'" (1949, p. 39).

sometimes means acting; but in the case of imperatives directed to others, it must mean willing that they do what is commanded (including any effects on oneself). Hare made this addendum explicit in a late essay: "A speech act is prescriptive if to subscribe to it is to be committed, on pain of being accused of insincerity, to doing the action specified in the speech act, or, if it requires someone else to do it, to willing that he do it" (1991, p. 458).

§2.5(B). IMPERATIVE INFERENCES

Hare's early work focused on showing how there could be a logic of imperatives that largely parallels that of indicatives. He did this by showing how both indicatives and imperatives could be recast in a way that separates out the mood-indicating part. In English, this is embedded in the conjugation of the verb, but when the mood-indicating part of the sentence—which Hare called the "neustic"—is separated out, it becomes clear how there could be a logic of inferences among imperatives that is isomorphic with the logic of inferences among indicatives. For instance, these two sentences:

Indicative: You will use a saw.
Imperative: Use a saw.

Hare restates as follows, in order to separate out the neustic (after the comma) from what he calls the "phrastic" (before the comma):

Indicative: Use of a saw by you shortly, yes.
Imperative: Use of a saw by you shortly, please.

"Use of a saw by you shortly" is not a complete sentence; it is neither an indicative nor an imperative. But it expresses what the indicative and imperative sentences have in common—which Hare calls the "phrastic" element of the two sentences (*LM*, p. 18; 1970, pp. 20–21). Separating the neustic and the phrastic in this artificial way makes it clear how parallel logic can apply both to indicatives (this example is from Hare 1949, p. 31):

Use of axe or saw by you shortly, yes.
No use of axe by you shortly, yes.
____
Use of a saw by you shortly, yes.

and to imperatives:

Use of axe or saw by you shortly, please.
No use of axe by you shortly, please.
____
Use of a saw by you shortly, please.

Hare goes on to discuss some ways in which the parallels seem to break down, but this suffices to make his core point in response to the emotivists, who were the prominent non-cognitivists of the time.

The emotivists held that moral "judgments" are actually expressions of the speaker's feelings, and they therefore denied that standards of logical argumentation could be applied to moral judgments. If moral judgments entail imperatives, however, then Hare's work on imperatives showed that a non-cognitivist need not abandon reasoning in ethics.

> Logic is primarily concerned, not with the truth of propositions, but with the validity of inferences; and it has long been a commonplace of traditional logic that it makes no difference to the validity of an inference whether its premises and conclusion are true or whether they are false. The argument is valid if the conclusion follows from the premisses, whether true or false, or, we might add, neither. (1949, pp. 36–37)

Indicatives are either true or false. Imperatives are neither. And yet, as the examples above show, valid reasoning can be applied to both. Indeed, Hare points out: "It is even possible to mix [imperatives and indicatives] within the same syllogism; Aristotle does it in his practical syllogism, where the major premise is a universal imperative, the minor an indicative, and the conclusion either a further imperative, or an action which, so to speak, elides an imperative" (1949, p. 34). According to the *Oxford English Dictionary*, "elide" is derived from a Latin verb meaning "to crush out," so that by "elides an imperative," Hare presumably meant something like "annuls or absorbs the motive force of an imperative."

§2.5(C). THE PRESCRIPTIVITY OF MORAL JUDGMENTS

According to Hare, the difference between imperatives in general and "normative statements" (*OP*, p. 51), or "value-judgments" (*LM*, p. 129) in particular, is that while the former need not be universalizable, the latter always are. And this means that normative statements or value judgments always "refer to, and express acceptance of, a standard which has an application to other similar cases" (*LM*, p. 129). In a late (1996) essay, Hare illustrates this point with a military example:

> A serjeant [*sic*] on the parade ground, who wishes merely to see whether a recruit will, or even can, obey him, may give the order "Slope arms," but on a precisely similar occasion with a precisely similar recruit he might say "Shoulder arms." The serjeant [*sic*] cannot be accused of having changed his mind. Contrast the case where an officer doing a tactical exercise with a cadet says "The situation being as it is, you ought to attack." If in a future identical exercise with identical cadets he says "The situation being as it is,

> you ought not to attack," he must have changed his mind. This is because the normative "ought" has to be used in accordance with a rule; and, the situation being the same, he has, unless he has changed his mind about the rule, to apply it and say the same as he said before. If he understands what "ought" means he will know this. (*OP*, p. 51)

One can consistently issue random imperatives, but to make a normative or value judgment is to commit oneself to a rule entailing similar imperatives in similar cases. Thus, in *The Language of Morals*, Hare describes moral judgments as expressing "universal imperatives" (pp. 175ff) and in "Imperative Sentences" he says that "a rule is a universal imperative" (1949, p. 23).

So if we agree that, as stated above, to assent to an imperative is to act in accordance with it, and if Hare is right that moral judgments are universal imperatives, then we must conclude that moral judgments are prescriptive in Hare's sense. This means that one sincerely assents to a moral judgment only if one acts accordingly, but further, as noted above, one must also *will* the consequences of others acting according to the judgment.

§2.5(D). HISTORICAL ANTECEDENTS

Although this prescriptivist doctrine has a queer sound to the modern lay ear, it has deep roots in Western moral philosophy. For, as Hare notes, there is "a prescriptivist strand" in Aristotle's treatment of weakness of the will (*OP*, p. 20). Like Socrates before him, Aristotle appeared to hold that it is impossible to "really" believe that one should not do something while doing it. He compared the weak-willed person's saying "I shouldn't do this" to the mutterings of a drunk or an actor reading lines in a play (*Nicomachean Ethics* 1147a10ff).

In modern times, Kant thought about moral obligations in terms of imperatives, and what he says about them presages Hare's prescriptivism. In Kant's most widely discussed work on moral philosophy, the *Groundwork of the Metaphysics of Morals*, he writes:

> The conception of an objective principle so far as this principle is necessitating for a will is called a command (of reason), and the formula of this command is called an imperative.
>
> All imperatives are expressed by an *"ought."* By this they mark the relation of an objective law of reason to a will which is not necessarily determined by this law in virtue of its subjective constitution . . .
>
> A perfectly good will would thus stand quite as much under objective laws (laws of the good), but it could not on this account be conceived as *necessitated* to act in conformity with law, since of itself, in accordance with

> its subjective constitution, it can be determined only by the concept of the good. Hence for the *divine* will, and in general for a *holy* will, there are no imperatives: "*I ought*" is here out of place, because "*I will*" is already of itself necessarily in harmony with the law. ([1785] 1948, p. 81)

"A perfectly good will" would conform perfectly to any "objective principle." Imperatives—"oughts"—are directed to human beings only because we are imperfectly rational and may therefore fail to conform to an "objective principle." Kant makes the same point early in his other main work in ethical theory, the *Critique of Practical Reason*:

> The practical rule is always a product of reason, because it prescribes action as a means to an effect which is its purpose. This rule, however, is an imperative for a being whose reason is not the sole determinant of the will. It is a rule characterized by an "ought," which expresses the objective necessitation of the act and indicates that, if reason completely determined the will, the action would without exception take place according to the rule. ([1788] 1956, p. 18)

The same point is also made in the introduction to his *Metaphysics of Morals* ([1797] 1956, p. 23). So Kant, like Hare, understands the moral "ought" as expressing an imperative, and Kant endorses a kind of prescriptivism: if one sincerely assents to an ought claim, then, at least to the extent that one is rational, one acts accordingly.

Among the classical utilitarians, Sidgwick also thought that the moral words are prescriptive, at least when they are being used in their distinctively moral sense. Unlike Hare, however, he denies that to sincerely assent to a moral judgment, one must act accordingly: "when I speak of the cognition or judgment that 'X ought to be done'—in the stricter ethical sense of the term ought. . . . I imply that in rational beings as such this cognition gives an impulse or motive to action: though in human beings, of course, this is only one motive among others which are liable to conflict with it, and is not always—perhaps not usually—a predominant motive" (1907, p. 34). And later he compares making a judgment about what is "good on the whole for me" to "assuming myself to have an equal concern for *all* existence" (p. 112).

§2.6. A Formal Reconstruction of Hare's Argument

I can now provide a formal reconstruction of Hare's argument that the logic of moral discourse entails utilitarianism. The argument has four premises.

1. All and only genuine moral judgments express prescriptions that are treated as both universal and overriding.

This first premise expresses Hare's basic claim that all genuine moral judgments have these three logical features. The second premise explains what Hare means by a prescription:

2. One sincerely assents to a prescription only if one both acts as required by it and wills all actions required or authorized by it.

Universalizability can be explained generically as the requirement to "judge similar cases similarly," but if Hare is right that moral judgments are also prescriptive, then the universalizability requirement can be restated thus:

3. To treat a prescription as universal is to acquire the preferences of everyone affected by one's action in the sense of willing whatever option would maximize the satisfaction of all one's preferences, under the assumption that one will have to undergo, in random order, the experiences of everyone affected by one's action.

This leads Hare to talk in Kantian terms about a "contradiction in the will" arising when I contemplate the universalization of a prescription (*MT*, p. 109).

Hare explains the overridingness of moral judgments in terms of both "principles" (or rules) and prescriptions:

4. To treat a prescription as overriding "is to let it always override other principles when they conflict with it and, in the same way, let it override all other prescriptions, including non-universalizable ones (e.g. plain desires)" (*MT*, p. 56).

Treating these four numbered statements as premises, we can now formulate an explicit argument in defense of Hare's claim that utilitarianism follows from the logic of moral discourse.

The argument takes the form of a conditional proof, in which we begin with a simple, abstractly described case:

5. Suppose that Deon prefers option *A* to option *B* by some amount *X*, and that the only other individuals affected by the option, Eddy and Flo, prefer option *B* to option *A* by amounts *Y* and *Z* that together are greater than *X*.

We now show why, given the premises of the argument, sound critical thinking leads to the utilitarian conclusion in any case that fits the above description:

6. On utilitarian grounds, Deon ought to choose option *B* and he ought not to choose option *A*. (This follows from [5] under the standard definition of utilitarianism.)
7. Deon cannot sincerely assent to the moral judgment "I ought to choose option *A*" (or to the weaker "It is false that I ought not to choose option

A"—that is, "It is morally permissible for me to choose option *A*"), for the following reasons:

A. To treat it as a moral judgment is to treat it as an overriding prescription (from #1), and this means that it overrides any prescription expressing Deon's preference for option *A* over option *B* (from #4).

B. If Deon treats it as a universal prescription, then he acquires the preferences of Eddy and Flo in the sense of willing that option *B* be chosen (from #3), and then it follows (from #2) that he does not sincerely assent to the prescription.

8. But Deon can sincerely assent to the judgment "I ought to choose option *B*" (or "I ought not to choose option *A*"), for precisely parallel reasons:

A. To treat it as a moral judgment is to treat it as an overriding prescription (from #1) and this means that it overrides any prescription expressing Deon's preference for option *A* over option *B* (from #4).

B. If Deon treats it as a universal prescription, then he acquires the preferences of Eddy and Flo in the sense of willing that option *B* be chosen (from #3), and then it follows (from #2) that he does sincerely assent to the prescription.

9. So Deon can only sincerely assent to the same moral judgments about this case that a utilitarian would make (from #6–8).

Now, since the case described in step 5 can be adjusted to fit any case in which the aggregate utility of option *B* is greater than that of option *A*, in ways that do not affect the outcome of the reasoning in steps 6 through 9:

10. The reasoning in steps ##5–9 can be replicated for any case in which one's moral judgment diverges from the conclusion that a utilitarian would reach.

Thus, given Hare's account of the logic of moral judgment, all clear-thinking people can and will sincerely assent to the same moral judgments, assuming that they are doing sound critical thinking and that they are working with the same detailed descriptions of the situations in question.

To illustrate this point, Hare asks you to imagine that in order to park your car, someone's bicycle must be moved, and that while it would be "highly inconvenient" not to be able to park one's car, having one's bicycle moved would be only mildly inconvenient. If the driver also owns the bicycle, then obviously the prudent thing to do is move it. But, according to Hare, the logic of moral discourse requires us to judge as if we ourselves had to experience all of the consequences of our actions. This, in effect, transforms an interpersonal case (in which the driver does not own the bicycle) into an intrapersonal case (in which the driver owns both); and this ensures that both driver and cyclist will reach the same conclusion:

> We see here in miniature how the requirement to universalize our prescriptions generates utilitarianism. And we see how in principle unanimity can be reached by our method of reasoning, once each fully represents to himself the situation of the other. And there is in principle no difficulty in extending the method to multilateral cases; the difficulties are all practical ones of acquiring the necessary knowledge and correctly performing some very complex thought-processes. In difficult cases it would take an archangel to do it. (*MT*, p. 111)

That is: "archangels, at the end of their critical thinking, will all say the same thing, on all questions on which moral argument is possible" (*MT*, p. 46—reference omitted). This is reminiscent of Kant's vision of a "Kingdom of Ends," where everyone's wills would be in systematic harmony because each moral agent follows the logic of moral discourse to precisely the same conclusions and allows the resulting moral imperatives to override all contrary norms and desires.

Unfortunately, as detailed in the following chapter, the best that real-world human beings can hope to achieve is a crude approximation of archangelic thinking; and to do that, they need to train themselves to rely heavily on what Hare calls "intuitive level rules."

{3}

The Nature of Intuitive-Level System (ILS) Rules

§3.1. The Need for Intuitive-Level Rules

As detailed in the preceding chapter, Hare argues that the logic of moral discourse forces us to think like utilitarians, at least when we are doing what he calls "critical thinking." For two reasons, however, Hare holds that real-world humans need to rely most of the time on what he calls "intuitive-level rules." The first reason is epistemic: extremely detailed information would be needed to apply the principle of utility correctly, but human beings are usually lacking some or much of the relevant information. An extreme example (in which you couldn't have known that the child drowning in the pond was the young Adolf Hitler) was given in §1.5, but in myriad, less dramatic ways real world humans must act in ignorance of many relevant details. An animal-related example is the debate, which is discussed in more detail in §10.6, over how many field animals perish under different agricultural regimes. Steven Davis (2003) argues that because row crop agriculture probably kills more field animals than does the practice of pasturing livestock, adopting a vegan diet actually results in more animal deaths than a diet that includes meat from pasture-raised livestock. Davis admits, however, that we do not have even close to all of the relevant data, and his critics (including Matheny 2003, Schedler 2005, and Lamy 2007) have accused him of piling assumption upon questionable assumption.

The second reason that real-world humans need non-utilitarian rules for day-to-day decision making is more cognitive and motivational. Humans are imperfectly rational, both (a) because we have limited data-processing abilities and make mistakes, and (b) because we are inclined to "cook the data" and deceive ourselves into thinking that we are justified in doing things that suit our individual interests, when clear utilitarian thinking would require us to sacrifice for the greater good of the whole.

For purposes of moral education, these rules need to be fairly simple, like "Don't lie," and given humans' tendency to self-deceit, they need to be (as I like to put it) "deontological in flavor." That is, they need to be internalized in a way that causes us to be diffident about violating them, even when, after careful critical thinking, we believe that we are justified in doing so. For this reason, Hare, a utilitarian, even finds rights claims—which have been characterized as "trump cards against utilitarian arguments"—useful in formulating some intuitive-level principles (*MT*, pp. 147–56; *EB*, pp. 24–30). In sum:

> Intuitive thinking has the function of yielding a working approximation to [archangelic thinking] for those of us who cannot think like archangels on a particular occasion. If we wish to ensure the greatest possible conformity to what an archangel would pronounce, we have to try to implant in ourselves and in others whom we influence a set of dispositions, motivations, intuitions, prima facie principles (call them what we will) which will have this effect. We are on the whole more likely to succeed in this way than by aiming to think like archangels on occasions when we have neither the time nor the capacity for it. The prima facie principles themselves, however, have to be selected by critical thinking; if not by our own critical thinking, by that of people whom we trust to be able to do it. (*MT*, pp. 46–47)

This chapter elaborates on the notion of intuitive-level rules, extending Hare's treatment of the subject in several ways.

In §3.2 I will discuss how Hare's theory distinguishes between the common morality of a culture or society, the laws of that society, codes of ethics for individuals who occupy various professions and roles, and the personal moral commitments of individuals. In §3.3 I will discuss some related, general characteristics of intuitive-level rules. From §§3.2 and 3.3 a number of important implications emerge. All of these will become important in part III of this book. One is that technology and economic conditions can affect what standards are appropriate from a utilitarian perspective in different societies and across generations. So "showing appropriate respect for animals" may mean something very different in a pre-contact Inuit society than it means in a wealthy industrialized state, and laws governing animal slaughter should change as new technologies become available. Relatedly, while utilitarian moral thinking has a conservative bias (because costs of change must be counted), it can simultaneously help to change the background cultural attitudes that contribute to that conservative bias. For instance, individuals inspired by utilitarian ideals might internalize intuitive-level rules in their personal morality that demand greater respect for animals than is demanded by the rules of their culture or profession. To the extent that these "heros" capture the imagination of the public, the goal posts of common morality can shift across generations. Another complication is that the Harean distinction between common morality on the one hand, and law and professional ethics on the other hand, may lead us to conclude that appropriate ways of incorporating animal welfare considerations in the thinking of people in various roles and professions might differ. For instance, the code of ethics of the Ecological Society of America might incorporate concerns about animal welfare very differently from the code of ethics of the American Veterinary Medical Association (see Varner 2008b).

In §3.4 a core point about the difference between critical thinking and intuitive-level thinking will be discussed in detail, adding another, more general kind of complication. Although what is called the "classical model" of moral reasoning applies to Harean critical thinking, recent research on moral judgment suggests that intuitive-level thinking proceeds according to a prototype-based model. On that model, explicitly stated rules and principles play multiple important roles, without necessarily articulating fully the intuitive moral beliefs of the individuals who champion them. This will make it difficult, in part III of this book, to say precisely which intuitive-level rules accurately characterize a society's common morality, or various individuals' personal moralities. Nevertheless, explicitly stated rules have several functions, and one of these involves furthering public debate about how the rules of common morality, law, and professional ethics ought to change over time.

§3.2. Four Types of "Good" ILS Rules

In the rest of this book, I will speak of "sets of ILS rules," with the embedded acronym standing for "intuitive-level system." The acronym (which is mine, not Hare's) is appropriate because it is used in aviation to stand for "instrument landing system," a system for finding the right path when you cannot clearly see it for yourself and you could easily drift off course or be blown off course. In Hare's theory, a set of ILS rules has a similar function. A set of ILS rules is designed to cover a range of ethically charged situations that are encountered by the target population in the normal course of their affairs. Internalizing the rules properly produces dispositions to judge, react emotionally, and act accordingly. It also makes the individual diffident about violating them, even when clear critical thinking indicates that doing so will maximize aggregate happiness.

A set of ILS rules is "good" to the extent that it approximates the set that an archangel would prescribe for the target population. Although Hare never notes this, one ILS rule can also be said to be "more important" than another, to the extent that good utilitarian reasons can be given for making people relatively more diffident about violating it. For instance, it may be a good thing for people to be more diffident about violating rules of non-maleficence than rules of beneficence. Relatedly, rights-language might be appropriate for formulating such relatively important ILS rules.

Since target populations differ in various ways, there will be at least four categories of ILS rules. Hare mentions three of them when he says that there will be "at least three sub-levels" of ILS rules, "the ones common to all, those common to particular roles, and those personal to individuals" (*MT*, p. 203).

These correspond, at least roughly, to what one engineering ethics textbook (Harris et al. 2000, pp. 13–14) calls "common morality," "professional ethics," and "personal morality."

§3.2(A). COMMON MORALITY

If Hare's theory is correct, then insofar as members of a society face similar ethically charged situations, we would expect agreement to emerge on basic standards that everyone in the society will be expected to live up to. This is what Harris and colleagues call a "common morality." As will be emphasized in part III of this book, to the extent that societies face different background ecological, technological, and economic conditions, there should be variations across cultures, but here I want to emphasize the similarities.

To the extent that there are certain universal features of the human condition, we would expect there to be many similarities between the common moralities of various cultures at different times and places. I think that we tend to miss how many points of cross-cultural agreement there are, because we tend to focus on controversial issues. As James Rachels pointed out in his fine treatment of cultural relativism in ethics, however, "there are some moral rules that all societies will have in common, because those rules are necessary for society to exist" (1999, p. 30, italics removed). Rachels gave as examples: caring for infants, "a presumption in favor of truthfulness," and a prohibition on murder. In *Human Universals* (a survey of the surprising number of institutions and practices that are universal across all known human cultures), anthropologist Donald Brown includes standards governing promise keeping and reciprocity, the division of labor and rites of passage, property inheritance, marriage and "sexual modesty," etiquette and hospitality, and aesthetics and storytelling in his catalog of characteristics of "the Universal People" (1991, chapter 6).

As noted in §9.6, cannibalism is also universally condemned (at least when practiced on members of one's own culture), except under the most extreme conditions. Four other candidates for cultural universals that can be understood as norms of common morality will be discussed in §4.2 and §9.2. These are: the golden rule; a right to equal respect and concern; what Hare calls "the formal principle of justice"; and "Don't kill sentient animals unnecessarily." Each of these requires some discussion to make them seem plausible candidates for cultural universals, however.

§3.2(B). PERSONAL MORALITY

Harris and colleagues describe "personal morality" as "the set of one's own ethical commitments, usually given in early home or religious training and often modified by later reflection" (2000, p. 13). In Hare's theory, personal morality has two important roles.

First, he says that some ILS rules will be "personal to individuals" (*MT*, p. 203), because the content of good ILS rules depends not only on "circumstances" but also on the "temperaments of individuals" (*MT*, p. 41) and their "capacities" (*MT*, p. 199). Insofar as individuals within a given society differ in their abilities to reason critically under various circumstances, critical thinking will lead different individuals to train themselves to adhere to different sets of intuitive-level rules, including what we might call "meta-rules" for deciding when to engage in utilitarian critical thinking and when to stick more unquestioningly to one's intuitive-level rules. As Hare puts it: "A person with any deep experience . . . will have acquired some *methodological* prima facie principles that tell him when to launch into critical thinking and when not; they too would be justified by critical thinking in a cool hour" (*MT*, p. 52). In this way, Hare's theory also has a place for what Harris and colleagues call "personal morality."

That there is a second role for personal morality in Hare's theory is implicit in his treatment of "supererogation." Generally, an act is said to be supererogatory if it is good, but more than duty requires. Utilitarianism is often criticized for having no place for this notion of "going above and beyond the call of duty." This is true of act utilitarian reasoning, in which one applies the principle of utility directly to the evaluation of a particular case. Since the principle defines the right action as that which maximizes aggregate happiness, there cannot be any supererogatory action, according to act utilitarianism. There might be more than one option from which to choose (if more than one option would maximize aggregate happiness), but there is no way to say, from this act utilitarian perspective, that what one did was *better* than what duty required.

In Hare's two level utilitarianism, however, one does not think like an act utilitarian most of the time. Hare's account of supererogation focuses on the third of the four situations (listed in §1.5) in which one should rely on critical thinking: when selecting and amending ILS rules. Then it will be appropriate for some people to set their sights higher and train themselves to do more than the ILS rules of common morality or professional ethics require: "Although it would probably be disastrous if moral education were devoted to the sole purpose of producing saints, it would be detrimental to nearly everyone if those who are able to be saints failed to become such, according to their capacity. Each of us, therefore, has to ask himself what is the level of saintliness of which he is likely to be capable, and strive for that" (*MT*, p. 201). So individuals who adopt personalized ILS rules that result in them habitually doing more to further aggregate happiness than is required by the relevant ILS rules of their common morality or professional ethics are in this way "going above and beyond the call of duty." The rationale for not requiring this level of sacrifice of everyone is that it would be requiring more than most people are capable of, but it is good to encourage those who are up to the task, and so it makes sense to praise those who internalize especially demanding ILS rules in their system of personal morality. So Hare characterizes "The saint or hero" as someone who "will do more and will be praised for it" (*MT*, p. 203).

Everyone does something supererogatory on occasion, without qualifying as a hero or a saint, and maybe sometimes this results from clear critical thinking rather than the sort of habitual response that results from internalizing an ILS rule in the way described below in §3.3(D). Note this related qualification, however: tragic situations can arise in which doing what critical thinking requires will involve actions that ought to be *condemned* rather than praised, as discussed in §4.6.

§3.2(C). PROFESSIONAL ETHICS

Insofar as those in certain roles face special situations in the normal conduct of their work, we would expect agreement to emerge on basic standards for the conduct of various professionals and others in special roles. These ILS rules, when explicitly formulated for a given profession, would constitute a code of professional ethics.

Hare tends to talk about ILS rules being formulated to cover the kinds of situations that members of the target audience "commonly" encounter, but when it comes to professional ethics, it would be more accurate to describe them as formulated to cover situations that are "normally" encountered. In this respect, professional ethics is importantly different from both common morality and personal morality. In the latter two kinds of ILS rules, as Hare always says, the focus is on cases that are commonly encountered in day-to-day life. In professional ethics, however, the focus is on both commonly encountered and on *normal but uncommonly encountered* cases. Soldiers, for instance, do not *commonly* find themselves in combat; indeed, many career soldiers never encounter it. Nevertheless, combat is *normal* for soldiers, in the sense that it is the reason for having them—if there were no need to fight wars, there would be no need for soldiers—so their training is designed to make them internalize rules regarding conduct in battle. Another relevant example would be police officers: they do not *commonly* get involved in high-speed chases, but conducting them is every bit as much a part of their normal duties as is soldiers' engaging in battle. Part of the goal is to produce unreflective dispositions to behave in certain ways in such cases, and this is what it means to have "internalized" a set of ILS rules in the proper way, as described in §3.3(D).

As anyone who has reviewed various codes of professional ethics will realize, there is a lot of overlap among them. From a Harean perspective, it is obvious why this should be so: a wide range of professionals normally encounter the same kinds of ethically charged situations that members of the laity do not. For instance, a wide range of professionals engage in research and publication, and so many codes of professional ethics discuss things like plagiarism, coauthorship, and so on. Similarly, codes of closely related professions, for example those of doctors and nurses, will share provisions that are not found in unrelated professions.

On the other hand, some professions are such that their members normally encounter ethically charged situations of a unique kind, and this is reflected in the fact that their codes of ethics include rules that are found in few or no other professions' codes. For instance, police officers' codes might contain provisions about high-speed chases, and codes of military ethics should contain provisions about targeting non-combatants. Such provisions are found in no other professions' codes, and they reflect the fact that members of these professions will, in the normal course of their work, encounter situations that members of other professions never encounter in the normal course of their work. Including related provisions in their codes of ethics ensures that properly socialized members of these professions will have developed strong dispositions to behave in certain ways in these situations, even if they do not commonly encounter them.

§3.2(D). LAWS

Hare does not mention laws as a fourth "level" of ILS rules, but they clearly are. In Harean terms, laws are a category of ILS rules that compare and contrast with common morality and professional ethics in two ways. On one hand, laws are typically written down or otherwise explicitly codified like codes of professional ethics, whereas the rules of common morality may not be codified anywhere (and, for reasons given in §3.4, they may be strictly uncodifiable). On the other hand, laws are more comprehensive in scope than professional ethics, since the latter are binding only on members of a given profession.

Laws should also be less broad in scope than common morality, however. For as Mill famously emphasized, good utilitarian reasons can be given for attaching legal penalties only to certain offenses against common morality. In *On Liberty* ([1859] 1956) he defended, on utilitarian grounds, the recognition of a private sphere that is immune to legal regulation. There he adopted the now standard liberal definition of "the private sphere" as including all actions that threaten no harm to others. From a utilitarian perspective, there is a prima facie reason for regulating behavior even in this private sphere, since "aggregate happiness" includes the agent's own happiness. For this reason, utilitarianism is sometimes said to be a busybody doctrine, one that would justify paternalistic interventions, "forcing people to be happier" than they might make themselves if left to their own devices. Mill argued, however, that allowing individuals to live their lives by their own lights (at least as long as their choices don't threaten harm to others) actually pays off better in the long haul. His main reason was that while doing so allows individuals to make mistakes, it also allows them to make lifestyle experiments that, when successful, will be adopted generally.

In chapter 5 of *Utilitarianism*, Mill offers the following related account of "justice." He characterizes "rights" as correlated with certain duties and with "valid claims" on society for protection from infringement of those duties ([1861] 1957, pp. 61, 66). The duties in question are defined by rules, the enforcement of which is particularly important to maintaining social order (pp. 73ff). Mill argues that people should be "compelled" by legal sanctions to conform to these duties, rather than leaving their enforcement to the informal mechanisms of persuasion and exhortation (p. 61). "Justice is a name for certain classes of moral rules, which concern the essentials of human well-being more nearly, and are therefore of more absolute obligation, than any other rules for the guidance of life; and the notion which we have found to be of the essence of the idea of justice—that of a right residing in an individual—implies and testifies to this more binding obligation" (p. 73). As long as the duties and rules in question all involve preventing harm to others, this account is consistent with his defense of the private sphere in *On Liberty*.

Similarly, Peter Singer holds that "It is a mistake to assume that the law should always enforce morality" (Singer 1993, p. 144); and Sidgwick says: "[N]ormally in a well-organised [*sic*] society the most important and indispensable rules of social behaviour will be legally enforced and the less important left to be maintained by Positive Morality. Law will constitute, as it were, the skeleton of social order, clothed upon by the flesh and blood of Morality" (1907, pp. 458–459). Holding that while the enforcement of some rules should be left to informal suasion, others should be enforced by a police/legal system underscores a difference that I have already noted between common morality and law: laws are formally codified in a way that common morality is not. Inasmuch as the police/legal system has a monopoly on sanctioned violence, it is important that the rules it is enforcing be fully spelled out in a public, written code.

§3.3. Some General Characteristics of ILS Rules

In this section, I discuss several general characteristics of ILS rules. Combined with the fact that there are, as described in the preceding section, at least four different types of ILS rules, these characteristics will significantly complicate (but also *enrich*) the discussion when we turn to evaluating ILS rules governing our treatment of animals in part III of this book.

§3.3(A). A Benign Form of Relativism

Hare's theory allows for what looks like a kind of relativism, as illustrated by a famous anthropological case study. At the time of their first contact with Europeans, the common morality of Arctic peoples permitted infanticide, and it was practiced far more often on newborn girls than on newborn boys. Knud Rasmussen, who studied Canadian Inuits just after the turn of the twentieth century, reported that in one village 18 marriages had yielded a total of 96 children, 38 of whom were females killed at birth. He does not say how many of the 96 children were female, but assuming that 50 percent were, that would mean that 38 out of 48, or 79.2 percent, of the female babies had been killed. Similarly, he reports that one woman from another village told him that of 20 children, she had 15 girls, 10 of whom (66 percent) were killed as infants (Rasmussen 1927, p. 226). Another anthropologist, E. Adamson Hoebel, cites several reasons that the Inuits needed to exercise sex selection. First, males did the hunting that provided a large majority of the community's food. Second, the death rate was high among hunters. Third, married couples settled near the husband's family, making it less likely that a girl child would support her parents in old age. And fourth, Inuit women nursed their children for a large number of years, which further increased the amount of effort "wasted" in raising a girl (Hoebel 1954, pp. 74–76).

A Harean archangel could cite the same reasons for crafting the ILS rules of pre-modern Inuit society differently from the ILS rules of modern Inuit society, that can, thanks to various technological advances, achieve family planning by various means that don't involve causing suffering to newborns. This is, however, a benign form of relativism. Full-blown cultural ethical relativism is the view that what justifies the members of society in believing an ethical rule and acting accordingly is the bare fact that it is accepted in their society. In Hare's theory, however, members of a society are justified in believing and acting according to the rules of their common morality to the extent that those rules approximate the ones that clear critical thinking would lead them to choose, and as the society's circumstances change, reformers can use critical thinking to urge changes in the common morality.

This complicates discussion of our treatment of animals in the following ways. Just as technological advances may have made it inappropriate for Inuits to continue practicing infanticide, technological advances may make it inappropriate for people in affluent, developed nations to practice forms of animal agriculture that would have been appropriate in earlier times. In chapter 10, changes in slaughter and husbandry practices will be discussed in detail to illustrate how this may be so.

§3.3(B). A DEGREE OF CONSERVATIVE BIAS OR INERTIA

There is also a degree of conservative bias or inertia in utilitarian thinking about ethics, because various costs must be taken into account when considering reforms to common morality, laws, or professional ethics. These would include, for instance, the economic losses that slave owners suffer during abolition, as well as the feelings of offense that racists suffer in the process and afterward.

Reformers find this feature of the theory repulsive. Convinced that the practices and attitudes in question are wrong, they argue that the feelings of slaveholders and racists should no more be counted than the pleasures of rapists. Tom Regan has pressed this point about rape in various places. Of course, in any realistic scenario the suffering of the victim, which can last a lifetime, outweighs the momentary pleasures of the rapist, and the costs to slaves outweigh the benefits to slaveholders. Still, many find it scandalous that the pleasures of rapists or the costs to slaveholders must be counted at all. In response to this kind of concern, Hare wrote:

> It has been thought to be a defect of utilitarianism that it is in this way at the mercy of the facts. Likewise, it shows the lack of contact with reality of a system based on moral intuitions without critical thought, that it can go on churning out the same defences [sic] of liberty and democracy *whatever* assumptions are made about the state of the world or the preferences of its inhabitants. This should be remembered whenever some critic of utilitarianism, or of my own view, produces some bizarre example in which the doctrine he is attacking could condone slavery or condemn democracy. What we should be trying to find are moral principles which are acceptable for general use in the world as it actually is. (*MT*, pp. 167–168)

And in the real world, people don't tend to prosper under tyranny and slavery, and the pleasures of rapists don't tend to outweigh the suffering of their victims.

Although Hare does not mention this, in his two-level version of utilitarianism we can make the following distinction. While the pleasures of the rapist must be counted when we are doing critical thinking, they might be ruled out entirely when we are doing intuitive-level thinking. It is a safe assumption that, in any realistic scenario, the pleasures of the rapist are outweighed by the suffering of the victim, and this means that we should only engage in critical thinking about rape when thinking about how to formulate ILS rules regarding sexual relations. The same assumption also ensures that rape (at least clear cases of it) will always be prohibited by good ILS rules.

That helps to blunt the reformers' criticism of utilitarianism, but it is still true that, when considering changes in ILS rules—including common morality, laws, and professional ethics—the costs to those who benefit from the status quo must be considered, along with the benefits to those whose interests would be better protected by the contemplated reforms. This does give Hare's theory a conservative bias or inertia, especially in the short term, and in the specific case of animal agriculture, this explains why proposals for "humane sustainable agriculture" must take into account not only technological limitations but also economic ones.

§3.3(C). A POTENTIALLY REVOLUTIONARY BENT

At the same time, utilitarianism can call for radical change, at least in the long haul. Famously, Mill argued on utilitarian grounds against slavery and in favor of extending voting rights to women, improved and universal public education, and universal medical coverage and a kind of welfare system. At the time Mill wrote, race-based slavery seemed natural, and women's suffrage unthinkable to many people. But across time, educational efforts (including Mill's writings) changed background cultural attitudes in ways that reduced the costs of changes in common morality and laws.

Also, note that Hare held that terrorist tactics could be justified under certain circumstances. Certainly it would be good to internalize an ILS rule against harming innocents, but like all ILS rules, if you "have a pretty cast-iron reason, based on firm knowledge that the case is an unusual one, for breaking them," then critical thinking will lead you to do so. Hare concluded that such cases do arise in practice: "I have no doubt that *some* of the acts of members of the Resistance against the Germans could be so justified" (*EPM*, chapter 4, "Terrorism," pp. 41–42).[1]

The definition of terrorist tactics and their role in the animal rights movement will be taken up in the sequel (Varner forthcoming) when discussing scientific research on animals, but chapter 10 of this book provides a brief illustration of the role of utopian writings in visions of humane sustainable agriculture. There I will describe how, like Mill's writings on women, utopian writings like Ernest Callenbach's *Bring Back the Buffalo: A Sustainable Future for America's Great Plains* (1996) and various authors' descriptions of "demi-vegetarianism" or "critical omnivory" can help to shift the goal posts of common morality, making feasible in future generations reforms that remain utopian today. Certainly this is true of Singer's *Animal Liberation*, as described at the beginning of §10.5.

§3.3(D). THE SIMPLICITY AND DISPOSITIONAL NATURE OF ILS RULES

Hare emphasized that, for purposes of moral education, ILS rules must be kept fairly simple, but he also said that they are not so much taught as "inculcated" (*MT*, p. 197) or "absorbed" (*EB*, p. 18), and that the rules one has internalized are often more complex than one can express in words (see §3.4 below). This led him to speak interchangeably of "dispositions" and "motivations" on the

[1] Relatedly, the United States' use of nuclear weapons in World War II is commonly defended in utilitarian terms. From a Harean perspective, the choice that American war planners faced can be understood as a conflict between a good ILS rule against knowingly killing non-combatants, and what critical thinking indicated about how long the war would have gone on and how many more would have died had the United States invaded mainland Japan instead of using nuclear weapons.

one hand, and "habits of thought," "moral intuitions," and "prima facie principles" on the other (*MT*, p. 47; *EB*, p.18).

To see how ILS rules would need to be kept simple in the beginning but would, when properly internalized, constitute dispositions and would be more complex than one could articulate, consider the way that rules function in a very different form of education. A good rule for beginning mountain bikers is:

> MBR-1: Don't use your front brake on unstable road surfaces.

The front brake is much more effective at braking than the rear, so it is important to learn how to rely heavily on the front brake in emergency braking conditions on good surfaces. On unstable surfaces, however, it is easy to cause a wheel skid, and it is much harder to recover from a front wheel skid than from a rear wheel skid. So MBR-1 is a good rule for mountain bikers to learn, but this might seem to be a better version of it:

> MBR-2: On unstable road surfaces (like loose dirt, gravel, or snow), don't rely primarily or heavily on your front brake.

This version is better insofar as it doesn't overgeneralize the way the first one does; experienced mountain bikers will certainly use their front brakes on some unstable surfaces, but they will rely primarily on their rear brake. Putting the relevant condition at the beginning suggests something about this rule's place in an organized, comprehensive statement of rules, and perhaps it's helpful to give some examples of the unstable road surfaces in question. Then again, perhaps the rule would be better stated as:

> MBR-3: On unstable road surfaces (like loose dirt, gravel, or snow), *and especially when going downhill*, don't rely primarily or heavily on your front brake.

The "especially" clause is important because wheels skid more easily when going downhill. Obviously, the rule could be formulated in other ways, and there are good reasons for adding many other qualifications.

This illustrates Hare's points that sometimes—indeed often or typically—there is no canonical way of stating the rule in question, but it also reveals how, when it comes to education, one generally uses oversimplified versions. When introducing a young child to mountain biking, for instance, one might begin by teaching the overgeneralized rule MBR-1 and switch to more complicated rules like MBR-2 or MBR-3 as the child becomes a more experienced rider, acknowledging all the while that no articulated version of the rule could ever cover all of the kinds of situations that a very experienced mountain biker would have learned to deal with. Hare claims that the rules of common morality are like this. With small children, we begin with overgeneralized rules like "Don't lie," and then, as the child becomes a more experienced moral thinker, we try more complicated versions like "Don't lie except in order to prevent needlessly upsetting people," or "Don't lie except in special circumstances, for example when lying is necessary to avoid serious harm coming to people, etc., etc. . . ." The "etc., etc. . . ." points to the fact that no articulated version of the rule could ever capture all and only the exceptions to the first (clearly overgeneralized) rule that would be allowed by the common morality of a modern, Western society.

Mountain biking can also be used to illustrate what it means to "internalize" such a rule. Regardless of how the rule has been articulated during training, the mountain biker who avoids a crash by reflexively releasing the front brake in a scary situation on an unstable surface is the one who has properly internalized the rule. The mountain biker who crashes because he reacted to the same situation by squeezing harder on both brakes has not properly internalized the rule, even though he stands up immediately afterward and says, "I know, I know: 'On unstable road surfaces, don't rely heavily on your front brake'!" According to Hare, intuitive-level moral rules are like this. Consider, for instance, the rules that soldiers learn about conducting combat operations around non-combatants. The rules are spelled out in some particular form or other in manuals, but the soldiers have not learned the rules in the relevant way unless they can do more than recite the versions in their books; the goal is to get them to internalize the rules as ingrained dispositions to behave (and to judge both themselves and others) accordingly. (As noted in §4.4, in this way Hare's theory incorporates virtue theory.)

These features of ILS rules introduce several complications when it comes to understanding and assessing attitudes toward animals. I once knew a vegetarian who said that she followed the rule: "Don't eat anything that tries to get away from you when you go to kill it." Insofar as her goal was to avoid causing suffering, this may be an over-generalized rule, but it would be a perfectly good rule for vegetarian parents to teach their young children. For it is easier to teach a young child to avoid eating things that move on their own than it is to teach them about arguments by analogy and what this implies about sentience in various kinds of animals (as discussed in chapter 5). And it might still be a perfectly good rule for an adult who remains uncertain about the scope of sentience in the animal kingdom.

A related complication is that from a Harean perspective, it makes perfect sense to internalize an ILS rule that will lead one to have intuitive reactions to cases in which clear critical thinking would lead one to different conclusions. So survey results, insofar as they report the target population's intuitive-level thinking, may misrepresent the subjects' underlying values, and that can go for both animal rights activists and their opponents in fields targeted by the animal rights movement. Indeed, individuals on either side may not be able to articulate any principle that accurately captures the ILS rules they have internalized, as is further emphasized in §3.4. (In this subsection I have been talking about the ILS rules of common morality, but the same things may be true of certain rules of personal morality.)

§3.3(E). COMPARISON TO "INTUITION" AND "ANALYSIS" IN THE EXPERT JUDGMENT LITERATURE

As summarized in Hammond and colleagues (1987), the literature on expert judgment recognizes a distinction between "intuition" and "analysis" that parallels, in some but not all ways, Hare's distinction between intuitive and critical moral thinking (see also the discussion of "intuitive judgment" and "deliberate reasoning" in §4.5).

One interesting parallel to Hare's distinction is the description of "confidence" in the two types of judgment. When relying on "intuition," experts tend to have high confidence in the answer, but low confidence in the method. When using analysis, the situation is reversed: the expert has high confidence in the method but low confidence in the answer. This corresponds to Hare's theory of moral judgment insofar as critical thinking, if done correctly, will always give the correct answer (high confidence in the method), although the answer will sometimes be counterintuitive (low confidence in the answer). There is also little "cognitive control" or "conscious awareness" in intuitive thinking insofar as one gives a habitual reaction rather than thinking about the case, and for the same reason, intuitive thinking is fast. Critical-level moral thinking can be fast too, however, if the situation is clear-cut (if it is obvious what would maximize utility under the circumstances).

Another parallel regards errors. When using intuition, experts make more errors, but the errors are distributed "normally," meaning that more serious errors are less likely than minor ones. When using analysis, on the other hand, experts make fewer errors, but there are a significant number of doozies. The parallel to Hare's account of moral thinking would be that while overreliance on ILS rules will produce many errors, they will tend to be relatively minor, whereas overreliance on critical thinking tends to produce serious misjudgments. For instance, a man in a failing marriage who continues to obey an ILS prohibition on adultery may fail to maximize aggregate happiness, but not as seriously as a man in a happy marriage who reasons that a clandestine tryst will maximize aggregate happiness and ends up ruining his marriage.

An important difference between Hare's intuitive/critical moral thinking distinction and the expert judgment literature's intuition/analysis distinction concerns the "organizing principles" at the two levels. According to Hare, in moral thinking the "organizing principle" is always the same, whether one is engaged in intuitive or in critical thinking. Ultimately, the principle of utility is always doing the work. The organizing principles that experts use when they do analysis are "task specific," however. For instance, there are various different algorithms or formulas to use in estimating highway traffic capacity, calculating how much weight a bridge can carry, or how much thrust a rocket engine needs to produce. Relatedly, Harean intuitive moral thinking employs lots of different organizing principles insofar as we internalize many different ILS rules.

Another important difference concerns how the presentation of a problem influences which kind of thinking will be used. Hammond and colleagues (1987) show that "if a) the task presents *many redundant* cues (attributes), b) the cue values are *continuous*, c) the cues are displayed *simultaneously*, d) the cues are measured *perceptually*, and e) the subject has available *no explicit* principle, scientific theory, or method for organizing cues into a judgment, then the subject will employ intuitive cognition" (pp. 755–756, italics in original). Experts tend to use analysis when, by contrast, they are presented with numerical values to plug into various variables in familiar algorithms. Perhaps sometimes critical moral thinking is triggered because we have relevant numeric values to plug into a utilitarian calculus. For instance, during legislative debate over a state's annual budget, having available figures on how many deaths could be prevented by installing traffic lights at intersections along country roads might inspire legislators to use explicitly utilitarian reasoning rather than relying on their intuitive judgments. Usually, however, we don't have any (meaningful) figures available, and human moral thinkers almost always have incomplete knowledge of relevant facts. Nevertheless, we need to engage in critical thinking when ILS rules conflict, in novel situations, and when considering altering the intuitive-level rules we teach ourselves and others. So when it comes to *moral* expert judgment, whether we engage in intuitive or critical thinking isn't tied to how the details of the problem are presented.

§3.4. Connectionism and ILS Rules

As noted earlier, Hare speaks interchangeably of "principles" or "rules" on the one hand, and "dispositions," "motivations," or "habits of thought" on the other, and he speaks of them all being "absorbed" or "inculcated" as much as "taught." He also states several times that one may have internalized ILS rules that one cannot formulate in words. For instance, he says:

> If principles reach more than a certain degree of complexity, it will be impossible to formulate them verbally in sentences of manageable length; but it might still be possible to learn them—i.e., to come to know them in some more Rylean sense which does not involve being able to recite them. Assuredly there are many things we know without being able to say in words what we know. All the same, there is a degree of complexity, higher than this, beyond which we are unable to learn principles even in this other sense which does not require that we be able to recite them. So principles which

> are to be learnt for use on subsequent occasions have to be of a certain degree of simplicity, although the degree has been exaggerated by some people. (*MT*, p. 35, reference omitted)

And elsewhere he says that "In a morally developed person the exceptions [that one assumes are built into the ILS rules that one has internalized] may get too complex to be formulated in words" (1991, p. 457).

In some of the recent literature on moral reasoning, I have found a discussion that sheds considerable light on Hare's tendency to conflate motivations, intuitions, and knowledge of rules or prima facie principles, and on his claims that "there are many things we know without being able to say in words what we know," and that this can include a list of exceptions to a moral rule that is too complex to be formulated in words. In this section I will argue that while a classical, deductivist model of moral reasoning captures the kind of thinking we are doing when we explicitly apply the principle of utility, what Hare calls "intuitive-level thinking" better fits a new, connectionist model of moral decision making. On this model, the individual may or may not be able to articulate a rule or principle that he is following, and learning these rules proceeds by examining paradigm cases rather than being taught a rule. This makes it more appropriate to speak indifferently of "dispositions, motivations, intuitions, [and] prima facie principles" as Hare does in the above passages.

§3.4(A). CLASSICAL AND CONNECTIONIST MODELS OF MORAL REASONING

According to what is called "the classical model," moral reasoning proceeds deductively, and the reasoning is either conscious or can be consciously formulated after the fact: the agent knows a rule of the form "In all situations with characteristics *X*, *Y*, and *Z*, the right thing to do is *A*," recognizes that "Situation *S* has characteristics *X*, *Y*, and *Z*," and concludes that "*A* is the right thing to do in situation *S*." On the classical model, drawing moral conclusions is in this way analogous to applying a definition or criterion that can be stated in terms of necessary and sufficient conditions.

Unfortunately, the classical account fails to accurately model the actual performance of people. With regard to definitions, for example, research shows that people do not work with necessary and sufficient conditions, but rather with "prototypical cases." The prototypes in question are not concrete examples, but rather statistically defined "central tendencies":

> Such [a] central tendency is calculated by treating each concrete example as a set of co-occurring features and generating (as the prototype) a kind of artificial exemplar that combines the statistically most salient features. . . . Novel cases are then judged to fall under a concept (such as "pet" or "crime")

> according to the distance separating the set of features they exhibit from the prototypical feature complex . . . (Clark 1998, p. 111)

To the extent that moral reasoning fits the classical model, relying on explicitly stated principles and deductive reasoning about their application, traditional computer programming might provide a good model for moral reasoning. However, to the extent that moral reasoning is more like the reasoning about prototypical cases that substitute for explicit definitions or criteria, another model is called for.

§3.4(B). CONNECTIONIST SYSTEMS

Artificial intelligence experts generally believe that connectionist systems, which do not rely on pre-programmed principles, provide a better model of such reasoning. Figures 3.1, 3.2, and 3.3 can be used to illustrate why connectionist systems have two properties relevant to the present discussion of moral reasoning. First, they are not pre-programmed with principles for making the distinctions they do; rather, they are trained using lots of examples. Second, even after training, when they make the same distinctions that human beings do, there may be no principle that one could articulate that would provide necessary and sufficient conditions for making the distinctions; rather, they have embedded a statistically defined "central tendency" that provides an abstract prototype.

Figure 3.1 illustrates the structure of a very simple system that could be trained up to distinguish "2" from "6" as projected on a five by nine pixel grid as illustrated in figures 3.2 and 3.3. Although only three inputs are shown in figure 3.1, the system actually has 45 input nodes, one for each pixel on the five by nine presentation grid. It does have just two output nodes, however: one corresponding to "2" and one to "6." It has two layers of so-called "hidden nodes." Each of the nodes in the first hidden layer is connected to each input node and to each node in the second hidden layer. The nodes in the second hidden layer are in turn connected to each of the two output nodes. There doesn't need to be any specific number of nodes in the hidden layers, although having more than a small number of nodes generally allows the system to make finer discriminations.

Every line in the drawing represents a flow of excitation from nodes on the left to nodes on the right. At the input level in this example, the excitation is either on or off, based on whether or not the corresponding pixel in the five by nine grid is lit. Which output is given depends on which output node ("2" or "6") receives more excitation from the nodes in the second hidden layer. The excitation level of the hidden nodes is initially set at random levels, and the amount of a node's excitation that is passed on to the second hidden layer and on to the output nodes (the strength of the connections between the nodes) is also initially randomly determined.

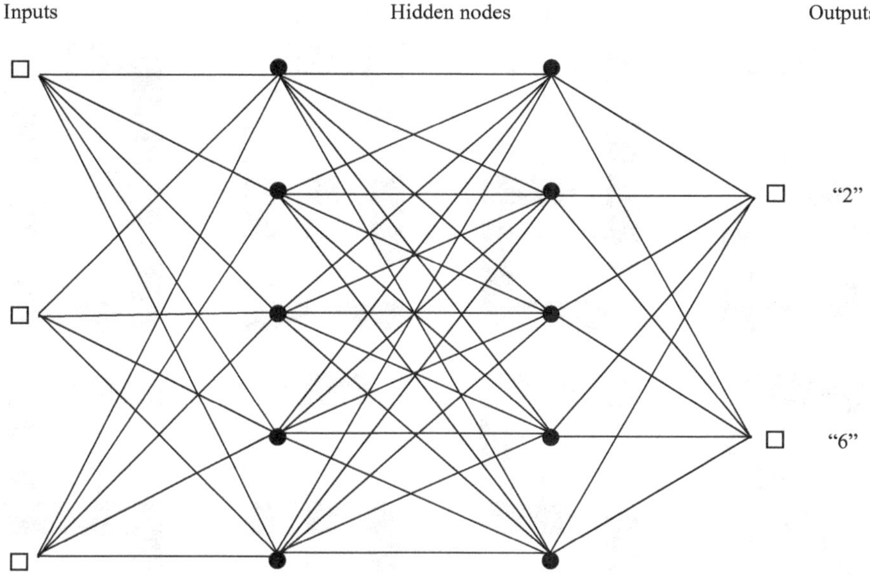

FIGURE 3.1. *A Simple Connectionist System*

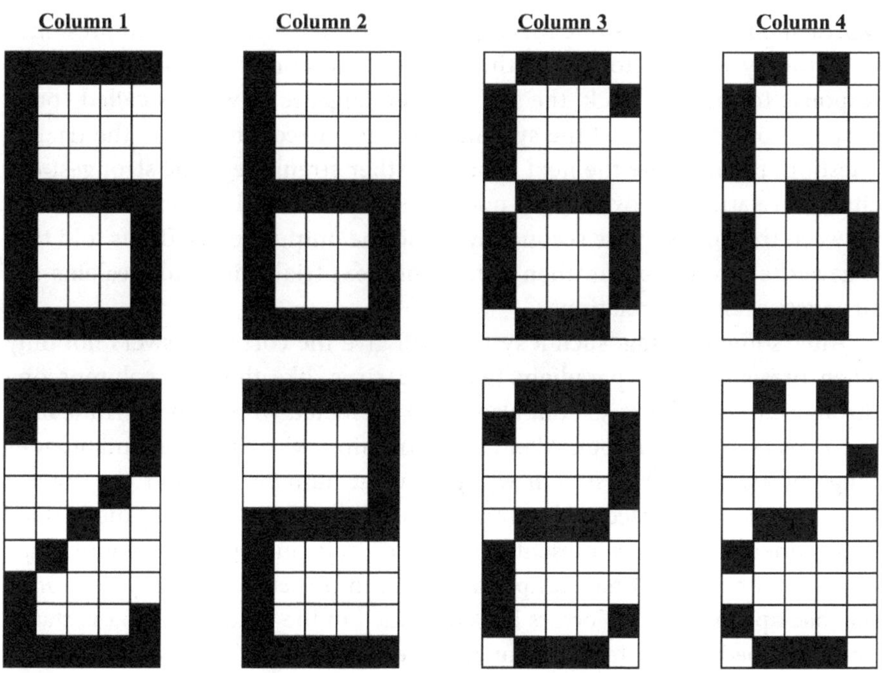

FIGURE 3.2. *Distinguishing "2"s and "6"s Projected in a Five by Nine Pixel Grid*

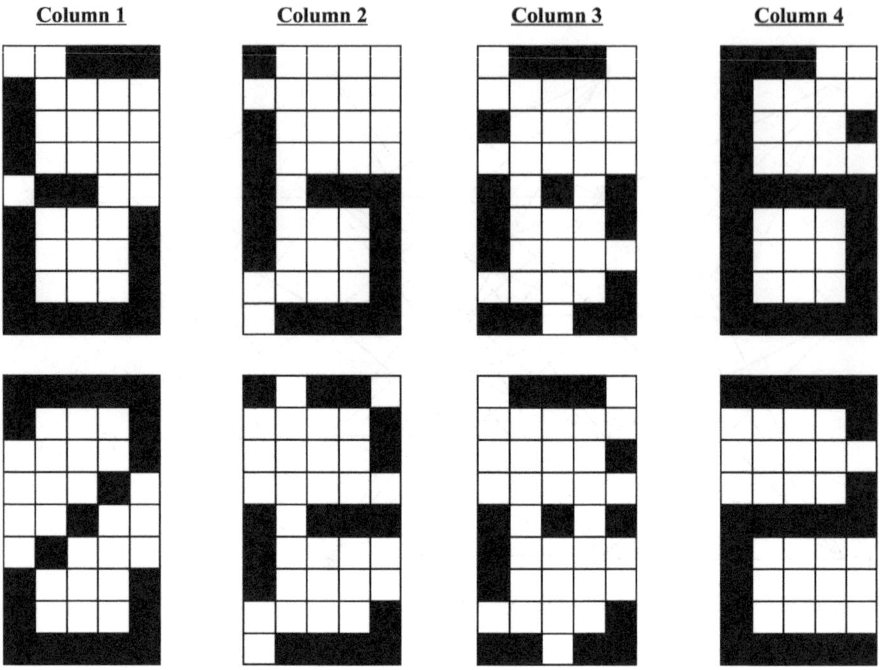

FIGURE 3.3. *Distinguishing "2"s and "6"s Projected in a Five by Nine Pixel Grid*

The system is then trained by exposing it to various inputs of the types represented in columns one through three of figure 3.2. After each presentation, the system is "told" whether it gave the correct output or not. In response to that feedback, the system then engages in what is called "back propagation of errors": if the system gave the correct answer on the trial in question, then, before the next trial, it further strengthens the strongest excitation relations between the nodes in the first and second layers and between the nodes in the second layer and the output nodes. If it is told that it gave the wrong output, then before the next trial it instead weakens the strongest excitation relations.

After some training, such a system will give the correct answers not only when presented with paradigm twos and sixes, like those in columns one through three of figure 3.2, but also when presented with such aberrant sixes and twos as in column four. After a lot of training, with a broader and broader range of examples, the system will give what humans consider the correct answers when presented with a more or less broad range of even more aberrant numerals like those presented in figure 3.3, including an indefinite range of cases with which it has not previously been presented. The explanation is that back propagation of errors has embedded in the hidden nodes the statistically defined central tendency of all the examples of "2" and "6" to which it has been exposed, thus creating artificial prototypes of each.

Connectionist systems with more complex inputs and outputs and more layers of hidden nodes can be trained to do things like decipher the handwriting on your tax returns, distinguish dogs from cats, and recognize individual human faces. Those are all tasks that would be difficult or impossible to accomplish with conventional programming. To illustrate why, imagine how complicated the instructions would have to be to program a computer to distinguish all of the twos and sixes in figures 3.2 and 3.3. In all, there are over 35 *trillion* different possible input states in a 45-pixel grid. If the system is only being presented with what a human would judge to be either a two or a six, only a subset of those input states would be relevant, but those would still be in the tens or hundreds of thousands at least. Enumerating them would still be impractical, and there is no simple algorithm that will work well. For instance, in all of figure 3.2, and in columns one through three of figure 3.3, the last pixel in the third line is on for each of the twos and off for each of the sixes. But the instruction "If the last pixel in the third line is on, output two, otherwise output six" would yield the wrong answer with the figures in the fourth column of figure 3.3. Even much more complex algorithms will yield the wrong answers on some range of cases in which the trained connectionist system would give the right answer.

For such reasons, people working in artificial intelligence think that connectionist models are more promising than conventional programming for various perceptual tasks, and as I will emphasize in the next subsection, one role of ILS rules may be to shape one's perception of situations, training one to see (at least in the sense of notice) various aspects of them. Insofar as the many interconnections among nodes in a connectionist system are analogous to the many interconnections among neurons in a nervous system, connectionist systems are also a more plausible physical model of how brains work than traditional computers with their central processing units that execute one programmed command at a time (although at present, connectionist systems are still almost always run on traditional computers that model what an actual network of nodes could be made to do). Back propagation of errors also has a known analogue in nervous systems, insofar as "neurons that fire together wire together" and "neurons that fire out of sync lose their link." That is, the degree to which excitation of one neuron is passed on to the neurons to which it is connected increases when those neurons are used together while solving problems, and it decreases when those neurons are not used together while solving problems.

§3.4(C). CONNECTIONISM AND THE ROLE OF RULES IN INTUITIVE-LEVEL THINKING

This simple example illustrates the two features of connectionist systems mentioned earlier. First, connectionist models do not work by being programmed with principles that are then applied in deductive fashion; rather, they are trained using lots of examples. "Literature," Andy Clark observes, "by depicting complex moral situations, may be seen as another kind of concrete case—virtual moral reality, if you will" (p. 114). The same can be said for other kinds of storytelling, in films and television, for instance, but also for rituals and myths, which anthropologist Lyle Steadman (1993 and 1994) characterizes as "transmitting behavior vicariously."

Second, even after a connectionist system is trained and makes the same discriminations that a human being will make, there is no such principle programmed into the system, and neither the system nor the computer scientist who created and trained it can state any such principle upon which it is relying; at least, there is no principle that exhaustively specifies, in terms of necessary and sufficient conditions, the concepts it is applying. Insofar as moral reasoning fits this connectionist model, what we know cannot be adequately captured in language; or, as Hare put it in the passage quoted at the beginning of this section, there are "things we know without being able to say in words what we know." Clark describes several functions that explicitly stated principles would still have, however.

(i) Education

One is early moral education, where simple rules are used to sensitize young children to morally salient features of situations. As soon as a child can understand what it means to tell a lie, he can pick out cases of lying, and the earliest phase of moral education would consist of learning to draw simple inferences of the kind that the classical model of moral reasoning describes. This role in early moral education is recognized by other authors who agree with Clark that what the moral expert knows cannot be stated in principles (Clark mentions Flanagan 1991, Johnson 1993, and Dreyfus and Dreyfus 1998). But these other authors write as if principles are to be discarded as we become moral experts. Clark thinks that this is a mistake, however, and he describes two ways in which rules and principles still function in the lives of mature moral experts.

(ii) "Context Fixing"

One way is by "context fixing," or drawing persons' attention to features of a situation that will trigger the application of one prototype or another. "Thus, talk of 'unborn children' may bias prototype-activation one way, while talk of 'unwanted pregnancy' may bias it another. Moral rules and principles, on this account, are nothing more than one possible kind of context-fixing input, among many. Others could include well-chosen images or non-rule-invoking discourse" (p. 118). Clark describes work by Kirsh and Maglio (1992) on the role that explicitly stated rules seem to have in expert Tetris play and their related speculation about how this context fixing could be designed into a connectionist system. Tetris is a computer game in which the object is to rotate falling geometrical objects so that they form complete lines as they pile up at the bottom of the screen. Completed lines disappear, allowing the game to continue, but incomplete lines pile up until the screen is filled and the game ends. Kirsh and Maglio found that expert Tetris players attend, consciously, to certain rules as they play, for example, "don't cluster in the center, but try to keep the contour flat" (p. 119). Kirsh and Maglio speculate that the rules function by somehow altering the salience of the inputs to a connectionist system, which in effect changes the player's focus of attention. Clark writes: "Just how the shift of attention is accomplished is left uncomfortably vague. But they speculate that it could work by 'biasing certain board regions' or by 'increasing the precision of [certain] values being returned by visual routines'" (p. 119, quoting Kirsh and Maglio 1992, p. 10). Although Kirsh and Maglio talk as if this shift in the salience of inputs results from conscious attention to rules, research on chess players suggests that rules can also perform this function unconsciously.

In a fascinating 1985 paper, Eugene Hargrove compares ethical decision making to decision making in chess and argues that rules can function unconsciously by altering one's perceptions of problematic situations. Hargrove describes a study by Adriaan de Groot ([de Groote] 1965) in which players of various skill levels were asked to "think out loud" when presented with a specific board situation. Hargrove reports that players only consciously applied a rule when they studied the situation and could "detect no favorable or unfavorable consequences resulting from any possible move available." Only then did "they apply a rule that they learned as a beginner for want of anything better to do" (p. 14). Nevertheless, Hargrove claims, the rules were still functioning by unconsciously shaping how they *perceived* the problem presented by the board situation.

> One of the situations involved the possibility of exchanging a bishop for a knight. Since it is generally believed that the bishop becomes more valuable as the game progresses while the knight becomes less valuable (the change resulting from the increased maneuverability of the bishop as the board assumes a more open character), beginning chess players are taught in their first set of rules that it is good to exchange a knight for a bishop in most cases, but not a bishop for a knight. In de Groote's experiment the exchange of a bishop for a knight—thereby breaking the beginner's rule—led to the winning of a piece; the exchange of a knight for a bishop did not, on the other hand, improve the player's circumstances in any way. The weaker players were affected by this rule to the extent that they did not "see" the exchange and missed the combination. The stronger players in contrast were not affected at all and immediately began considering the exchange. None of the weaker players mentioned the rule in their mumblings—which de Groote calls protocol statements or protocols—and were not therefore consciously thinking about the rule, but all of them were unconsciously being guided, or more correctly, *limited* by it. (p. 14)

That chess players do not consciously apply rules is surprising, because there is, Hargrove reports, universal agreement that the way to learn to play chess is to study rules of increasing complexity. Hargrove argues that it still makes sense to learn rules and apply them, however, as a way of shaping perception:

> The players who failed to find the correct continuation were guided or limited by a rule, although they were not specifically thinking about it while under the rule's influence. In this case, the players were not trying to apply rules, but rather were simply trying to perceive the board situation correctly. The rule entered the decision process as an element or factor in the players' ability to perceive specific states of affairs . . . If based on this evidence one is willing to make a leap of faith, then it could be concluded, as I wish to do, that the immense amount of activity in learning how to play chess involving the study of rules and their application is not a bizarre joke on the beginning chess player—even though rule application virtually never takes place—since that study translates into elements of the player's perception which contribute significantly to his ability to find correct moves in actual games. (pp. 18–19)

Hargrove suggests that rules might function similarly in ethical decision making, by improving persons' perceptions of problems, rather than giving them a set of rules that are applied in deductive fashion.[2]

In the context of the connectionist model of moral reasoning, Hargrove's suggestion could be rephrased as follows. Like Kirsh and Maglio, he is speculating that the rules function by somehow altering the salience of the inputs to the connectionist system, which in effect changes the player's focus of attention, but Hargrove is speculating that rules perform this function only. So one function of relatively simple, explicitly stated rules is to train people to perceive situations differently, to notice certain morally salient features of situations that they might have missed had they not "internalized" the rules, and the rules may perform this function either consciously or unconsciously.

Note that the foregoing discussion also gives us another reason for speaking of "internalizing" ILS rules rather than "learning" them. Earlier I spoke this way to emphasize that "learning" the rule involves developing a disposition to

[2] This is a valuable insight, but Hargrove's paper includes a fairly long critique of Hare's views (pp. 22–29) that misfires two ways. First, although Hargrove understands that Hare distinguishes between ILS rules and critical thinking, he describes Hare as recommending "that simple prima facie rules of the kind represented by non-constitutive rules in chess be replaced by rules which are both universalizable and prescriptive" (p. 24). Hargrove then argues that strictly universalizable rules, which would specify all categories of exceptions, are both unnecessary and undesirable in moral thinking. But, as has already been detailed in the text, Hare is quite explicit that this should not be our goal in formulating ILS rules. Relatedly, Hargrove characterizes Hare as holding that rules "should always be *consciously* involved in the thought process which constitutes decision making" (pp. 22–23, italics in original). But, while this may be true of critical thinking, this is not true of intuitive-level thinking, as has been made clear in the text.

act accordingly, without reflection, and diffidence about violating it, even when clear critical thinking says that you ought to. But now another reason is apparent. To say that a child has "learned" a simple ILS rule suggests that the child is able to articulate it and to employ it in classical, deductive moral reasoning. To say that someone has "internalized" an ILS rule suggests something much more subtle and comprehensive. The explicitly stated ILS rule (e.g., "Don't lie") does not carry the statistical information that is embedded in the abstract prototype that is activated when a trained moral expert engages in connectionist moral reasoning. To say that "Raj has internalized the ILS rule against lying," then, may mean that Raj will judge a broad range of cases according to an abstract prototype that cannot be given in any one example or expressed in any explicitly stated moral rule. It also means, as Hargrove's discussion emphasizes, that the rule will function unconsciously by focusing the moral expert's attention on morally salient features of the situations that he subsequently encounters. Along with developing relevant dispositions to act and to be diffident about violations, these are two more things that can be conveyed by saying that someone has "internalized" an ILS rule. None of these things is as clearly conveyed by saying that they have "learned" it.

(iii) "Collaborative Problem Solving"

Clark argues that explicitly stated rules have one other, distinct function in the thinking of mature moral experts. Given that "many moral problems basically consist in the need to find some practical way of accommodating multiple perspectives, . . . attempts by each party to articulate the basic principles and moral maxims that inform their perspective provide the only real hope of a negotiated solution . . . they provide the essential starting point of informed moral debate" (pp. 121–122). In this way, trying to articulate principles that will help others to understand one's moral concerns can contribute to what Clark calls "collaborative problem solving."

§3.4(D). SUMMARY

All of this helps us to understand Hare's statement that it may be possible to know intuitive-level principles without being able to "recite them," and this provides us with a plausible view of how intuitive-level moral thinking proceeds. It is important to notice, however, that when it comes to critical thinking, the classical model always applies. For in every case, what we know (or at least believe that we know) can be completely stated. The relevant principle is always the principle of utility, and the reasoning takes the classical form:

1. The right thing to do is maximize aggregate happiness.
2. In these circumstances doing *A* would maximize aggregate happiness.

C. Therefore, the right thing to do in these circumstances is *A*.

There is just one case in which it initially seems plausible to say that critical thinking is engaged but the classical model could not apply: when one encounters a novel situation that is not covered by the ILS rules that one has internalized and there is no time for reflection. If we stipulate that there is literally "no time for reflection," then we cannot possibly engage in critical thinking, but the above picture of how we internalize ILS rules does suggest that intuitive-level thinking could still function in such a situation. For even if the "novel situation" was not explicitly captured by any ILS rule that one had ever articulated, the situation would in fact have various features in common with some internalized paradigms, and this could trigger paradigm activation under the connectionist model.

To sum up: there are good reasons for believing that some or many of the ILS rules that adults use most of the time cannot be adequately stated in language. At the same time, however, explicitly stated ILS rules have several functions in both the moral education of children and in adult discussions of ethics. In the case of the child, we have to be able to state the rules, and state them in very simple form, in order to train the child to recognize various morally salient features of situations. ILS rules have a similar function in adult moral thinking, insofar as they serve to focus attention on morally salient features of situations, and they can also function in attempts at collaborative moral problem solving, by helping others to understand what your moral concerns are—what you are seeing as the morally salient aspects of a situation.

In discussions of animal ethics, the attempt to articulate the competing ethical principles of various interest groups is a necessary starting point for debate over how our shared common morality, law, and professional ethics should treat animals. Even such obviously oversimplified principles as "Animals are not ours to eat, wear, experiment on, or use for entertainment" and "Medical progress would be impossible without animal experimentation" can help participants in the debate understand what ingrained, habitual reactions each side brings with them at the beginning. However, adopting a Harean perspective on such debates allows us to simultaneously acknowledge that each side has internalized ILS rules that cause them to have strong intuitive reactions to the cases under discussion, while also recognizing that collective critical thinking must be used to modify various relevant ILS rules of common morality, law, and professional ethics.

{4}

Assessing Hare's Theory

§4.1. Assessing the Descriptive Adequacy of Hare's Theory

According to universal prescriptivism, "the moral words" have (at least when used in their genuinely or distinctively moral sense) three logical properties: they are universalizable, overriding, and (in Hare's special sense) prescriptive; and these three properties together force us to think like utilitarians, at least when doing what Hare calls "critical thinking." As emphasized in §1.6, this is offered both as an accurate description of what logical properties moral judgments have and as a theory of normative ethics.

In terms of evaluating universal prescriptivism as a theory of *descriptive* ethics, the most obvious objection is that people do not, in fact, think like utilitarians most of the time. Of course, Hare's theory predicts just that, since he holds that real-world human beings must (for the reasons given in §1.5 and chapter 3) use intuitive-level thinking most of the time, and ILS rules are internalized in a way that gives them a "deontological flavor" and makes their "application" unreflective and automatic. In this way, utilitarianism is "self-effacing." This mirrors an old argument about egoism: an individual who does not always think egoistically will benefit in diverse ways, so a consistent egoist has a good reason to train himself not to think egoistically all of the time (see Butler [1729] 2006, sermons I and XI). Similarly, if Hare's theory is true, then we should expect people not to think like utilitarians most of the time, even though the reasoning that shows why they should not is itself ultimately based on the principle of utility.

Only in the four kinds of cases described in §1.5 must we rely on explicitly utilitarian thinking, according to Hare:

1. In novel cases (which intuitive-level rules are not designed to handle, and on which they give little or no guidance),
2. When intuitive-level rules conflict,
3. To select and amend intuitive-level rules over time and in light of new information and experiences, and

4. When something is ruled out by the intuitive-level rules one has internalized but both (a) it is clear that the violation will maximize aggregate happiness, and (b) one can trust one's judgment that this is so.

But how often do people think like utilitarians in these kinds of cases?

In a range of cases that can plausibly be construed as fitting into the first, second, or fourth categories, people clearly do, at least sometimes, rely on explicitly utilitarian thinking. Consider, in addition to the captain in the wartime convoy who leaves the man overboard behind (§1.5), the pilots who were ordered to shoot down a hijacked airliner on 9/11/2001, and the crews that delivered the atomic bombs to Hiroshima and Nagasaki. Leaving their professional training to one side, and looking at such cases from the perspective of common morality alone, each of these seems like either a novel case (category 1), a case in which various ILS rules are in conflict (category 2), or a case in which the ILS rules that one has internalized would rule out the action in question, even though it seems clear that the action would maximize aggregate happiness (category 4). In each of these cases, it does seem likely that the agent would invoke utilitarian reasons for doing what they did. It also seems likely that they would experience compunction afterward, just as Hare's theory would predict (to the extent that violations of important ILS rules are involved).

With regard to the third kind of case, I don't know how many parents would report that utilitarian thinking guided their decisions about what kinds of rules to employ in the moral education of their children. Some social reformers report basing their suggested reforms on utilitarian considerations, but the most obvious examples—Bentham, Mill, and Singer—are themselves champions of utilitarianism in ethical theory, and many social reformers report basing their suggested reforms on purely deontological considerations, such as God's word or respect for rights.

So even within the limited range of cases in which Hare's theory predicts that they will employ explicitly utilitarian thinking, some people do, but many don't. Where does that leave Hare's theory, considered as a theory of descriptive ethics? In the first half of this chapter (§§4.2–4.6), I aim to defend—as far as I can—the descriptive adequacy of Hare's theory. That is a hard row to hoe, and I will not be able to establish its descriptive adequacy to any degree of certainty. The best that I can hope to do is to give reasons for thinking that Hare's theory is not as inconsistent with peoples' actual behavior as it at first appears to be. That is enough, however, to make Hare's theory worthy of careful study. For, as I will argue in the second half of this chapter (§§4.7–4.12), Hare's two-level version of the theory provides quite convincing replies to the standard "test cases for utilitarianism." Given that utilitarianism is as old as Western ethical theory, and given that (as noted in §1.7) it plays a very large role in our thinking about our treatment of animals, Hare's theory is therefore worthy of serious study—especially given that Peter Singer embraces Hare's theory, at least the broad strokes (§11.1).

§4.2. The Ubiquity of the Golden Rule

One reason for thinking that Hare's theory is not as inconsistent with peoples' actual moral thinking as at first appears is that the golden rule has been claimed to be a cultural universal (Shermer 2004, p. 25; and Hauser 2006, p. 357), and Hare claims that the golden rule, properly understood, provides a commonsense expression of what is required by the three logical properties of all genuine moral judgments.

Although most Americans think of the golden rule in Christian terms, the same precept is found in Hebrew scripture: "You shall not hate your brother in your heart, but you shall reason with your neighbor, lest you bear sin because of him. You shall not take vengeance or bear any grudge against the sons of your own people, but you shall love your neighbor as yourself: I am the Lord" (Leviticus 19:17–18). "The stranger who sojourns with you shall be to you as the native among you, and you shall love him as yourself; for you were strangers in the land of Egypt" (Leviticus 19:34). And as I noted in §1.1, in the Talmud story of the impatient convert, Rabbi Hillel characterizes the golden rule as capturing the entire message of the Torah.

In Christian scripture, both Matthew and Luke report that Jesus repeated that precept of Hebrew scripture during the Sermon on the Mount: "So whatever you wish that men would do to you, do so to them" (Matthew 7:12); "As you wish that men would do to you, do so to them" (Luke 6:31). And, like the Talmud, both report that Jesus characterized the golden rule as capturing the entirety of his teaching. When asked "which is the great commandment in the law?" Matthew reports that Jesus replied: "You shall love the Lord your God with all your heart, and with all your soul, and with all your mind. This is the great and first commandment. And a second is like it, You shall love your neighbor as yourself. On these two commandments depend all the law and the prophets" (Matthew 22:37). Similarly, Luke reports that when Jesus was asked what one must do to "inherit eternal life," he responds: "You shall love your God with all your heart, and with all your soul, and with all your strength, and with all your mind; and your neighbor as yourself . . . do this, and you will live" (Luke 10:25–28). When asked "And who is my neighbor?" Jesus responded with the story of the Good Samaritan, indicating that your "neighbor" includes any human you encounter (Luke 10:29–37).

Given the common origins of Judaism, Christianity, and Islam, it is not surprising that what Christians call the golden rule also appears in Muslim scriptures: "None of you [truly] believes until he wishes for his brother what he wishes for himself" (Hadith 71–72); "Seek for mankind that of which you are desirous for yourself. . . . The most righteous of men is the one who is glad that men should have what is pleasing to himself, and who dislikes for them what is for him disagreeable" (Sukhanan-i-Muhammad 21 and 306). But the precept is also claimed to be at the heart of Confucianism: "Tzu-kung asked, 'Is there a single word which can serve as the guiding principle for conduct throughout one's life?' Confucius said, 'It is the word "consideration" [*shu*]. Do not impose on others what you do not desire others to impose upon you" (Analects 15.23). And versions of it can be found in Buddhism: "Hurt not others in ways that you yourself would find hurtful" (Udana-Varga 5.18); Hinduism: "Let no man do to another that which would be repugnant to himself" (Mahabharata bk. 5, chap. 49, v. 57); Jainism: "treat all beings as [one]self would be treated" (Sutrakritanga 1.10.13); Zoroastrianism: "That nature only is good which shall not do unto another whatever is not good for its own self" (Dadistan-i-Dinik 94.5); and among the Bantu: "O man, what you do not like do not to your fellow,"[1] and "If you see a jackal in your neighbour's garden drive it out, one may get into yours one day and you would like the same done for you" (Claridge [1922] 1969, pp. 254, 255).

In a 1934 article on the golden rule, J. O. Hertzler claimed that it had the advantage of *not* being based on knowledge of "what his fellow's desires may be, which incidentally may be faulty, but his own desire transferred in imagination to him" (p. 429). Thus the golden rule "does not require any great individual intelligence. . . . It uses the clever psychological principle of starting with the individual actor, who, on the basis of his own concrete experience . . . has fairly definite and secure feelings, if not ideas, about what is good or not for *him*" (p. 428, emphasis added).

---

[1] The preceding quotations in this paragraph are all in Wattles 1996, pp. 4 and 191–192. Wattles includes a statement attributed to an Inca ruler in his list of examples of the golden rule in world religions (p. 192), but I think this example is suspect for the following reasons.

The statement appears in Garcilaso de la Vega's account of the first Inca ruler, Manco Capac, an account based on an Inca legend that his uncle told him as a teenager. The uncle described how the sun god sent a son and a daughter to teach the Incas "how to live, how to clothe and feed themselves like men, instead of like animals" (Vega [1609] 1961, p. 6). Vega describes Capac as "assemb[ling] his first subjects together" in order to teach them

> the natural and reasonable laws that were to govern their relationships to one another.
>
> In order that peace and concord might reign, and that neither anger nor passion should come between them, he succeeded in convincing them that each one should do unto others as he would have others do unto him, so true it is that the law must be the same for all. (p. 11)

But while Vega's mother was descended from Inca royalty and his first language was the Incas' Quechua, his father was a Spaniard and he was given a Spanish-style education. Also, in his forward to the 1961 English edition of Vega's *Royal Commentaries*, Alain Gheerbrant describes Vega as having a "fervent Catholic faith" (Gheerbrant 1961, p. xxvii), and as being convinced that the Spanish conquest of the Incas had "served humanity by extending belief in Christ's redemption to the Peruvian Indians" (p. xxviii). So Vega's choice of words in the above description of Manco Capac's teachings may have been influenced by his immersion in Christian culture.

As explained in §2.2, this is (according to Hare) an incorrect way of understanding the golden rule. As I said there, getting children to imagine what *they* would want if they were in the other person's position *with their own preferences intact* is the first step toward understanding the golden rule. Hertzler is correct that this "does not require any great individual intelligence," and this is part of what makes the golden rule a useful tool for helping children learn to see their actions from a moral point of view. But it is also easy to use the golden rule to improve on a child's moral thinking from a utilitarian point of view. For when another child's preferences are different, one can ask the child: "What if you didn't like rough and tumble play, the way Harold here doesn't like rough and tumble play, how would you then feel about this situation?" This application of the golden rule also does not require any great individual intelligence and it helps move the child in the direction of thinking like a utilitarian.

Although Hare himself never appealed to the ubiquity of the golden rule in support of his theory, if he was right that the golden rule, "properly understood," expresses the three logical requirements on critical-level moral thinking, then its ubiquity supports the claim that people really do think like utilitarians, "at least ultimately." For the golden rule is commonly used as a critical check on personal and social standards, and in using it this way people appear, from a Harean perspective, to be using utilitarian critical thinking to correct standards of personal morality, common morality, and the law.

§4.3. Prescriptivism and the Evolution of Utilitarianism

I have already emphasized (in §2.1) that Hare does not claim that "the moral words" ("right" and "wrong," "good" and "bad," "must," "desirable," etc.) always have the three logical properties ascribed to them by universal prescriptivism. He claims only that there is a distinctively moral sense of the words in which they have these three properties, and that the English "must" comes closest to always having them. Similarly, Hare did not himself claim that all natural languages have a universal prescriptivist component. He claimed only that English and closely cognate languages have a universal prescriptivist component (*MT*, pp. 18, 20, 55). In this section, I offer reasons for making the stronger claim, that all natural languages may have a universal prescriptivist component, because evolution would have favored communities of early humans who thought like utilitarians.

The "classical utilitarians" (Jeremy Bentham, John Stuart Mill, and Henry Sidgwick) all claimed that even people who explicitly reject the principle of utility tacitly accept it. Mill claimed that "the principle of utility . . . has had a large share in forming the moral doctrines even of those who most scornfully reject its authority" (¶1.4). Bentham claimed that "By the natural constitution of the human frame, on most occasions of their lives men in general embrace this principle, without thinking of it," and that "When a man attempts to combat the principle of utility, it is with reasons drawn, without his being aware of it, from that very principle itself" ([1780] 1948, chapter 1, ¶¶12, 13). For his part, Sidgwick argued that "the Morality of Common Sense may be truly represented as at least unconsciously Utilitarian" insofar as:

1. "existing moral rules" have "in each case some manifest felicific tendency,"
2. "their generally received limitations and qualifications" correspond to the exceptions that utilitarian thinking would recommend, and
3. "the remarkable discrepancies found in comparing the moral codes of different ages and countries are for the most part strikingly correlated to differences in the effects of actions on happiness, or in men's foresight of, or concern for, such effects" (1907, pp. 424–426).

Sidgwick concluded that "the intuitions of Common Sense" are "inchoately and imperfectly Utilitarian" (p. 427), "unconsciously or 'instinctively' utilitarian" (p. 463).

Sidgwick went on to claim that utilitarian thinking seems to be guiding the evolution of common morality, arguing that "when we compare the different moral codes of different ages and countries, we see that the discrepancies among them correspond, at least to a great extent, to differences either in the actual effects of actions on happiness, or in the extent to which such effects are generally foreseen—or regarded as important—by the men among whom the codes are maintained" (p. 454). For specific examples, Sidgwick relied on Scottish philosopher Dugald Stewart:

> Thus Dugald Stewart lays stress on the extent to which the moral judgments of mankind have been modified by "the diversity of their physical circumstances," the "unequal degrees of civilization which they have attained," and "their unequal measures of knowledge or of capacity." He points out, for instance, that theft is regarded as a very venial offence in the South Sea Islanders, because little or no labour is there required to support life; that the lending of money for interest is commonly reprehended in societies where commerce is imperfectly developed, because the 'usurer' in such communities is commonly in the odious position of wringing a gain out of the hard necessities of his fellows; and that where the legal arrangements for punishing crime are imperfect, private murder is either justified or regarded very leniently. Many other examples might be added to these if it were needful. But I conceive that few persons who have studied the subject will deny that there is a certain degree of correlation between the variations in the moral code from age to age, and the variations in the real or perceived effects on general happiness of actions prescribed or forbidden by the code. And in proportion as the apprehension of consequences becomes more comprehensive and exact, we may

> trace not only change in the moral code handed down from age to age, but progress in the direction of a closer approximation to a perfectly enlightened Utilitarianism. (Sidgwick 1907, pp. 454–455)

In the remainder of this section, I want to suggest, in a similar vein, that Hare's universal prescriptivism helps explain how moral thinking could have evolved among humans.

Hare himself never addressed the question of *why* moral judgments have the specific logical properties they do. In defense of his linguistic hypothesis, he appeals to his readers' linguistic intuitions, and he argues that his hypothesis being true would help explain certain features of moral controversies and even the development of various competing ethical theories (as described below in §4.6 and §4.4, respectively). He did, however, say some isolated things that suggest he would have welcomed an evolutionary explanation. For instance, he said: "Because we are human beings and not angels we have adopted *or inherited* what I called the intuitive level of moral thinking with its prima facie principles, backed up by powerful moral feelings, and attached to rather general characteristics of actions and situations" (*MT*, p. 59—emphasis added). And elsewhere he said that "Evolution (if we may personify her) has had the same idea" as utilitarian critical thinking regarding parental "loyalties and affections which are genetically transmitted, and have no doubt favoured the survival of the genes which transmit them" (*MT*, p. 137). But Hare never directly addressed the question of whether and how the three logical features of moral discourse might be explained in evolutionary terms.

In the remainder of this section, I provide a sketch of such an account. My argument is that group selection would tend to favor the evolution of utilitarian thinking in ethics under the following two assumptions.

(A1) If the members of a group talk about ethics and have as their ultimate ethical standard the principle of utility, then they will be willing to sacrifice their individual interests for the aggregate happiness of the group.

(A2) The members of a group with higher aggregate happiness will tend to be more fit than members of a group with lower aggregate happiness.

In support of (A1), I offer the following argument:

1. Utilitarianism is the view that (at least ultimately) right actions and institutions maximize aggregate happiness.
2. On Hare's prescriptivist account of moral language, one sincerely assents to a moral judgment only if one acts accordingly and welcomes others acting accordingly.
3. And a community of rational users of such language will tend toward agreement on conclusions that are—at least ultimately—utilitarian.

C. So members of a group who use moral language with the three logical features in Hare's account will tend to both think and act like utilitarians.

The first premise of this argument is just a standard definition of utilitarianism, and the second premise just expresses the claim that moral judgments are prescriptive in Hare's special sense. So what needs defending here is the third premise.

Hare clearly believed that sincere debate about ethics would tend to produce consensus on what are good intuitive-level rules to abide by and utilitarian conclusions about unusual cases. But why think that a community of rational users of universal prescriptivist language will "tend toward agreement on conclusions that are—at least ultimately—utilitarian"? Sometimes we can point to real-world examples, such as the changes in Inuit attitudes toward infanticide discussed in §3.3(A), but Hare did not claim that changes in intuitive-level rules always track archangelic thinking so well. Citing Singer's (1972) "Famine, Affluence and Morality" essay, Hare wrote: "[our intuitions about] whether we ought to give away our goods to feed the poor throughout the world . . . are suspect. They date from a time when it was not possible to do very much about poverty in remote parts of the world, and they may owe something to a natural desire to insulate our morality from these problems" (*MT*, p. 199). But there are reasons to think that the *general tendency* will be in this direction.

Assuming that Hare's account of the logic of moral discourse is correct, the intuitive-level rules characteristic of a group's common morality will tend to track the recommendations of a Harean archangel to the extent that the following things are true.[2] First, members of the group engage in debate about when they are justified in breaking whatever intuitive-level rules they now endorse or how they ought to amend their current intuitive-level rules. Engaging in such debates involves them in *using* or *applying* the universal prescriptivist component of their language. Second, members of the group are *sincere* in their use of this component. By this I mean that they make a serious, genuine attempt to "stand in others' shoes," as the golden rule requires. This requires, as Hare emphasizes, that one imagine being in others' situations with *their* preferences. This sincerity requirement ensures that critical thinkers will reach the same conclusions to the extent that they accurately represent the sufferings and enjoyments of others to themselves. "For I am to imagine myself

---

[2] Students of Hare will recognize these two conditions as eliminating what Hare calls, respectively, "amoralists"—those who have a universal prescriptivist component in their language, but refuse to use it (*MT*, chapter 10, sections 7–8)—and the "impure" fanatic—those who engage in critical thinking but reach different conclusions from a utilitarian only because they fail to adequately represent the sufferings and enjoyments of various affected individuals (*MT*, chapter 10, sections 1–6). Thus the two conditions in this paragraph pick out societies with few amoralists and few impure fanatics.

in his situation with *his* preferences. Unless I have an equal aversion to myself suffering, forthwith, what he is suffering or going to suffer, I cannot really be knowing, or even believing, that being in his situation with his preferences would be like *that*" (*MT*, pp. 94–95). To the extent that these two things are true of the general population, a group's system of intuitive-level rules will tend to change in the directions that a perfect Harean critical thinker would send it.

Relatedly, this might explain why the common moralities of some cultures change more slowly. To the extent that the society discourages debate about the content of their intuitive-level rules and when exceptions to them are justified, those rules will tend to change more slowly, and to the extent that such debate is encouraged, the rules will tend to change more quickly. In more rigidly conformist societies, then, change would occur more slowly, and adaptation to changing circumstances would be slower.

The second assumption (A2) that supports my argument in this section is that the members of a group with higher aggregate happiness will tend to be more fit than the members of a group with lower aggregate happiness. There is an obvious problem with this assumption, however: maximizing aggregate happiness and maximizing average fitness are not the same thing. For it might be that the happiest people don't reproduce as well, and people might reproduce best under relatively miserable conditions.

This is not the place for an extended discussion of the concept of happiness, which has been unpacked in very different ways by utilitarians. Mill defined "happiness" as "pleasure and the absence of pain" ([1861] 1957, p. 10), Hare (*MT*) and Richard Brandt (1979) both define it in terms of achieving some kind of intelligently integrated satisfaction of one's preferences, and there are further variations (e.g., living a good life story, as described in chapter 6). The important thing, for present purposes, is that however the details are spelled out, happiness is a function of positive mental states. Fitness, on the other hand, is a function of one's potential for reproductive success, and some things that improve an individual's chances of reproductive success don't seem to maximize the happiness that they experience in life. A dramatic example is provided by salmon. Fish are now known to feel pain (for the reasons given in chapter 5), and presumably salmon returning to spawn in the shallow streams where they hatched suffer greatly. But a salmon that does not endure that terrible and always terminal trip has no chance of spawning at all. So, in the case of salmon, individual happiness ends where fitness begins.

People are not salmon, however. In modern industrialized nations, where the "demographic transition" has occurred and average reproductive rates are falling to below replacement level, people do happily squander the higher reproductive capacity provided by good nutrition and health care, but in pre-industrialized societies it seems likely that there was at least a rough correlation between reproductive success and living a happy life.

So under assumptions (A1) and (A2) above, it is plausible to think that group selection could have driven the evolution of universal prescriptivist language in humans. This increases the plausibility of Hare's theory in the following way. His claim that genuine moral judgments are prescriptive in his special sense is the most controversial feature of his account of moral language. In particular, it makes it difficult to account for moral backsliding. It is this same logical feature, however, that provides a key part of a Harean account of how moral discourse might have evolved. For Harean prescriptivism provides a tight link between thinking like a utilitarian (at least ultimately) and acting like one, and only if thinking like a utilitarian leads one to act like a utilitarian would thinking like a utilitarian raise the fitness level of one's group. Adopting a form of moral discourse that entailed utilitarianism but was not prescriptivist would not necessarily raise the aggregate happiness level of the population, but adopting a form of moral discourse that both entails utilitarianism and is prescriptivist would.

§4.4. The Existence of Other Theories of Normative Ethics and Metaethics

The preceding two sections described two reasons for thinking that universal prescriptivism might be true, despite the fact that people do not use explicitly utilitarian thinking in a range of cases where Hare's theory says that they should. One (§4.2) is that the golden rule is universally recognized and, according to Hare, the golden rule embodies the three logical requirements of critical-level thinking. The other (§4.3) is that incorporating a universal prescripitivist component in their language would have been adaptive for our ancestors and would help explain how moral thinking and the general practice of ethical debate evolved.

Hare did not consider either of the preceding arguments in defense of his theory's descriptive adequacy, but he did devote all of chapter 4 of *Moral Thinking* to showing how his theory would explain the emergence of various schools of thought in ethical theory. Since Hare claims that his universal prescriptivism tells us what Kant should have said about the logic of moral discourse, and that all thinking is at least ultimately utilitarian, if Hare's theory is correct then we would expect moral philosophers to develop variations on both of these theories. But we would also expect philosophers to defend versions of moral relativism, rights theory, and virtue ethics, and (when it comes to metaethics rather than normative ethics) to embrace such diverse views about meaning and justification in ethics as intuitionism, naturalism and nonnaturalism, and emotivism. For while each of these latter theories is ultimately mistaken, according to Hare's theory, each of them also emphasizes at least some aspect(s) of the truth. In this way, his theory explains why even careful moral philosophers would deny the truth of utilitarianism and/or universal prescriptivism.

Relativism would be suggested by the fact that what are the best ILS rules to adopt will vary across background ecological, sociological, economic, and technological conditions. Other philosophers would tend to put rights, conceived of as "trump cards against utilitarian arguments," at the heart of their theory. This is because rights have the deontological flavor of certain important ILS rules that are best expressed in non-utilitarian terms.

In virtue theory, the focus is not on principles or rights, but on *good character traits*. A virtue, generally conceived, is a character trait, acquired by habituation, which makes one more likely to do the right thing or live a good life. Hare's theory incorporates the virtues via the way ILS rules are "internalized." For to properly internalize an ILS rule is to acquire a character trait, embodied in dispositions to behave and judge accordingly; and that disposition remains in place when one "acts out of character" (see also §§4.9, 4.11, 6.3).

Heavy reliance on intuitive-level thinking would incline some toward intuitionism in metaethics. "Metaethics" refers to study of the meanings of moral words, of logical requirements on moral judgments, and of how to go about justifying an ethical principle or theory of normative ethics. A society of people who used universal prescriptivist language without understanding it and who relied heavily on intuitive moral thinking, the way Hare claims that humans can and must, would be expected to produce philosophers who thought that various principles are self-evident or "known by intuition." As Hare puts it, "Intuitions are certainly used in our moral thinking, and this is highly desirable; how easy it is, therefore, to attribute to intuitions an epistemological status which they do not have!" (*MT*, p. 75).

Hare also says that heavy reliance on intuitive-level moral thinking would incline some to embrace naturalism, the view that key moral words—like "good" and "bad"—can be defined in terms of natural properties, such as enjoyment and suffering. At the same time, the fact that any good set of ILS rules will refer to many properties with little in common would incline others to hold that terms like "good" and "bad" must refer to some "non-natural property" of actions, events, and institutions.

Finally, a philosopher who noticed that prescriptivity (in Hare's special sense) is a logical feature of all genuine moral judgments would tend toward emotivism, if he failed to also notice the other logical features of moral judgments. For if all the philosopher noticed was that people who sincerely assent to a moral judgment are inclined to act and judge accordingly, he could imagine people having no way to use reason to reach consensus on ethical questions and being instead condemned to endless disagreements because they have different "ultimate attitudes" (cf. Stevenson 1944). But this would only happen if he fails to see—as Hare emphasizes—that when combined with the other two logical features of moral judgments, *universal* prescriptivism ensures that all careful critical thinkers will arrive at the same conclusions if they are working with the same set of facts.

In these ways, Hare's universal prescriptivism allows him to explain why other philosophers would endorse other theories. I know of no other ethical theory that does this in the comprehensive way that Hare's does. Relatedly, Hare's theory allows him to explain why the objections of critics have the influence that they do. For as noted above (§1.4), the critics of utilitarianism rely on the method of reflective equilibrium, and if Hare's theory is correct, people's intuitive moral judgments will conflict with critical thinking in a range of cases. So, when doing intuitive-level thinking, people feel that it is inherently wrong to break the promise to the dying man, to punish the innocent, to enslave people, and so on. They think that these things would be wrong, even if doing so clearly would, under certain specified circumstances, maximize aggregate happiness. The critics of utilitarianism appeal to such strongly held intuitions to turn people against the theory.

§4.5. The Compartmentalization Problem

A related doubt about the descriptive adequacy of Hare's theory focuses on the division of moral thinking into two levels with such different characteristics. One version of this objection claims that one's intuitive-level rules cannot continue to have a non-utilitarian "flavor" once one understands that the ILS rules are justified by explicitly utilitarian thinking; once one understands that this is a "noble lie," it ceases to be effective. This is what I call "the compartmentalization problem." In response, Hare simply says, "I do my own moral thinking in the way described in this book . . . To say that it is impossible to keep intuitive and critical thinking going in the same thought-process is like saying that in a battle a commander cannot at the same time be thinking of the details of tactics, the overall aim of victory, and the principles . . . which he has learnt when learning his trade" (*MT* p. 52). For what it's worth, since experiencing a conversion to Hare's theory over the last decade, I myself also think, very explicitly, in terms of what ILS rules I have internalized or should internalize, on the one hand, and, on the other, in terms of what act utilitarian thinking would require. I do this very explicitly and sincerely. I do not think for a moment that surveys of the general populace would in any direct way reveal that they think the way that Hare and I do.

There is, however, good evidence that people use, in a broad range of situations, two very different kinds of cognitive systems, with features that correspond to those of Harean critical thinking and intuitive-level thinking. In his 2002 Nobel Prize lecture, psychologist Daniel Kahneman described how "intuitive judgment" and "deliberate reasoning" differ in similar ways across diverse domains, including sense perception, probability, and statistics. In each of these domains, two very different kinds of systems operate. The operation of what Kahneman calls "system 1" is fast and automatic, while the application of "system 2" is slow and requires effort. Kahneman characterizes system 1 as "emotional" and "associative," system 2 as "neutral" and "rule-governed" (Kahneman 2003, p. 698, figure 1). Kahneman details how, across various domains of judgment, system 2 can, with effort, be used to override the automatic outputs of system 1. In terms of the distinction between "intuition" and "analysis" in the expert judgment literature discussed in §3.3(E), this could be described as using analysis to override intuition.

Kahneman does not discuss moral judgment specifically, but Joshua Greene and colleagues have used brain-imaging techniques to show that when subjects are asked to consider certain kinds of moral dilemmas, brain activity indicates a primarily emotional response when the subjects do what Hare would interpret as following an intuitive-level rule, whereas the subjects' response primarily involves reasoning when they reach the utilitarian conclusion that the intuitive-level rule should be violated. Greene and colleagues presented subjects with moral dilemmas like the following, in which the utilitarian conclusion would involve what Hare would interpret as overriding an ILS rule:

> Enemy soldiers have taken over your village. They have orders to kill all remaining civilians. You and some of your townspeople have sought refuge in the cellar of a large house. Outside, you hear the voices of soldiers who have come to search the house for valuables.
>
> Your baby begins to cry loudly. You cover his mouth to block the sound. If you remove your hand from his mouth, his crying will summon the attention of the soldiers who will kill you, your child, and the others hiding out in the cellar. To save yourself and the others, you must smother your child to death.
>
> Is it appropriate for you to smother your child in order to save yourself and the other townspeople? (Greene et al. 2004, p. 390)

Subjects who answered "yes" exhibited longer reaction times and greater activity in the dorsolateral prefrontal cortex, an area associated with "engagement of abstract reasoning processes and cognitive control" (ibid.), than subjects who answered "no." Greene and colleagues interpret these results as confirming "the hypothesis that utilitarian moral judgments engage brain areas associated with 'cognitive' processes" (p. 392) to override what they characterize as "prepotent social-emotional responses elicited by these dilemmas" (p. 390).

On Hare's theory, internalizing an ILS rule involves developing just such "prepotent social-emotional responses," which could be characterized, in Kahneman's terms, as a moral "system 1." Greene and colleagues' research suggests that overriding the output of this "system 1" involves conscious application of a utilitarian moral "system 2." Note also that in various situations, the output of the automatic, unreflective "system 1" does not end when it is cognitively overridden by the effortful, reflective "system 2." In the corridor illusion, for instance, two figures of identical height are displayed so that one appears to be farther down a corridor from the viewer. This causes one figure to appear taller than the other, and the optical illusion remains even after subjects measure the heights of the two figures and reason through to the conclusion that the figures are indeed the same height. From a Harean perspective, the same will be true in ethics: after one uses careful critical thinking to convince oneself to violate an ILS rule that one has internalized, the action will still "look" wrong (see the man overboard case in §1.5, and the discussion of "test cases for utilitarianism" in §§4.7–4.9).

For over a decade, Jonathan Haidt, an undergraduate philosophy major turned psychology PhD, has championed a widely discussed "social intuitionist model" of moral judgment. Haidt's model corresponds in significant ways to things said earlier in this book about Harean intuitive-level thinking. It is *intuitionist* because it holds that "intuitions (including moral emotions) come first and directly cause moral judgements" without any causal role for reasoning. The same was said in §3.4(A) about Harean intuitive-level thinking in contrast to the classical model of moral reasoning; to have properly internalized an ILS rule is to have developed a set of dispositions to judge and act accordingly. Haidt's model is *social* insofar as it holds that "Moral reasoning is usually an ex post facto process used to influence the intuitions (and hence judgments) of other people" (2001, p. 814). Similar things were said in §3.4(C) about the "context fixing" and "collaborative problem solving" roles of ILS rules: ILS rules often don't precisely express the internalized paradigm, but invoking them is nevertheless useful as a way of shaping the moral perceptions of others (or even oneself), and invoking ILS rules can be a way of contextualizing disagreements among people who have internalized different sets of ILS rules.

So far, Haidt's social intuitionist model corresponds to half of the story from a Harean perspective, but in a 2001 article titled "The Emotional Dog and Its Rational Tail," Haidt went further and suggested that the model could account for all moral reasoning. Haidt acknowledged that various lines of research had shown a correlation between moral judgments of various practices on the one hand, and belief in certain moral principles or rules and related factual claims on the other. He emphasized, however, that correlation does not establish causation: "An intuitionist interpretation is just as plausible: The anti-abortion judgment (a gut feeling that abortion is bad) causes the belief that life begins at conception (an ex post facto rationalization of the gut feeling)" (p. 817).

In a subsequent study prompted by Haidt's argument, Greene and colleagues (2008) designed an experiment that could, they thought, support the *causal* connection between reasoning and moral judgment when subjects reach a utilitarian judgment about dilemmas like the crying child case described above. They designed a variation on their earlier experiment; in this variation, the subjects' "cognitive load" was increased during the same time period that they were asked to read and reach a judgment about a dilemma. To do this, they presented the dilemmas as text scrolling across a video screen and either asked subjects just to respond to each dilemma by pushing a button (the no-load condition), or to do so while simultaneously watching a string of numbers scroll across the same screen, pressing another button each time the number five appeared in the string (the load condition). If reasoning is causally involved when subjects overcome "prepotent social-emotional responses elicited by these dilemmas" in order to reach a utilitarian conclusion in such dilemmas, then increasing the subjects' "cognitive load" should either reduce the rate at which they give utilitarian moral judgments or increase the reaction time on utilitarian judgments without similarly increasing the reaction time for non-utilitarian judgments.

Although the load condition did *not* decrease the rate at which subjects gave utilitarian versus non-utilitarian judgments, the load condition *did* increase the average reaction time on utilitarian judgments by about 11.5% (from about 5.75 seconds to 6.5 seconds), without at all affecting reaction times on non-utilitarian judgments (pp. 1149–1150). They concluded that "the present data may provide the strongest evidence yet that controlled cognitive processes play a causal role in *ex ante* moral judgment" (p. 1151).

Regarding the fact that the frequency of utilitarian judgments did not decrease under load, they speculate: "One explanation for this is that participants were keenly aware of the interference created by the load manipulation and were determined to push through it. Like motorists facing highway construction, they may have been delayed, but not ultimately prevented from reaching their destinations" (p. 1151). So additional manipulations of cognitive load might decrease the rate of utilitarian judgments in subjects.

They also note that, unlike their 2001 study cited above, there was only "a marginally significant" difference in this study between average reaction times for utilitarian and non-utilitarian judgments when made under no load: 5.736 versus 6.130 seconds, respectively (p. 1149). When participants were divided into high- and low-utilitarian groups based on the frequency with which they made utilitarian judgments (averages of 80% versus 42%, respectively), however, the average response times of the low-utilitarian group *were* significantly higher when giving utilitarian responses (p. 1150, figure 2). Relatedly, under the no-load condition there was "a robust correlation" between lower response times and a subject giving the utilitarian response in a high percentage of dilemmas, but no correlation to response time under load: "This suggests that there is an additional process that drives down RT in high-utilitarians in the absence of load, although this process still remains susceptible to cognitive interference" (p. 1152).

Greene and colleagues do not speculate about what that additional process might be, but from a Harean perspective, it could be that subjects in the high- and low-utilitarian groups have internalized different ILS rules of personal morality. For, as noted in §3.2(B), these include "meta-rules" for deciding when to engage in utilitarian critical thinking and when to stick more unquestioningly to one's intuitive-level rules. This suggests that while the moral system 2 (critical thinking) can be used, with cognitive effort, to override an output of the moral system 1 (internalized ILS rules), the moral system 1 also meaningfully restricts the operation of the moral system 2. That is a complication of Harean utilitarianism that could, perhaps, be explored in subsequent empirical research by Greene and colleagues, although I confess that at this point I have no suggestions about how to design an appropriate study.

While the research discussed in this section hardly constitutes empirical proof of Hare's two-level utilitarianism, it does address the objection that two types of moral thinking with such different characteristics as Harean ILS rules and Harean critical thinking could not continue to coexist while the output of the former is periodically overridden by the latter.

§4.6. Shedding Light on Some Real-World Controversies

Hare's theory is highly controversial, but I know of no other ethical theory that does such a good job of explaining why its own critics' tactics work so well, while shedding significant light on the history of philosophical thought about ethics. I also find that Hare's theory sheds light on various real-world controversies. All of part III of this book will be devoted to showing how this is so in some animal-related issues. Thus part III of this book can be read as an extended argument that if shedding significant light on real-world issues helps justify a theory of descriptive and normative ethics, then Hare's theory is well justified.

In this section, I will offer just one isolated example of how Hare's theory does this. In the currently booming controversy over the use of torture in "the war on terror," those who invoke "ticking time bomb" scenarios can be seen as appealing to our need for explicitly utilitarian, critical thinking in unusual cases, while those who insist that the United States should adopt a rule against torture are invoking what would be a perfectly good ILS rule that should be followed in almost all "normal" cases. If Hare's theory is correct, you would expect just such a disagreement to arise regarding the use of torture.

Identifying this *cause* of the controversy doesn't tell us how to *resolve* it, of course. This particular controversy is, I think, *unresolvable*, because it is a fundamental manifestation of "the problem of dirty hands," and a Harean analysis of the controversy shows why both sides are clearly right. Those who argue for publicly adopting a ban on torture have (as Hare would sometimes put it) "gotten hold of a good intuitive level rule." For if we publicly commit to torturing terrorists, even if only in "ticking time bomb" cases, then U.S. soldiers are more likely to be tortured when taken prisoner in the future, and information extracted through torture is generally unreliable anyway. So we should adopt an official, public policy of not resorting to torture under any circumstances. At the same time, however, those who raise the ticking time bomb case are surely correct that, given the specified circumstances (that, in the instant case, the information *would* be reliable, and by torturing this single terrorist, one *would* prevent many deaths and injuries in an imminent attack), this is clearly a case where a good intuitive-level rule should be overridden.

The controversy will never go away because this is a case where we need leaders who are willing to "get their hands dirty," but, at the same time, we have good reason to fear people who would do what critical thinking demands in such a case. We fear that they will be prone to misdiagnose a ticking time bomb case and that they will be too ready to override other good intuitive-level principles. There is also just something scary about such people. The right person for the job would be one who would both (a) internalize the ILS rule against torture in such a way that violations of it genuinely revolted him, and (b) torture *con brio* in the ticking time bomb case. Because such people are impossible to identify with certainty, neither side in this controversy can trust the other: to Rush Limbaugh (who is on the side of critical thinking in this case, although on other issues he often gets stuck thinking on the intuitive level), the advocates of a torture ban are "limp-wristed cowards," unwilling to do what "must" be done; but the advocates of a ban fear that any outspoken opponent of the ban will be as generally untrustworthy as they take Limbaugh to be.

In the end, I think that this controversy nicely illustrates how complicated utilitarian thinking can get. Among the classical utilitarians, Sidgwick best appreciated these complications, and discussed them at length in Book IV of *The Methods of Ethics*. In part III of this book, some of the complications will be discussed in the context of advocating changes to ILS rules governing our treatment of animals, but the present controversy over the use of torture in "the war on terrorism" illustrates Sidgwick's statement that "it may conduce most to the general happiness that *A* should do a certain act, and at the same time that *B*, *C*, [and] *D* should blame it" (1907, p. 491). That is, this may be a case where we should adopt a ban on torture, and enforce it, hoping all the while that if a genuine ticking time bomb case ever actually arises, someone will be willing to get their hands dirty and face punishment for their actions.

§4.7. Philosophical "Test Cases for Utilitarianism"

Critics of utilitarianism claim that the theory implies the opposite of deeply held intuitions and moral common sense in a broad range of what Rawls called "test cases for utilitarianism" (1955, p. 30). Although Hare eschews the method of reflective equilibrium for the reasons given in §1.4, it is natural to ask what his theory has to say about the test cases. In the remainder of this chapter, I will describe what Hare says in response to such objections and add some further remarks about differences among the various cases that critics use in such objections.

To understand how varied Hare's responses could be, it will be useful to consider a range of different "test cases" that various philosophers have used in formulating objections to utilitarianism in general or to Hare's theory in particular. The first case was introduced earlier, in §1.5:

1. **The Bloggs transplant case**: "Suppose that you are a famous transplant surgeon, and that your transplants always work. You have five patients, each of whom needs a transplant. One needs a heart, one a brain, two need one lung apiece, and one needs a liver. One of your patients, Bloggs, has come in today to find out the results from some lab work. You know from the results of the lab work that Bloggs would be a perfect donor for each of your five other patients, and you know that there are no other available donors. So you ask Bloggs if he would be willing to be cut up and have his organs distributed. He declines your kind offer, but you realize that you could easily overpower Bloggs and cut him up without his consent, without your crime ever being discovered."

The next four are quoted from two famous critiques of utilitarianism:

2. **Punishing the innocent man**: "[I]f some kind of very cruel crime becomes common, and none of the criminals can be caught, it might be highly expedient, as an example, to hang an innocent man, if a charge against him could be so framed that he were universally thought guilty; indeed this would only fail to be an ideal instance of utilitarian 'punishment' because the victim himself would not have been so likely as a real felon to commit such a crime in the future; in all other respects it would be perfectly deterrent and therefore felicific" (Carritt 1947, p. 65).
3. **Jim in the jungle**: "Jim finds himself in the central square of a small South American town. Tied up against the wall are a row of twenty Indians, most terrified, a few defiant, in front of them several armed men in uniform. A heavy man in a sweat-stained khaki shirt turns out to be the captain in charge and, after a good deal of questioning of Jim which establishes that he got there by accident while on a botanical expedition, explains that the Indians are a random group of the inhabitants who, after recent acts of protest against the government, are just about to be killed to remind other possible protestors of the advantages of not protesting. However, since Jim is an honoured visitor from another land, the captain is happy to offer him a guest's privilege of killing one of the Indians himself. If Jim accepts, then as a special mark

of the occasion, the other Indians will be let off. Of course, if Jim refuses, then there is no special occasion, and Pedro here will do what he was about to do when Jim arrived, and kill them all. Jim, with some desperate recollection of schoolboy fiction, wonders whether if he got hold of a gun, he could hold the captain, Pedro and the rest of the soldiers to threat, but it is quite clear from the set-up that nothing of that kind is going to work: any attempt at that sort of thing will mean that all the Indians will be killed, and himself. The men against the wall, and the other villagers, understand the situation, and are obviously begging him to accept. What should he do?" (Williams 1973, p. 98).

4. **The promise to a dying man**: "Suppose that two explorers in the Arctic have only enough food to keep one alive till he can reach the base, and one offers to die if the other will promise to educate his children. No other person can know that such a promise was made, and the breaking or keeping of it cannot influence the future keeping of promises. On the utilitarian theory, then, it is the duty of the returned traveler to act precisely as he ought to have acted if no bargain had been made: to consider how he can spend his money most expediently for the happiness of mankind, and, if he thinks his own child is a genius, to spend it upon him" (Carritt 1947, pp. 64–65).
5. **George the conscientious job seeker**: "George, who has taken his Ph.D. in chemistry, finds it extremely difficult to get a job. He is not very robust in health, which cuts down the number of jobs he might be able to do satisfactorily. His wife has to go out to work to keep them which itself causes a great deal of strain, since they have small children and there are severe problems about looking after them. The results of all this, especially on the children, are damaging. An older chemist, who knows about this situation, says that he can get George a decently paid job in a certain laboratory, which pursues research into chemical and biological warfare. George says that he cannot accept this, since he is opposed to chemical and biological warfare. The older man replies that he is not too keen on it himself, come to that, but after all George's refusal is not going to make the job or the laboratory go away; what is more, he happens to know that if George refuses the job, it will certainly go to a contemporary of George's who is not inhibited by any such scruples and is likely if appointed to push along the research with greater zeal than George would" (Williams 1973, pp. 97–98).

This final one is my own description of a real case:

6. **Lying to attend a funeral**: Maude has worked in a professional capacity at a facility for four years. She is on the verge of quitting, however, because the management chronically understaffs the facility, resulting in frequent, last-minute calls for additional staff. On the last

> day of a scheduled, two-week vacation, Maude's favorite aunt dies unexpectedly and the funeral is planned for three days later, when Maude is scheduled to work. She believes that, having just taken two weeks off, and given the chronic understaffing of the facility, the manager will not grant her request for a day off to attend the funeral, so instead she calls in sick that day.

These are fairly representative of the range of cases in which, according to its critics, utilitarianism gets the wrong answer. In each of them, critics claim that the implications of utilitarianism fail to match our pre-theoretic intuitions. Using the method of reflective equilibrium (§1.4), the critics dismiss utilitarianism because it fails to match these intuitions in such a broad range of cases. But Harean, two-level utilitarianism has several interrelated responses to this line of criticism.

§4.8. Hare's "Template" Response to Unrealistic Test Cases

In chapter 8 of *Moral Thinking*, Hare offers a sort of template for responding to those who criticize his two-level form of utilitarianism for having counterintuitive implications. He illustrates the template using a version of the Bloggs transplant case, so that is what I will use here (although, as I will stress in the remainder of this chapter, there are other important considerations raised by some of the other cases).

In Hare's template response, you begin by asking the objector which kind of thinking we are supposed to be doing about the case. Since the objection is that what Hare's theory implies about the case runs counter to our intuitions, and what Hare's theory implies about the case depends on whether we are doing intuitive or critical thinking about it, this is a reasonable question to ask.

If we apply intuitive-level thinking to the Bloggs case, then the answer will obviously be that it is wrong to cut up Bloggs, since good ILS rules would prohibit such involuntary organ harvesting. This is because ILS rules are designed to handle cases that are commonly encountered (or, in the case of professional ethics, *normally* encountered—see §3.2[C]), and in such cases, it would obviously be good to require informed consent.

So the objection only gets traction if the critic expects us to engage in *critical* thinking about the case. But then, Hare emphasizes, it will only be clear that one is justified in breaking the ILS rules that one has internalized if the case can be made quite realistic:

> Your opponent will now object that although on the utilitarian view it is a good thing for people to have these intuitions or feelings, it also follows from that view that they ought to overcome and act contrary to them in cases like this, in which, *ex hypothesi*, it is for the best to do so. Let us ask, then,

> whether the doctors in the hospital ought to do this if they are utilitarians. It will turn upon their estimate of the *probability* of hitting off the act which is for the best by so doing. The crucial words are, of course, '*ex hypothesi*'; for your opponent has constructed his example with the express purpose of making the murder the act which will have the best consequences. You must not allow him simply to *assume* that this is so; he has to convince the audience, not just that it really could be so in a real-life situation, but that it could be known to be so by the doctors with a high degree of probability. . . .
>
> It is fairly obvious that this high degree of probability will not be forthcoming in many actual situations, if any at all. Have the doctors checked on the down-and-out's connexions [*sic*] or lack of them? (How? By consulting the police records, perhaps! But a colleague of my psychiatrist sister once wrote in his notes, about a dishevelled [*sic*] individual brought in off the streets very late at night by the police, "Has delusion that he is a high-ranking civil servant," and it turned out that he was in fact a *very* high-ranking civil servant.) Have they absolute confidence in the discretion and support of all the nurses, porters, mortuarists, etc., who will know what has happened? Add to this the extreme unlikelihood of there being no other way of saving these patients, if they can be saved at all, and it will be evident that your opponent is not going to get much help out of this example, once it is insisted that it has to be fleshed out and given verisimilitude. (*MT*, pp. 133–134)

So the Bloggs case isn't an instance of the fourth kind of case listed in §1.5, in which "something is ruled out by the intuitive level rules one has internalized but both (a) it is clear that the violation will maximize aggregate happiness, and (b) one can trust one's judgment that this is so."

On the other hand, Hare continues, if the critic "claims the right to introduce any *logically possible* example" (that is, if the critic insists that we treat the example as what I called "strictly hypothetical" in §2.3[A], by assuming that, unless uncertainty about some things is stipulated in the description of it, we know everything there is to know about it, and with absolute certainty),

> then he is exposed to the other prong of your attack. For then he has put himself beyond the range of intuition and cannot appeal to it. Critical thinking can certainly deal with such cases, and will give a utilitarian answer. If he tailors the case so that the utilitarian answer is that murder is the right solution, then that is the answer he will get. What you have to say to the audience is that this does not in the least matter, because such cases are not going to occur. This has two important consequences for the argument. The first is that allowing that in such a case murder would be justified commits us to no prescription to murder in the actual world, in which we have to live our lives. The second is a generalization of the first: the prima facie principles which the critical thinker will select for use in this world can also, and will,

> include a ban on murder, because for the selection of these principles this peculiar case, since it will not occur, is irrelevant. (*MT*, pp. 134–135)

That is, Hare believes that the standard "test cases" will all turn out to be either (a) cases in which we couldn't trust ourselves to do critical thinking, because the stipulated understanding of the facts is incomplete, unbelievable, or otherwise unreliable; or (b) so unrealistic that they would never arise in the real world and thus have no role to play in choosing good ILS rules.

Thus, according to Hare, Bloggs transplant-type "test cases" for utilitarianism do not succeed in showing how his theory would "get the wrong answer" in any real-world case, and his theory explains why the conclusions that critical thinking would reach in the unrealistic (strictly hypothetical) cases should seem counterintuitive. I agree, and the same goes for the second case on the above list, in which the authorities propose to frame and execute an innocent man. If we treat the case as strictly hypothetical—if we stipulate that doing this will cause the many people who are committing the "very cruel crime" to stop, that the injustice will never be discovered, and that there are no other relevant facts—then yes, applying critical thinking to the case yields the conclusion that the man should be framed and hanged. In the real world, however, one should never trust oneself to apply critical thinking based on that description of a case. For how on earth could the authorities be sure that executing one man will stop the wave of cruel crimes, and that their injustice will never be discovered? The answer, obviously, is that they can't. Anyone who seriously believed that he faced a situation that was accurately described by #1 or #2 above would be delusional, and that is why no one should trust themselves to do critical thinking about a case that is described in such unrealistic terms. And such unrealistic cases will never be taken into account when framing good ILS rules for people to follow in their intuitive-level thinking.

§4.9. Responding to More Realistic Test Cases

As we continue through the above list of cases, however, we encounter complications that go beyond Hare's template response. For one thing, the cases later in my list become increasingly realistic. In the third case, in which Jim the botanist is offered the chance to kill one Indian himself and thereby prevent all 20 of them from being killed by Pedro, the stipulated details are highly unrealistic, but—tragically—in its broad strokes the case parallels the real-world cases referred to above in §4.1: the pilots who were ordered to shoot down a hijacked airliner on 9/11/01, and the crews who delivered the atomic bombs to Hiroshima and Nagasaki. In all three cases, people believe that by killing some number of people they can prevent some larger number of deaths at the hands of others. Similarly, while the promise to a dying man case involves an incredibly unlikely specific scenario, there really are cases in which one has made a promise to someone and it is true both that no one else knows about the promise and that the promisee is no longer alive. The case of George the conscientious job seeker is quite realistic: a lot of people have job opportunities arise in industries or companies that offend their personal moralities. And the lying to attend a funeral case, as I said when introducing it, is an accurate description of a decision that a friend of mine once made. Of the last four cases, then, it will not do to say either that they are cases in which we couldn't trust ourselves to do critical thinking because the stipulated understanding of the facts is incomplete, unbelievable, or otherwise unreliable, or that they are so unrealistic that they would never arise in the real world.

As the test cases get more realistic, however, it becomes less and less clear that what Harean critical thinking would imply about them conflicts with common sense, and at the same time Hare's theory explains why, if we sided with critical thinking, we would still have the contrary intuition. In this way, for those who rely on the method of reflective equilibrium, discussion of the "test cases" ends up *strengthening* rather than weakening the case for Harean, two-level utilitarianism.

Take the case of Jim in the jungle, trying to decide whether or not to shoot one Indian in order to save 19 others. Common sense is at least divided on shooting down hijacked airliners and the United States' use of nuclear weapons to end World War II, so to the extent that Jim's situation is analogous to those cases, consistency requires that we take seriously the claim that he should shoot one Indian. Indeed, Bernard Williams (who made up the case) himself says that "the utilitarian is probably right in this case" (1973, p. 117). I believe that common sense is similarly divided over the questions of whether George should turn down the job in the weapons lab and whether Maude should call in sick in order to attend her aunt's funeral. So, in each of these cases, it is no longer clear that the contemplated decision (kill the one yourself in order to prevent the deaths of many at the hands of others, take a job in what one considers a morally reprehensible industry, or lie about being sick in order to attend your aunt's funeral) conflicts with common sense.

At the same time, Hare's theory can explain, in each of these cases, why the agent who goes ahead and makes the choice that critical thinking requires would still have an intuition that what he or she is doing is wrong. For if Jim has been well brought up, in a community with good ILS rules of common morality, then he will have internalized an ILS rule against killing. This is because the kind of case he faces is not one that people commonly face, and thus will not figure into the ILS rules of common morality. This in turn means that he will have developed a disposition to act and judge accordingly, a disposition that persists even as he "acts out of character" (remember that, as noted in §4.4, internalizing ILS rules is a part of character development). For his part, George has apparently internalized, with its associated dispositions, some kind of ILS rule against supporting chemical and biological warfare. The associated dispositions are expressed, perhaps, when he contributes to organizations that oppose such weapons, but those dispositions will persist even if he "acts out of character" and takes the job.

A related complication is that some of these cases involve questions of professional ethics and, as noted in §3.2(C), this means that more than "commonly encountered" cases should be taken into consideration in crafting the relevant ILS rules. When selecting ILS rules of common morality, the focus is always on "commonly encountered" cases, but when selecting ILS rules for a code of professional ethics, the focus is on both commonly encountered cases and "normal" but uncommonly encountered cases. So a well-trained soldier is less likely than Jim to experience an intuitive revulsion when offered the chance to kill some number of people himself in order to prevent some larger number of people from being killed by others. Well-trained professionals are also members of the larger community, however, and so they have also internalized the ILS rules of their community's common morality. That is the Harean explanation of why soldiers who instinctively "do what is necessary" in time of war nevertheless feel something like regret after the fact: apparently professional ethics and common morality can become as compartmentalized as the intuitive and critical levels of thinking (§4.5). Similarly, any code of professional ethics would rule out manipulating the sick leave system the way that Maude did, and a professional who has internalized that code will have qualms about doing so.

The real-world Maude felt no compunction at all about calling in sick, however. In fact, that was my only misgiving about her decision. That is, I had no doubt that what she ought to do, all things considered, was call in sick, but I was somewhat disturbed by the fact that she felt no compunction at all about doing so. At the same time, the real-world Maude has a reputation for being a top-notch professional. So what are we to say about the real-world Maude case? We might think that she had not internalized the ILS rules of her profession very thoroughly. But remember that Maude did not think that the facility was well run. As a result, she did not think that the management properly respected the employees. So I think that Maude's lack of compunction in this case reflected what she would claim to be a kind of justified indifference to rules to which she would otherwise have felt more compelled to conform.

§4.10. Promises and Rawls on the "Practice Conception" of Rules

Something more needs to be said about the case of the promise to a dying man. As described above by E. F. Carritt, this test case raises a fundamental objection that is rooted in the forward-looking nature of utilitarian reasoning. Utilitarianism evaluates actions in terms of their consequences, and it has been claimed that this makes the fact that a promise has been made irrelevant to how the keeping of the promise gets evaluated. Admittedly, in a case in which others know about the promise, people may take note and the example may influence them, but these are still effects in the future. The fact that the promise was made, per se, seems to be, strictly speaking, irrelevant, and Carritt's case of the promise to the dying man ensures that these side effects that would normally accompany the breaking of a promise are irrelevant. Carritt concludes that, according to utilitarianism, the promiser should, in this case, do whatever he should have done had the promise never been made. This is a different kind of objection that goes beyond noting that peoples' intuitive judgments about the case conflict with what the principle of utility implies about it.

Although he was an ardent critic of utilitarianism, in 1955 John Rawls published a paper in which he defended utilitarianism against this specific objection. Rawls' argument involved distinguishing between a "summary conception" of rules and a "practice conception." Critics like Carritt assume that utilitarians must always think of rules as summarizing the tendencies of past actions, as when Mill writes that during "the whole past duration of the human species . . . mankind have been learning by experience the tendencies of actions" ([1861] 1957, p. 30). On this "summary conception," decisions on particular cases are logically prior to the rules, and individuals are entitled to reconsider the correctness of the rule on each occasion falling under it.

Rawls argues that the rules governing promises are of a fundamentally different type, however:

> It would be absurd to interpret the rules about promising in accordance with the summary conception. It is absurd to say, for example, that the rule that promises should be kept could have arisen from its being found in past cases to be best on the whole to keep one's promise; for unless there were already the understanding that one keeps one's promises as part of the practice itself there couldn't have been any cases of promising.

This is because saying "I promise" is "a performative utterance": saying those words *does* something. Specifically, it places one under an obligation, the nature of which is in part determined by what kinds of excuses will, and what kinds will not, absolve one of the obligation. This is demonstrated, according to Rawls, by the fact that if someone explains that they broke a promise "simply because breaking it was best on the whole," we would say that they don't understand what it means to make a promise.

> Is this to say that in particular cases one cannot deliberate whether or not to keep one's promise? Of course not. But to do so is to deliberate whether the various excuses, exceptions and defenses, which are understood by, and which

> constitute an important part of, the practice, apply to one's own case. Various defenses for not keeping one's promise are allowed, but among them there isn't the one that, on general utilitarian grounds, the promisor (truly) thought his action best on the whole, even though there may be the defense that the consequences of keeping one's promise would have been extremely severe.[3]

In this way, "practice conception" rules are logically prior to decisions about particular cases, and it is not open to individuals to reconsider the rules of the practice whenever they face a decision falling under the rules. As Rawls puts it, agents working within an existing practice hold the "moral office" of judges rather than legislators, and it is not up to them to redesign the practice, any more than it is up to baseball players to redesign the game.

At the same time, the principle of utility can be applied to the question of what practice rules to adopt, or how to design various practices, and from this perspective the practice of promising looks like a better one to adopt than the practice of what we might call *telosing*. In *telosing*, the excuse that "it was just better on the whole not to do it" would be permitted, but a society in which there was no practice of promising, but only that of *telosing*, would be much less efficient. For every time I had *telosed* to do something for you, I would be immune from blame if I (truly) found that, when it came time to do what I had *telosed* to do, doing it would inconvenience me more than my not doing it would inconvenience you. People just couldn't count on each other the way we need to in a society that adopted the practice of *telosing* rather than the practice of promising.

Rawls claims that this allows the utilitarian to give an effective response to Carritt's objection (which he had quoted earlier in his essay):

> It must, of course, be granted that the rules defining promising are not codified, and that one's conception of what they are necessarily depends on one's moral training. Therefore it is likely that there is considerable variation in the way people understand the practice, and room for argument as to how it is best set up. For example, differences as to how strictly various defenses are to be taken, or just what defenses are available, are likely to arise amongst

[3] Note that Rawls does acknowledge that "the consequences of keeping one's promise would have been extremely severe" is, nevertheless, an allowed excuse:

> It is no doubt part of the utilitarian view that every practice should admit the defense that the consequences of abiding by it would have been extremely severe; and utilitarians would be inclined to hold that some reliance on people's good sense and some concession to hard cases is necessary. They would hold that a practice is justified by serving the interests of those who take part in it; and as with any set of rules there is understood a background of circumstances under which it is expected to be applied and which need not—indeed which cannot—be fully stated. Should these circumstances change, then even if there is no rule which provides for the case, it may still be in accordance with the practice that one be released from one's obligation. But this sort of defense allowed by a practice must not be confused with the general option to weigh each particular case on utilitarian grounds which critics of utilitarianism have thought it necessarily to involve. (1955, pp. 17–18)

> persons with different backgrounds. But irrespective of these variations it belongs to the concept of the practice of promising that the general utilitarian defense is not available to the promisor. That this is so accounts for the force of the traditional objection which I have discussed. And the point I wish to make is that when one fits the utilitarian view together with the practice conception of rules, as one must in the appropriate cases, then there is nothing in that view which entails that there must be such a defense, either in the practice of promising, or in any other practice. (pp. 30–31)

Rawls was skeptical that there are very many instances of practices outside legal contexts (i.e., in common morality); in the essay he claimed only that both promising and punishing are practices that are embedded in our common morality. But he held that Carritt's objection about promise keeping misfires as soon as the rules regarding promising are understood as practice conception rules, rather than summary conception rules.

This adds yet another complication to Hare's theory, for now we see that while most ILS rules are to be understood as summary conception rules, at least some are to be understood as practice conception rules. Unlike the rules of an individual's personal morality, these will always be social constructs, because they will always involve communal understandings about what sorts of exceptions will be allowed. Like all ILS rules, internalizing them properly will involve acquiring dispositions to act and judge accordingly, but internalizing practice conception rules always involves acquiring some specific understanding of what kinds of exceptions are allowed, and understanding that the general, utilitarian rationale that "it was just better on the whole not to do it" is disallowed. (In §3.2, I noted that while legal standards are codified in writing, common morality is not. When two parties execute a legal contract, the contract, together with the common law of contracts, codifies a promise by providing the kind of explicit, written-out version of the exceptions that will be allowed, which is, as Rawls notes, usually lacking in day-to-day talk about promises.)

§4.11. Williams on Integrity and Utilitarianism

Finally, I want to return to the Jim in the jungle and George the conscientious job seeker cases, which Bernard Williams uses to motivate one of his central objections to utilitarianism. Williams begins by noting how strained it is to say that if Jim refuses to shoot one of the Indians, then he is responsible for the deaths of the 20 whom Pedro will then shoot:

> While the deaths, and the killing may be the outcome of Jim's refusal, it is misleading to think, in such a case, of Jim having an *effect* on the world through the medium (as it happens) of Pedro's acts; for this is to leave Pedro out of the picture in his essential role of one who has intentions and projects, projects

> for realizing which Jim's refusal would leave an opportunity. Instead of thinking in terms of supposed effects of Jim's projects on Pedro, it is more revealing to think in terms of the effects of Pedro's projects on Jim's decision. This is the direction from which I want to criticize the notion of negative responsibility.

In particular, Williams argues, to rely on explicitly utilitarian thinking to decide such difficult cases would be to abandon commitments that are essential to one's identity as an individual person.

> The determination to an indefinite degree of my decisions by other people's projects is just another aspect of my unlimited responsibility to act for the best in a causal framework formed to a considerable extent by their projects.
>
> The decision so determined is, for utilitarianism, the right decision. But what if it conflicts with some project of mine? This, the utilitarian will say, has already been dealt with: the satisfaction to you of fulfilling your project, and any satisfactions to others of your so doing, have already been through the calculating device and have been found inadequate. Now in the case of many sorts of projects, that is a perfectly reasonable sort of answer. But in the case of projects of the sort I have called "commitments," those with which one is more deeply and extensively involved and identified, this cannot just by itself be an adequate answer, and there may be no adequate answer at all. For, to take the extreme sort of case, how can a man, as a utilitarian agent, come to regard as one satisfaction among others, and a dispensable one, a project or attitude round which he has built his life, just because someone else's projects have so structured the causal scene that that is how the utilitarian sum comes out? . . . It is absurd to demand of such a man, when the sums come in from the utility network which the projects of others have in part determined, that he should just step aside from his own project and decision and acknowledge the decision which utilitarian calculation requires. (pp. 115–116)

It is important to understand that William's point is not that it would be absurd to say that Jim should shoot the one Indian; in fact, Williams says that "the utilitarian is probably right in this case" (p. 117). His point is that an individual who sincerely took a purely utilitarian perspective on his own "projects" and "commitments" would have ceased to identify with them in a robust, normal way. Utilitarian thinking "is thus, in the most literal sense, an attack on [one's] integrity," and "it is not hard to see that in George's case, viewed from this perspective, the utilitarian solution would be wrong" (p. 117).

This concern cannot be adequately addressed until later, after the importance of complicated, long-term projects in humans' lives has been discussed at length in chapters 6 and 7, but here I can sketch the form that my reply will ultimately take at the end of §7.3(C).

In chapters 6 and 7, I will argue that persons have what Williams calls "ground projects" because they take a narrative perspective on their lives.

Persons are, quite literally, the authors of their own identities insofar as they adopt a self-narrative, and these narratives involve the pursuit of certain ground projects. For this reason, Williams is correct that it is in one sense a threat to one's integrity to conclude, as the result of Harean critical thinking, that one should abandon or severely compromise one's ground projects. Reaching this conclusion does threaten one's integrity insofar as acting on it dooms one's ground projects, and the pursuit of these projects is integral to one's identity as it is embedded in one's life story. But doing explicitly utilitarian critical thinking is not a threat to one's integrity as a moral agent. For taking the moral point of view requires us to adopt an attitude of indifference between our own and others' interests, and this is why morality can require sacrifice, up to and including everything that one values in life.

As I emphasized above, Williams himself admits that in Jim's case "the utilitarian is probably right," but he thinks that "it is not hard to see that in George's case . . . the utilitarian solution would be wrong." If ground projects are what give persons' lives special moral significance, however, then it is not so obvious what "the utilitarian solution" would be in George's case. Given that—as I will argue in chapters 6 and 7—ground projects define the most important preferences of persons, and their pursuit is what gives persons' lives special moral significance, something important must be at stake to justify dooming a ground project. The military-industrial complex will churn along just the same whether he takes the job or not, so if George's taking the job really would severely compromise his ground project, it is not so obvious that utilitarian critical thinking would lead him to conclude that he should take it.

For his part, Williams does not specify how one could make such judgments on non-utilitarian grounds; indeed, when he says (in the long quotation above) that "there may be no adequate answer at all" he seems to doubt that such a decision could be made on the grounds of *any* principle. How that is supposed to be preferable to utilitarianism escapes me. Williams is correct that taking an explicitly utilitarian perspective on one's own commitments is, in a sense, "alienating," but being indifferent between one's own and others' interests is precisely what taking the moral point of view requires, and the principle of utility embodies this. It is a disquieting feature of human life that we are both self-determining beings who choose the ground projects that determine our most important interests, and moral agents who are capable of the most profound self-denial.

§4.12. Summary and Transition to Part II

Descriptively, Hare's theory gains plausibility from the fact that the golden rule is universally acknowledged as a commonsense expression of what it means to take the moral point of view (§4.2), and from research indicating that moral agents employ both a more emotion-based and deontological system, and a more reason-based and utilitarian system (§4.5). It also explains the development of alternative moral theories and views in metaethics (§4.4), and it helps to explain how moral thinking might have evolved in humans (§4.3). Furthermore, as was illustrated briefly in this chapter (§4.6) and will be emphasized in part III of this book and the sequel (Varner forthcoming), Hare's theory also sheds considerable light on various real-world controversies.

As a theory of normative ethics, Hare's Kantian approach to justifying his theory (chapter 2) is a virtue, given the problems with the method of reflective equilibrium that were described in §1.4, and his two-level version of utilitarianism does a good job handling the standard "test cases" for utilitarianism (§§4.7–4.11). Chapter 3 emphasized how the nature of ILS rules introduces a number of complications, but in part III of this book I will emphasize how these complications result in a much more nuanced analysis of animal ethics issues than either the standard animal welfare stance or the standard animal rights stance. By incorporating elements of both (as described in §1.7), Hare's theory emphasizes rather than obscures the complexity of these issues, but where the issues are hard, a theory that gives clear-cut answers probably isn't a good theory. For if the right answers to the issues were really so clear-cut, then the best minds of our generation wouldn't find them hard.

For the foregoing reasons, Hare's theory deserves—on its own merits—the kind of detailed attention that I am giving it in this book. Even if it didn't deserve so much attention on its own merits, however, it would be a worthwhile exercise to apply it carefully to animal ethics issues for the reasons given in §1.1: Hare was Singer's dissertation advisor, Singer claims to endorse a Harean version of utilitarianism, and he invokes Hare's distinction between the intuitive and critical levels of moral thinking in response to certain objections. Neither philosopher has provided the kind of detailed application of two-level utilitarianism to animal issues that is provided in parts II and III of this book, however.

Having laid out and defended Harean utilitarianism here in part I, in part II I will turn to the place of personhood in the theory. In chapter 5 I first address the question of which animals are sentient, for this defines the outer limit of the moral community for a Harean utilitarian. I will then spell out, in chapters 6 and 7, the reasons that a Harean utilitarian has for recognizing a distinction between "merely sentient" individuals and "persons," defined as individuals with a biographical sense of self. I give parallel reasons for recognizing a third category of "near-persons," defined as individuals who have a fairly robust sense of their own past, present, and future, but who lack a biographical sense of self. At the end of part II, Hare's theory will be significantly supplemented by an account of what gives humans' lives special moral significance, and by the distinction between near-persons and the merely sentient, which can justify significantly different ILS rules to follow in our dealings with different animals, as will be illustrated in part III.

{ PART II }

Persons, Near-Persons, and the Merely Sentient

{5}

Which Animals Are Sentient?

§5.1. Sentience, Phenomenal Consciousness, and Pain

In his later works, Hare mentioned that sentient animals certainly have moral standing on his theory (*MT*, pp. 90–91; *EB*, p. 82), and he credited his former graduate student Peter Singer with drawing his attention to the moral standing of animals (*EB*, p. 221). Hare published only one paper that was focused on issues involving animals, however—"Why I am only a Demi-Vegetarian," first published in 1993 as chapter 15 of his *Essays on Bioethics*—and he nowhere systematically investigated the question of which animals are sentient. Yet this is a key question, because sentience defines the outer limit of the moral community for a Harean utilitarian. Informally stated, Hare argues that the logic of moral discourse requires you to choose as if you had to "stand in the shoes" of everyone affected by your actions. In those informal terms, the answer to the question "Which animals are sentient?" tells us who has shoes to stand in.

In this chapter, I consider how far sentience extends in the animal kingdom. The bulk of the chapter concerns what I call the "standard" argument by analogy concerning pain, which concludes that (with some exceptions perhaps) vertebrates probably are conscious, while invertebrates probably are not. After getting clear about the meanings of "sentience" and some other key terms in this first section, I will discuss animal consciousness and arguments by analogy consciousness in §5.2, and then present the standard argument in §5.3. In §5.4, I respond to two important criticisms of the standard argument. The final section of this chapter is a broader consideration of "what consciousness might do for an organism," apart from helping it to avoid noxious stimuli, as in the case of pain. The upshot of this first chapter of part II is that good ILS rules governing our treatment of animals will cover at least all vertebrates. The remaining three chapters in this second part of the book will complicate the picture by arguing that a Harean utilitarian should recognize a distinction among persons, near-persons, and the merely sentient, and that while we have no good evidence that any non-human animals are persons, at least some animals probably qualify as near-persons. The extended illustration of Harean utilitarianism in practice provided in part III of this book will include some illustrations of how good ILS rules might therefore distinguish between animals that are merely sentient and those that are near-persons.

According to the *Oxford English Dictionary*, the word "sentient" is derived from the Latin *sentient-em*, the present participle of *sentīre*, meaning "to feel." So, etymologically, "sentient" just means conscious of something or other. In the animal ethics literature, however, the term is usually assumed to mean capable of feeling pain, specifically, meaning physical, bodily pain. When Peter Singer adopted the term in *Animal Liberation*, he characterized it as "a convenient if not strictly accurate shorthand for the capacity to suffer and/or experience enjoyment" (1990, p. 8 [= 1975, pp. 8–9]). Singer went on to focus on pain when discussing the scope of sentience in the animal kingdom, but it would be a mistake to equate "suffering" with "pain." For one can suffer from depression without feeling pain, and a masochist is said to enjoy pain. Even normal people describe themselves as sometimes experiencing pain without *suffering* from it, and for this reason David DeGrazia defines "suffering" as "a highly unpleasant emotional state associated with more-than-minimal pain or distress" (DeGrazia 1996, p. 116).

It is also possible to suffer and enjoy things without being able to feel pain at all, as illustrated by the fact that approximately 100 people worldwide suffer from congenital insensitivity to pain (CIP). Patients with CIP lack functional nociceptors (the peripheral elements of the nervous system that register tissue damage) and are completely unable to feel pain. The condition is highly maladaptive, as illustrated in the 2005 film *A Life Without Pain*, directed by Melody Gilbert (http://alifewithoutpain.com), which documents the extraordinary measures that parents must take to avoid serious injury to their children with CIP. One child chewed her fingers bloody while teething until her parents forced her to wear arm restraints. Later, she began rubbing her eyes so hard that she did serious damage to one of them, so they forced her to wear goggles constantly. A second child once ironed her hand because she didn't want it to be wrinkled, and a third failed for two weeks to notice that she had broken her leg. The children in Gilbert's film describe feeling only pressure when pricked with a pin in a way that would cause a normal person to feel a sharp pain, but they are otherwise emotionally normal, expressing joy, humor, anger, fear, depression, disappointment, and so on. Clearly, they are capable of many kinds of suffering and enjoyment, despite being completely unable to experience pain.

Of course "pain" is sometimes used in a wider sense in which it is synonymous with suffering, for instance in "It pained him to see the Republicans take control of Congress." This broad usage is reflected in the etymology of the word. According to the *Oxford English Dictionary*, "pain" is derived from the classical Latin word *poena*, meaning "penalty" or "punishment." When I speak of "pain," however, I will always mean physical pain, specifically, rather than "pain" in this broader sense.

A typical clinical definition of "pain" in this physical sense would be: *an aversive bodily sensation normally associated with actual or potential tissue damage*. This is a slight variation on the formulation given by David DeGrazia: "an unpleasant or aversive sensory experience typically associated with actual or potential tissue damage" (1996, p. 107). I say "normally" rather than "typically" in order to highlight the evolutionary role of pain—it is abnormal to feel pain when there is no tissue damage, as in phantom limb pain and other forms of chronic pain. And I say "bodily sensation" rather than "sensory experience" because pain is normally a bodily sensation.

Two general kinds of pain have been identified that feel different and have different evolutionary functions. "Fast pain" is the first pain that arrives during injury or upon contact with a stimulus that will cause tissue damage if contact is allowed to continue. Typically, we respond to it by rapidly withdrawing from the harmful stimulus, often reflexively. "Slow pain" arrives after an injury has already happened. We typically respond to slow pain by favoring or massaging the injured limb or tissue. The two types of pain are also mediated by different kinds of nerve fibers. Fast pain depends on the activity of Aδ (A-delta) fibers, slow pain on the activity of C fibers.

On the opposite pole from pain, we need to be careful about the words "pleasure" and "pleasant." When "pleasure" is narrowly construed as physical pleasure, like those associated with eating, drinking, and sex, psychological hedonism (the view that human beings are, ultimately, only motivated by "pleasure") sounds implausible. But if "pleasure" is construed broadly, to include all "pleasant" states of mind, including those associated with intellectual endeavors, the sense of accomplishment, doing the right thing, and so on, then psychological hedonism sounds more plausible. As Sidgwick observed, in its broadest connotation, "pleasant" is synonymous with "desirable" (1907, pp. 44–45, 127–128).[1]

Finally, the term "conscious" is used in several discrete senses in day-to-day speech (see Allen & Bekoff 1997, p. 143; Carruthers 2000, chapter 1, section 3). In the question "Is he conscious yet?" the word means awake rather than asleep or drugged, or "responsive" versus "unresponsive." In "Were you conscious of that?" it is used to refer to a mental state that is available for conscious reflection. But when we ask, "Are animals conscious?" we use it to refer to the general capacity for what philosophers call "phenomenal" consciousness. Phenomenal consciousness is the subjective "feel" of our lives as we experience them, and as such, phenomenal consciousness cannot be studied directly.

---

[1] Compare Mill's stronger claim that "desiring a thing, and finding it pleasant" are "in strictness of language, two different modes of naming the same psychological fact" ([1861] 1957, p. 49). This is only true of "finding it pleasant" in the broadest sense, not of "getting pleasure from it," however, and this is why Mill's competent judge test and his distinction between "higher" and "lower" pleasures makes it such a stretch for him to define "happiness" as "pleasure and the absence of pain" (p. 10).

Phenomenal consciousness is simultaneously extremely important and deeply mysterious. It is deeply mysterious because it is so hard to define or characterize clearly. We use mysterious expressions like "the subjective 'feel' of our lives or experiences" to describe it, and, in a phrase made famous by Thomas Nagel (1974), we say that for a phenomenally conscious being "there is something it is like to be that being." At the same time, phenomenal consciousness is extremely important, because we each value our own lives in large measure as a function of the positive (and negative) phenomenally conscious experiences they contain. This is illustrated by the fact that most of us would be indifferent to the option of going on living for many years, even if we behaved in complicated ways, if we were to do so as "zombies," stripped of phenomenal consciousness.

In summary, as I use the term, "sentience" refers to the capacity for phenomenally conscious suffering and/or enjoyment. I assume that it is a contingent fact about life on earth that only a very few, abnormal individuals are sentient without also being able to feel pain (by which I always mean physical pain). In fact, given how maladaptive the condition is, it seems unlikely that any individual with CIP would survive to adulthood outside human society, and humans with CIP are the only individuals we know of that are capable of suffering and/or enjoyment who are not also capable of feeling pain. So while determining which animals can feel pain is a convenient way of determining which animals are sentient, strictly speaking the capacity for pain is a sufficient but not a necessary condition for being sentient.

§5.2. Animal Consciousness and Arguments by Analogy

Arguments by analogy have the following general form, where properties *a*, *b*, *c*, . . ., and *n* represent some number of observed comparisons between two things *P* and *Q*, and *x* represents the property in question, which has been observed in *P* but has not been (or cannot be) observed in *Q*:

1. We know that both *P* and *Q* have properties *a*, *b*, *c*, . . ., and *n*.
2. We know that *P* has property *x* as well.

C. So *Q* probably has property *x*, too.

Arguments by analogy are commonly used to argue about things that it is *impossible* to observe. This is true of the so-called "argument from design":

1. Both various artifacts (*P*) and the world (*Q*) show various signs of intelligent design (*a*, *b*, *c*, . . ., and *n*).
2. All artifacts are known to have been created by an intelligent designer (*x*).

C. So the universe probably was created by an intelligent designer too.

As Hume ([1779] 1947, pp. 149–151) famously observed, the creation of the universe was a unique and unobservable event; as worldly creatures, we could not, even in principle, observe the world's creation.

Arguments by analogy are also commonly used when the question is which animals are conscious, because consciousness—at least in the morally relevant sense—cannot be directly observed in others. Each of us directly experiences our own phenomenally conscious states, but we can never literally observe phenomenal consciousness in anyone else. When it comes to the question of which animals are conscious in this sense, we assume that we know that other human beings (at least normal, mature ones) are phenomenally conscious, and "human beings" becomes *P* in the above schema, with *Q* being animals of the species in question. It is important to realize, however, that even in the case of our fellow human beings, we never directly observe their phenomenally conscious mental states. We sometimes say things like "I feel your pain," but of course we don't mean that literally. We are very confident of our knowledge that other humans are phenomenally conscious, because the relevant analogies are so extensive. Our fellow human beings are, we presume, thoroughly analogous in terms of the relevant physiology and behavior, and we share the same evolutionary history. In the case of our fellow human beings we also have a particularly nuanced type of behavioral analogy: we can compare introspective descriptions of our phenomenally conscious states.

Part of the reason that the question of which animals are phenomenally conscious is thought to be so difficult stems from the fact that they do not talk. If an animal could describe the phenomenal character of its mental states, most of us would believe that it was conscious, unless its "speech" could be exposed as some kind of hoax. Perhaps some animals are capable of being taught to communicate in relevant ways using specially constructed artificial languages, or perhaps some animals already can communicate some aspects of their phenomenally conscious mental states to us if we "ask them the right questions" in an experimental setting or if we attend to their natural communicative expressions with adequate care. For the foreseeable future, however, language gives each of us an especially detailed form of behavior to plug into an argument by analogy for phenomenal consciousness in our fellow human beings. Only they can tell us, after we have explained to them what we mean by the terms, that they not only know what happened yesterday but that they also "consciously remember" the experience, that they are now "consciously aware of" the sound that the motor is making, that the pain in their toe today feels more like a "throb" than a "shooting" pain, and so on. Only human beings can describe to us in such detail their phenomenally conscious experiences, and because we directly experience phenomenally conscious states that we would describe in similar terms, we feel perfectly confident that our fellow human beings are having similar, phenomenally conscious experiences. So only in the arid environment of philosophy courses do people have any doubt that "there is something it is like to be" one of their fellow human beings, and much of the time, we think that we have a pretty good handle on what, specifically, our fellow human beings are feeling.

When it comes to non-human animals, however, opinions diverge. Some of us assume that various animals are physiologically similar to us in relevant ways because their behaviors seem so similar, at least in some range of cases in which we feel basic emotions or pain. Others sincerely doubt that animals "feel pain" in the same way that we do and confidently deny that animals can ever literally "suffer." Scientists tell us that the brains of fish, herps (reptiles and amphibians), birds, and mammals evolved in fundamentally different ways and that we rely heavily on a part of the brain that is entirely absent in non-mammals (the neocortex), while birds rely heavily on a part of their brain that is entirely absent in ours (the diencephalon). In light of these fundamentally different nervous systems, and lacking the detailed behavioral analogies that language can provide, there are sincere skeptics about animal consciousness outside the philosophy classroom. Even among those who know certain types of animals very well, there are some who deny that "there is anything it is like to be" those animals.

§5.3. The "Standard" Argument by Analogy Regarding Pain

Nevertheless, among panels of experts assembled to address animal welfare questions from a public policy perspective, there has emerged something of a consensus on what I call the "standard" argument by analogy. In the mid-1990s, I put together table 5.1. It lists six observable comparisons between normal humans and various categories of animals that were mentioned in four metastudies of pain in the animal kingdom that were available by the early 1990s. They were "metastudies" in that they did not report any newly conducted research; rather, they reviewed the available research that they deemed relevant to consciousness of pain in the animal kingdom and offered their assessment of what light the evidence then available shed on the question of which animals could feel pain. The plus marks ("+") indicate where the authors of the four metastudies concluded that there was good evidence supporting the comparison, the minus signs ("–") indicate where they concluded that there was good evidence against the comparison, and the question marks ("?") indicate where the evidence was judged inconclusive. My table was based on a similar one provided by a multidisciplinary working party of the British Institute of Medical Ethics (Smith & Boyd 1991, p. 63). I modified a few cells and qualified the entries in the cells with a tedium of footnotes, so that my finished chart incorporated the findings of the three other metastudies (Rose & Adams 1989, DeGrazia & Rowan 1991, and Bateson 1991).

The first comparison listed is the presence of nociceptors. These are specialized elements of the peripheral nervous system whose function is to respond to

TABLE 5.1. Comparisons Relevant to Consciousness of Pain in the Animal Kingdom, circa 1995

	Invertebrates			Vertebrates			
	Earthworms	Insects	Cephalopods	Fish	Herps[a]	Birds	Mammals
1) Nociceptors present	?	-	?[b]	-[c]	-[c]	+	+
2) Brain present	-	-	+[d]	+	+	+	+
3) Nociceptors connected to brain	-	-	?/+[e]	?/+[f]	?/+[f]	?/+[f]	+
4) Endogenous opiods present	+	+	?	+	+	+	+
5) Responses modified by analgesics	?	?	?	?	?	+	+
6) Response to damaging stimuli analogous to that of humans[g]	-	-	+[h]	+	+	+	+

Note: A "+" indicates a positive comparison between the animals in question and normal humans, a "-" indicates a negative comparison, and a "?" indicates inadequate data for making the comparison.

a Reptiles and amphibians have been collapsed into a single category. Smith and Boyd score both taxa identically.
b Notice, however, Rose and Adams's (1989) conclusion cited in note (c) below. With regard to cephalopods, they argue that responses to electric shock are evidence of nociception (p. 48). As noted in §5.5(A)(ii), however, even decapitated cockroaches can be habituated to avoid shocks. The more salient reason for expecting to find nociceptors in cephalopods is the overall similarity of their behavior to that of humans in situations that would cause us pain (note [h] below).
c Rose and Adams do conclude that "Evidence supports the existence of nociception in all vertebrates" (p. 49), but this is on the basis of avoidance behavior rather than identification of functioning nociceptors. Smith and Boyd's "-" implies that here, unlike in the case of earthworms and cephalopods, a disciplined (but still unsuccessful) search for nociceptors has been conducted.
d Changed from "?" to "+" in light of Wells 1978 and Young 1965.
e Changed from "?" to "?/+" in light of the fact that both Young's and Wells's "models of learning in octopi include a 'pain' pathway leading to the vertical lobe of the brain," cited in Smith and Boyd, p. 64.
f Changed from "?" to "?/+" in light of Rose and Adams's discussion of spinothalamic connections, pp. 50–51 (concluding that "All vertebrates posses neural connections between peripheral nociceptors and central nervous structures").
g This category collapses two of Smith and Boyd's: (6) "The animal's response to stimuli that would be painful for a human is functionally similar to the human response (that is, the animal responds so as to avoid or minimize damage to its body)"; and (7) "The animal's behavioral response persists and it shows an unwillingness to resubmit to a painful procedure; [and] the animal can learn to associate apparently nonpainful with apparently painful events" (p. 62). Smith and Boyd give fish and herps a "?" on category #6 and a "+" on category #7.
h Changed from "?" to "+" in light of Smith and Boyd's own discussion of the response of octopi to pain, pp. 64–65 (that octopi have "distinct and easy-to-recognize responses to noxious stimuli").

Source: Reprinted from Varner 1998, p. 53, table 2.3, which was adapted, with noted modifications and qualifications, from Smith and Boyd (1991, p. 63).

damaging or potentially damaging stimuli such as intense pressure, high temperatures, torn skin, and the presence of caustic chemicals. The second comparison listed is having a brain, because consciousness in humans is thought to depend on the general integrity of the brain. The third comparison, nociceptors being connected to the brain, is mentioned because nociceptors can trigger spinally mediated reactions without producing conscious pain, as when the limbs of patients with spinal cord injuries continue to exhibit withdrawal reflexes after they have lost all conscious sensation in the limbs. (My original version of this table described the third comparison as "nociceptors connected to central nervous system," but since the central nervous system includes the spinal cord, in this version I have changed the wording accordingly.) The presence of endogenous opioids is the fourth comparison because in humans these substances are known to block conscious pain (or at least the affective component of pain, as discussed in §5.4) during intense concentration or exertion. Finally, the metastudies considered whether animals' general responses to painful stimuli are modified by known analgesics and whether animals learn to avoid them, favor or rub damaged areas, and so on, that is, whether or not their responses to damaging stimuli are analogous to those of humans.

Today I would modify several entries in this table in light of evidence that has emerged since the mid-1990s. First, as explained in footnote (c) to table 5.1, one of the metastudies that I consulted (Rose & Adams 1989) claimed that "Evidence supports the existence of nociception in all vertebrates" (p. 49), but that was based on avoidance behavior rather than identification of functioning nociceptors. Smith and Boyd's table included a minus sign on nociceptors for fish and herps, implying that here, unlike in the case of earthworms and cephalopods, a disciplined but still unsuccessful search for nociceptors had been conducted. Nociceptors have now been positively identified in fish (Sneddon et al. 2003a) and frogs (Downes et al. 1999), however, so I would replace the minus signs in row one under fish and herps with plus signs. Second, in identifying nociceptors around the mouths of trout, Sneddon and colleagues (2003a) also verified that signals from them reached the brain via the trigeminal nerve, so the question mark can be removed from row three, at least under fish. Finally, Sneddon and colleagues have also shown that trout injected with irritants do not respond normally to novel objects (they normally avoid them) but their normal response is restored if they are treated with morphine (Sneddon et al. 2003b). So the question mark in row five can be removed, at least for fish. These findings alter my original 1998 chart as represented in table 5.2.

The standard argument does not hinge on any one comparison. Rather, the evidence taken as a whole suggests the conclusion, because there are almost all plus signs under the vertebrates, but mostly minus signs or question marks under the invertebrates (with the exception of the cephalopods). As one of the studies put it, "Much remains to be discovered, but taking the evidence as a whole, the most obvious divide is between the vertebrates and the invertebrates," with the caveat that "One invertebrate group stands out as being different from the others—the cephalopods" (Smith & Boyd 1991, p. 64).

Relatedly, at the time that Sneddon and colleagues published their paper confirming functional nociceptors in trout, a series of studies of nociceptor-like receptors in cartilaginous fishes (e.g. rays and sharks) had been done (Coggeshall et al. 1978, Leonard 1985, and Snow et al. 1993), but these had concluded that "the receptors were not responding in the way true nonciceptors do" (Braithwaite 2010, p. 48). So Sneddon and colleagues proposed that

TABLE 5.2. The "Standard" Argument by Analogy, Initial Update (see also table 5.3)

	Invertebrates			Vertebrates			
	Earthworms	Insects	Cephalopods	Fish	Herps	Birds	Mammals
1) Nociceptors present	?	-	?	+	+	+	+
2) Brain present	-	-	+	+	+	+	+
3) Nociceptors connected to brain	-	-	+	+	?/+	?/+	+
4) Endogenous opiods present	+	+	?	+	+	+	+
5) Responses modified by known analgesics	?	?	?	+	?	+	+
6) Response to damaging stimuli analogous to that of humans	-	-	+	+	+	+	+

Source: Modified from Varner 1998, p. 53, table 2.3, for the reasons given in the text.

consciousness of pain may have evolved in the early bony fishes (telosts), from which the herps, birds, and mammals all radiated (2003a, p. 1120). Of course, convergent evolution could have produced consciousness of pain in some or many invertebrates, too, but the fact that all the other vertebrates radiated from early telosts provides an evolutionary explanation for the fact that the observable comparisons listed in table 5.2 are uniformly positive throughout the vertebrates (other than the cartilaginous fishes).

So the conclusion of the standard argument is that probably all vertebrates can feel pain, while invertebrates (with the exception of cephalopods) probably cannot. It would be a mistake to conclude from this that we know that invertebrates are not sentient. Rather, the upshot of the standard argument is that good ILS rules regarding how we treat animals cannot exclude any vertebrates the way the U.S. Animal Welfare Act does. As amended in 1970, it protects only *warm-blooded* animals (Public Law 91–579, section 3[g]).

§5.4. Allen's Criticisms of the "Standard" Argument

That the "standard" argument by analogy has been widely embraced is reflected in the authorship of the four metastudies that I originally summarized in table 5.1. These include philosophers, but also various scientists, veterinarians, and medical researchers. The Working Party that wrote the report edited by Smith and Boyd included:

> basic biological and ethological scientists, toxicologists, physicians, and veterinary surgeons with particular experience of research using animals, a Home Office inspector, an expert in alternatives to the use of animals in research, and officers of animal welfare organizations. The Working Party,

> which was chaired by a moral theologian, also included philosophers and a lawyer, and its research staff comprised a biologist and an ethicist. (Smith & Boyd 1991, p. 1)

But Colin Allen (2004) claims that the standard argument is flawed in two fundamental ways: (1) it is prone to being undermined by new evidence, and (2) we cannot rank the presented analogies in terms of relevance without a "guiding theory" of what consciousness does for an organism.

Allen observes that after I had finished the manuscript of my 1998 book, nociceptors were found in frogs, and also "invertebrates such as the leech . . . and the marine snail" (Allen 2004, p. 617, citing Walters 1996). His second objection jumps off from that observation:

> Thus it seems rather likely that the indicators in Varner's table noting the lack of evidence for nociceptors among invertebrates, fish, and amphibians and reptiles represent just that—namely a lack of evidence, not a lack of nociceptors in those groups. The direction of discovery here seems uniformly towards identifying *more* similarities between diverse species, especially in the domains of anatomy and physiology. What's left, then, are behavioral differences. But to draw strong distinctions primarily on the basis of behavioral differences such as the tendency to vocalize when in pain seems like a chauvinistic response favoring forms of life that are similar to our own. Of course, there are other behavioral comparisons besides vocalizations, particularly in the domain of learning . . . and in behavioral responses to pharmacological interventions, but in the absence of a guiding theory, it is virtually impossible to decide how to weight these various dimensions of similarity for the purposes of constructing an argument by analogy to the human case. (p. 623)

That is, to decide how salient the various comparisons listed in a table like mine are, we need a theory about what consciousness of pain does for an organism.

Allen's point about needing such a "guiding theory" can be illustrated with this obviously bad argument by analogy:

1. Both turkeys (*P*) and cattle (*Q*) are animals, they are warm blooded, they have limited stereoscopic vision, and they are eaten by humans (*a*, *b*, *c*, . . ., and *n*).
2. Turkeys are known to hatch from eggs (*x*).

C. So probably cattle hatch from eggs, too.

One could come up with more and more analogies to list (e.g., turkeys and cattle both have hearts, they have lungs, they have bones, etc., etc.). The above argument is weak, not because of the number of analogies considered, but because it ignores a crucial *disanalogy*: that cattle are mammals, whereas turkeys are birds, and we have very different theories about how the two are conceived, and how they develop through to birth and hatching, respectively. Another way of putting the point would be to say that the listed analogies are irrelevant because we have a "guiding theory" about the various ways in which reproduction occurs, and within that theory the analogies listed above are all irrelevant. Another illustration of the role of "guiding theories" in assessing arguments by analogy is provided by the argument from design as represented earlier. That argument is weak insofar as we have, in the theory of evolution, a good theory of how the appearance of design can arise in living organisms without the intervention of an intelligent designer, but we have no good theory of how this can happen in the case of artifacts.

So in assessing an argument by analogy, we do not just look at the raw number of analogies cited. Rather, we look at both how salient are the various analogies cited and whether there are any relevant disanalogies, and we determine how salient various comparisons are by reference to a "guiding theory."

In the remainder of this section, I respond to Allen's criticisms. In the process, I will offer a speculation about what, in evolutionary terms, consciousness of pain "does for" an animal and a related, observable analogy between normal humans and other animals that can be added to the standard argument by analogy.

§5.4(A). "THE RUMSFELD RESPONSE" AND TWO CAVEATS

My first response to Allen's criticisms of the standard argument might be labeled "the Rumsfeld response." For it is a paraphrase of what U.S. Secretary of Defense Donald Rumsfeld said in response to criticism that the best body armor and "up-armored" Humvees were not available to all U.S. troops going into Iraq. Rumsfeld reportedly said: "As you know, you go to war with the Army you have. They're not the Army you might want or wish to have at a later time" (*Washington Post*, December 15, 2004, p. A33). Similarly, I say: *You have to make ethical decisions with the evidence you have, not the evidence you might want or wish to have.*

Perhaps a scientist or a philosopher specializing in the philosophy of mind can afford to suspend judgment indefinitely. Ethicists and legislators must make decisions on the best evidence available at the time of the decision, however, and everyone is an ethicist in day-to-day life. I usually describe myself an "ethical theorist" rather than an "ethicist," because the popular media portray "ethicists" as people who have an answer to every ethical question addressed to them, and (as you'll see later in this book) my answer usually is, "Well, that depends on what the facts are." When I say that in everyday life "everyone is an ethicist," however, I don't mean that everyone has an opinion to express in response to every ethical question. All I mean is that each of us makes countless ethically charged decisions, many of which cannot wait, and that we must make them without all of the relevant information we might want or need. This is most obvious in the case of legislators, who must frequently establish policies with widespread and weighty ramifications based on incomplete information, but it's true of the least influential among us as well.

I admit that Allen is correct that the picture regarding animal pain is still radically incomplete at two levels. First, there are new findings coming in fast, and lots of relevant research remains to be done. Second, there is no consensus at the theoretical level about what consciousness in general is, or about why phenomenal consciousness of pain in particular would be adaptive for an organism. This is because evolution acts on what an organism *does*, and phenomenal consciousness is as invisible to evolution as it is to us. Nevertheless, the question of which animals are capable of feeling pain is a pressing public policy issue, and, from a Harean perspective, we must take pain seriously wherever it occurs. So I think that we have no choice but to rely on arguments by analogy.

I want to emphasize two caveats, however. The first is that we should downplay any analogy that we have reason to believe should not generalize across taxa. The significance of this caveat is evident from the example of vocalization in the above quote from Allen. When it comes to the question of which behavioral comparisons are relevant to consciousness of pain, vocalization should not count if facts about a given species' evolutionary history give good reason for discounting it. For instance, in species that live solitary lives from birth, vocalizing when injured would not be adaptive—cephalopod species would probably be a good example (some fish do in fact vocalize in certain contexts, including the plainfin midshipman, the Atlantic croaker, and the African pompano; Richard Moccia, personal communication).

My second caveat relates to Allen's second objection, that we cannot rank presented analogies in terms of relevance without a "guiding theory" of what consciousness (or a specific form of it, like pain) does for an organism. My related caveat is that where we do have such a theory, the relevant analogies suggested by the theory should be given special weight in setting public policy. In the next subsection, I will discuss recent research on the medial and lateral pain pathways in the mammalian nervous system. After discussing that research, in the subsequent subsection I will be able to provide an answer, albeit a very speculative one, to the question of what evolutionary function consciousness of pain, specifically, might have, and that will allow me to illustrate my second caveat in action.

§5.4(B). TWO PAIN PATHWAYS IN MAMMALS

Scientists who study pain in humans conceive of it as having two components: affective and sensory. The sensory component consists of information about the location and tactile qualities of the pain. For instance, on the commonly used McGill Pain Questionnaire, the "sensory" categories (#1–10) include: "aching," "boring," "gnawing," "sharp," "shooting," "stabbing," "throbbing," and so on. The affective component, in contrast, consists of the "bad feel," it concerns how aversive the pain is, rather than its location and tactile qualities. For instance, the "evaluative" category on the McGill Pain Questionnaire (#16) consists of the words: "annoying," "troublesome," "miserable," "intense," and "unbearable." Equally aversive pains can have very different sensory components. For instance, patients normally rank both arthritis pain and toothaches as "annoying," but they describe them using very different terms from the sensory categories. Arthritis pain is described as "gnawing" and "aching," whereas toothaches are described as "throbbing," "boring," and "sharp." (This example is based on Melzack and Wall 1988, p. 42, table 1. The entire 20 categories on the McGill Pain Questionnaire are given in figure 3, p. 40.) The McGill Pain Questionnaire has come into frequent use because patients with similar diseases and injuries use similar terms to describe the sensory and affective dimensions of their pains. In fact, when it comes to distinguishing between various pairs of pain syndromes, computer programs can successfully diagnose the problem with near perfect accuracy just by asking the patients which words best describe their pains (Melzack & Wall, pp. 43–44).

In humans, it is well established that the affective component of pain can, under some circumstances, be attenuated without attenuating the sensory component to the same degree. A famous example is patients on certain doses of morphine, who are said to still "feel" the pain, because they describe it using the same set of sensory terms on the McGill Pain Questionnaire, but they describe it using less aversive terminology from the evaluative category. The explanation appears to be as follows. Pain signals ascend through our brains along both a lateral and a medial pathway, with the medial pathway strongly associated with the affective, and the lateral pathway strongly associated with the sensory component. The medial pathway has far more opiod receptors along it, and the medial pathway projects through the anterior cingulate cortex (ACC). Damage to the ACC attenuates the affective component of pain similarly to morphine, and PET scans show that activity in the ACC varies proportionally with the experienced unpleasantness of pains.[2] Taken together, these facts explain why patients on morphine would say that they still feel the same pain (they describe it as having the same sensory component), but that they no longer mind it so much (they describe it as less aversive in affective terms).

In "Minding Mammals," Adam Shriver describes two lines of research indicating that the affective and sensory components of pain are similarly dissociable along the same two pathways in non-human mammals. First, experiments by

[2] "Interestingly, excitation in the same areas of the ACC has been observed in a range of circumstances as diverse as social exclusion, human mothers hearing the distress cries of infants, and increased aversive sensation due to expectation of pain. Thus, a wide range of behavior described as negative affect states in humans appear to selectively activate regions of the ACC" (Shriver 2006, p. 437, references removed).

Perry Fuchs and colleagues (LaGraize et al. 2004) induced hyperalgesia (increased sensitivity to the affective component of pain) in the left paws of rats (by ligating an associated nerve) and then introduced them to cages with light and dark areas. The cages were rigged so that the rats received shocks in either area of the cage, but in the dark area the shock was administered to the left, hyperalgesic paw, while in the light area the shock was administered to the right, normal paw. Rats normally prefer darker areas of their cages, but under these conditions, the rats preferred the light area. After their ACCs were lesioned, however, they preferred the dark area again, even though their withdrawl reflexes from shocks were unaffected, "thus producing something of a functional equivalent to the reports of 'feeling pain but not minding it' by humans with ACC lesions" (2006, p. 437). He then describes a similar study of monkeys:

> In one task, the monkeys had to hold a button down while intense heat was applied to a section of their skin in order to obtain a food reward. The monkeys could escape the thermal pain by releasing the button, but by doing so they would lose their food reward. Using this task, the researchers found a threshold at which the monkeys were not willing to endure the pain in order to obtain the reward. In a separate task, the monkeys were rewarded for pressing a button when the temperature changed from noxious to non-noxious levels. After damaging a part of the posterior parietal cortex that provides input to the insular cortex, another important structure in the medial pathway . . ., the researchers found that the monkeys no longer utilized the escape responses but were able to discriminate noxious and non-noxious temperatures. Again, this research fits very well with the interpretation that the monkeys could still feel the pain but no longer found it unpleasant. (Shriver 2006, pp. 437–438, describing work by Dong et al. 1996)

These studies certainly suggest that the affective and sensory components of pain are processed along the same two pathways in both human and non-human mammals, making them dissociable in the same way that they are in humans.

§5.4(c). An Evolutionary Function for Phenomenally Conscious Pain?

The separate processing of the affective and sensory components of pain in the mammalian nervous system may point to an evolutionary function for phenomenal consciousness of pain. Late in his 2004 paper, Allen suggests a hypothesis about a learning-related ability that consciousness of pain might give an animal, specifically the ability "to discriminate among different pain experiences having different intensities." He describes the ability thus: "Since these different intensities are variably related to tissue damage, it might very well be adaptive to discriminate between the painful experiences *as such* so as to allow independent assessment of the severity of the underlying conditions that they purport to represent" (2004, p. 633). His point appears to be that the ability to think independently about the sensory and affective components of pain would allow both (1) different responses to two pains with different sensory components but similar affective/motivational components, and (2) similar responses to two pains with similar sensory components but different affective/motivational components.

To illustrate the first kind of case, suppose that (contrary to fact!) having the arch of your foot broken and picking up a splinter on the sole of your foot were equivalent on the affective side, that is, they "felt equally bad" and were equally aversive—maybe something like "intense" under the "evaluative" category of the McGill Pain Questionnaire. Presumably the pain of picking up a splinter in the sole of your foot would still have a different sensory component from the pain of having your arch broken. For instance, you might describe the splinter pain as "stabbing" and the pain of the broken arch as "throbbing." If you couldn't separately consider the affective and sensory components of the two pains, then you could not learn to differentially favor your foot in the two situations. For if they were affectively equivalent, then you would be equally motivated to favor the limb, whichever kind of injury you had sustained. Being able to think separately about the sensory and affective components of pain in this way would be adaptive, for if you were close to killing something for dinner tonight, continuing to run on the foot with the splinter in it would probably be "worth it," while continuing to run on the broken arch probably would not be.

To illustrate the second kind of case, suppose that once when I was running, I suffered a broken arch, that it was intensely painful on the affective dimension, that this inspired me to stop running right away, and that it quickly became evident that I was badly injured and that I would have injured myself much more severely had I kept running. Now suppose that I am running again, that I again suffer a broken arch, but that the affective dimension of the pain has been completely attenuated without the sensory dimension being attenuated at all. This is not completely unrealistic: opioids sometimes do something like this, although they do not completely dampen the aversive dimension and they tend to simultaneously dampen the sensory dimension. To make the illustration clear, however, imagine that I am experiencing a complete deadening of the affective dimension without any attenuation of the sensory dimension. If I keep on running, I will be much more severely injured, but in this case the affective dimension of the pain cannot inspire me to stop. This is a situation in which I can avoid doing myself significantly greater harm if I think about the affective and sensory components of my pain separately and reason as follows: "The last time I felt a pain that was similar to this one on the sensory component, I was badly hurt and I really needed to stop running. So even though this pain doesn't feel bad, I'm going to stop running to prevent more serious injury." By hypothesis, my currently experienced pain is not at all aversive, so I would only be motivated to stop running if I could think about the sensory and affective dimensions of the two pains independently.

That explains how the ability to process the affective and sensory components of pain separately could confer an evolutionary advantage. But now let me be explicit about how this relates to consciousness, specifically. There is, of course, no consensus on what consciousness *is*, but one family of functionalist accounts are "higher order thought theories," or "HOT theories." According to these theories, a mental state (e.g., a perception) is conscious if and only if it is available to be thought about, that is, if and only if it can become the object of a higher order thought. On such a theory, an animal that perceives things in its environment but acts unreflectively would not be conscious, but an animal that is able to think about its perceptions (e.g., by recognizing that it misperceives in some cases) would be conscious.

HOT theories come in several varieties. For instance, one could hold either that a thought, in order to be phenomenally conscious, must be *actually* targeted by a higher order thought or only that it must be *available for* targeting by higher order thoughts. There is also no consensus that HOT theories in general are the way to go, but here are a couple of considerations in favor of them.

We know that humans use representations of our environment that guide our behavior subconsciously, sometimes in fairly sophisticated ways, and sometimes these representations are not accessible to consciousness. Perhaps the best available example is blindsight, which results from damage to the striate cortex at the back of the brain. Blindsighted patients

> can detect stationary and moving stimuli, and discriminate their orientation, motion direction and wavelength. . . . [They] can voluntarily initiate responses to visual stimulation, and, under ideal conditions, can perform up to 100% correct. Nevertheless, when asked what they perceive, they insist that they do not see anything at all. . . . The patients not only state that they do not see anything, they also state that they are merely guessing; they have to be convinced by the experimenter to give it a try, and often show astonishment when told that they performed quite well. (Stoerig & Cowey 1995, pp. 149–150, 154)

HOT theories provide a plausible explanation of blindsight: damage to the striate cortex somehow prevents visual information from being targeted by higher order thoughts. So the patients are unable to think about that information, despite the fact that they use it, unconsciously, when "guessing" the answers to the experimenters' questions.

Although their interpretation is more controversial than blindsight, some quotidian examples may present cases in which we use representations that are not phenomenally conscious and that are not available for targeting by higher order thoughts. For instance, most people have had the experience of being at the wheel of the car and suddenly realizing that they have no memory of driving the last few miles of the highway, and most people have returned to the house to make sure that they locked the door. In such cases, we were obviously using visual information about the environment to pilot the car and to lock the door, but it may be that we were not phenomenally conscious of the information at the time, and according to HOT theories, what this means is that the information was not available for conscious reflection. The fact that we cannot later recall pointing the key into the lock or what we saw as we drove at least suggests that this is the case.

Finally, consider what might be the implications of endorsing a first order representational (FOR) theory of consciousness. If you want to say that first order representations of the environment are conscious, then how are you going to maintain that very simple organisms that react to representations of their environment are not phenomenally conscious? FOR theories might imply that all kinds of organisms are phenomenally conscious, which seems counterintuitive to many. You might say that it's a very short step to panpsychism from there.

If we do endorse a HOT theory of phenomenal consciousness, however, then the above story about the evolutionary advantage of being able to separately think about the affective and sensory components of pain provides us with an account of what consciousness of pain can do for an organism. On a HOT theory, pains would be conscious so long as we are able to think about them, whether we can separate the affective and sensory components or not, but the above story about the two pain pathways in mammals shows how the ability to think separately about the affective and sensory components of pain could be adaptive in a way that the ability to think about pain univocally would not. For only if you can think about them separately can you learn to continuing running in the case of the splinter but not in the case of the broken arch. If you could not separately target (i.e., be conscious of) the sensory components of the pains, then they would be equally motivating (because we have assumed that they are equivalent on the affective side, i.e., they "felt equally bad" and thus were equally aversive), and you would either stop or keep running in response to both. Similarly, in the second example, if you could not separately target (i.e., be conscious of) the sensory and affective components of the two pains associated with broken arches, then you would stop running in the first case but go right on running during the second break (because we have assumed that the second break did not "feel bad" at all, and thus was not aversive at all).

How plausible this story is depends on how often mammals encounter situations when the ability to make such choices would be adaptive. In the real world, the phenomena of hyperalgesia (increased reactivity to the affective component of pain) and anti-nociception (reduced reactivity to the affective component) suggest that there may indeed be a range of cases in which thinking about the two components of pain separately might be adaptive. Hyperalgesia can result from damage to nociceptors and withdrawal from drug addiction, but how often does an animal in the wild find itself in a hyperalgesic state when being less reactive to the affective component of pain would be adaptive? Anti-nociception can result from various stresses, but how often would an animal in the wild find itself in an anti-nociceptive condition when reacting normally would be more adaptive?

Although highly speculative, the above story about what consciousness of pain does for an organism allows me to illustrate the second caveat that I placed on the Rumsfeld response to Allen's criticisms of the standard argument by analogy in §5.4(A). The story provides us with a partial theory of what consciousness of pain could do for an organism, at least one that, like mammals, processes the affective and sensory components separately. My second caveat was that "where we do have such a theory, the relevant analogies suggested by the theory should be given special weight in setting public policy." On the assumption that the ability to separately target the affective and sensory components of pain depends on them being processed along separate ascending pathways in the nervous system, the theory suggests both a physiological and a behavioral analogy to add to the standard argument by analogy. The physiological analogy is: animals in the taxon have separate ascending pathways for the affective and sensory components of pain. The behavioral analogy is: the animals can separately attend to the affective and sensory components of pain.

The two experiments described by Shriver do suggest that in both human and non-human mammals, the affective and sensory components of pain are processed along separate ascending pathways in the nervous system. These experiments do not, however, show that the animals can separately attend to the two components of pain. For in both experiments, the animals treated pains that were similar along the sensory dimension differently when the affective dimension was manipulated by the experimenters in a way that reduced its valence to the animal, independent of the animal's voluntary attention. That is, they attended less to the affective component of a pain after structures along their medial pathways were lesioned, but that was a result of the lesions, which attenuated the affective component. But the postulated evolutionary advantage of separately processing the affective and sensory components of pain comes from the animal's ability to voluntarily attend to just one of the components while the other component remains unattenuated. How an experiment might be designed to test for that, specifically, I will leave to others.

§5.4(D). THE "STANDARD" ARGUMENT, UPDATED

The previous two subsections give reasons for adding two additional comparisons to the standard argument by analogy, as represented in table 5.3. The argument for adding those two comparisons turned on a theory about what consciousness of pain does for an animal—on the kind of "guiding theory" that Allen calls for—and that, as promised, allows me to illustrate my second caveat in action.

My second caveat on the Rumsfeld response to Allen was that where we do have such a theory, the relevant analogies suggested by the theory should be given special weight in setting public policy. As detailed in Shriver's essay, the separate processing of the affective and sensory components of pain has only been investigated to date in mammals. Taking my second caveat to heart thus means concluding that, at the present time, the argument by analogy for pain in non-human animals is strongest in the case of our fellow mammals, and weaker for all of the other taxa.

Note that my second caveat also suggests two important new directions for research. First, scientists should develop ways to adapt the research protocols

TABLE 5.3. The "Standard" Argument by Analogy, Further Updated

	Invertebrates				Vertebrates			
	Earthworms	Leeches & snails	Insects	Cephalopods	Fish	Herps	Birds	Mammals
1) Nociceptors present	?	+	-	?	+	+	+	+
2) Brain present	-	-	-	+	+	+	+	+
3) Nociceptors connected to brain	-	-	-	+	+	?/+	?/+	+
4) Endogenous opiods present	+	?	+	?	+	+	+	+
5) Responses modified by known analgesics	?	?	?	?	+	?	+	+
6) Response to damaging stimuli analogous to that of humans	-	?	-	+	+	+	+	+
7) Has separate ascending pathways for the affective and sensory components of pain	?	?	?	?	?	?	?	+
8) Can separately attend to the affective and sensory components of pain	?	?	?	?	?	?	?	?

Source: Modified from table 5.2 above, for the reasons given in the text.

described in Shriver's essay to other, non-mammalian animals. This would address the physiological comparison that is suggested by the guiding theory. Second, they should develop protocols that test for the behavioral comparison: the ability to separately consider the two components of pain by selectively attending to one when the other's salience has not been manipulated by the experimenters.

§5.5. What Might Consciousness of Things Other Than Pain Do for an Organism?

The foregoing sections considered the "standard" argument by analogy concerning animal consciousness. That version focuses on consciousness of pain, specifically. Pain is important for two reasons. First, it is (normally at least) an undesirable form of consciousness, and in more than minor amounts, pain constitutes suffering. Second, as noted in §3.2, it is a contingent fact about life on earth that only a very few, abnormal individuals are sentient without also being able to feel pain, specifically. So the capacity for pain can be expected to figure prominently in a good set of ILS rules governing our relations with nonhuman animals.

It is worth considering, however, what consciousness might do for an organism apart from helping to mediate its reactions to noxious stimuli in the ways described above. Recall the general form of an argument by analogy:

1. We know that both *P* and *Q* have properties *a*, *b*, *c*, . . ., and *n*.
2. We know that *P* has property *x* as well.
C. So *Q* probably has property *x*, too.

Apart from pain, a popular type of argument by analogy concerning consciousness in animals plugs in various kinds of learning or problem solving for observed analogies *a* through *n*. Given that human beings (*P*) report that they do those kinds of learning or problem solving consciously (*x*), it is probable that animals observed doing the same kinds of learning or problem solving (*Q*) are also doing it consciously.

For instance, in my first book I emphasized the fact that some mammals and birds are capable of certain kinds of learning/problem solving that fish and herps are not and, because humans report that they do these kinds of learning/problem solving consciously, I concluded that mammals and birds probably are conscious in ways that fish and herps are not. Specifically, I concluded that they probably have conscious desires that can be frustrated in unpleasant ways (1998, chapter 2). In this section, I consider several types of learning/problem solving that might be indicative of conscious desires. If these types of learning/problem solving are found in some invertebrates (other than cephalopods), that might be a reason for recognizing sentience more broadly in the animal kingdom than is suggested by the standard argument by analogy regarding pain, specifically. For even if they cannot feel physical pain, such organisms might be capable of suffering insofar as their problem-solving abilities are indicative of conscious desires that can be frustrated in unpleasant ways.

§5.5(A). DOES INSTRUMENTAL LEARNING REQUIRE CONSCIOUSNESS?

An initially plausible candidate for a general form of learning that requires consciousness is instrumental conditioning. If it turned out that instrumental conditioning requires consciousness but classical conditioning does not, then demonstrating that members of a given species are capable of instrumental learning would suffice to show that they are sentient.

"Classical" conditioning is what Pavlov studied; it consists of habituating an animal to react to what was previously a neutral stimulus, for example a bell, in the way that it reacts, in a reflex-like way, to an "unconditioned stimulus," for example appealing food. (I say "reflex-like" because to stipulate that the unconditioned stimulus produces a *reflexive* response would be too strong. Compare Holland, who characterizes the unconditioned stimulus as "a biologically more meaningful event"; 1990, p. 105.) When Pavlov's dogs began salivating in response to the bell prior to the arrival of food, the bell had become a "conditioned stimulus." Instrumental conditioning, by contrast, relies on getting the animal to do something in a situation in which its behavior causes it to receive a reward or avoid a punishment. It is called "instrumental" conditioning because the subject's doing something is instrumental to producing a desired result.

Psychologists describe instrumental conditioning as requiring the organism to "deduce the connection between its behavior and an outcome in the environment" (Grau 2002, p. 77, citing Maier & Seligman 1976), so it is tempting to think that instrumental conditioning requires consciousness. Thus an initially plausible hypothesis about what consciousness does for an organism is that while classical conditioning does not require consciousness, instrumental conditioning does, and instrumental conditioning allows an organism to learn things that an organism capable only of classical conditioning could not.

A very tidy picture would emerge if all and only organisms that are capable of instrumental conditioning turned out to be conscious. For two reasons, however, this initially plausible picture must be abandoned.

(i) Trace Conditioning

The first reason is that "trace conditioning," which is a type of classical conditioning, seems to require conscious attention to the association in question. Trace conditioning is closely related to what is called "delay conditioning." In the classic delay conditioning paradigm, a tone or a burst of static begins a fraction of a second before the onset of a puff of air to the eyeball. The puff triggers a blink reflex. Delay conditioning is acquired when the subject blinks in response to the tone or burst of static. In trace conditioning, the sound is again presented a fraction of a second before the onset of the puff of air, but whereas in delay conditioning the sound co-terminates with the puff of air, in trace conditioning the sound is terminated a fraction of a second before the beginning of the puff of air. Trace conditioning is acquired when the subject blinks in response to the tone or burst of static. Both of these are examples of classical conditioning; like Pavlov's dogs, subjects have acquired the association when they blink (a reflex) in response to what was previously a neutral stimulus (the associated sound).

Clark and Squire (1998) showed that, in human subjects, whereas delay conditioning is acquired whether or not the subjects are consciously aware of the association, trace conditioning is acquired only by those who are consciously aware of the association. With their heads restrained in an appropriate apparatus, volunteers watched a silent movie. As the movie played, the subjects were exposed to either a delay conditioning or a trace conditioning paradigm. All of the subjects repeatedly heard either a tone or a burst of static and all of them received a puff of air onto their eyes that was associated with just one of the two sounds. In the delay conditioning paradigm, the sound co-terminated with the puff. In the trace conditioning paradigm, the sound lasted 1/4 second, and the puff followed 1/2 second later. All subjects were then given a series of true/false questions, some of which related to the association between the sounds and the puffs of air. Treating subjects who answered these questions correctly at well above chance levels as having been consciously aware of the association, Clark and Squire found that whereas volunteers acquired the delay conditioning whether or not they were consciously aware of the association between the sounds and the puffs, only volunteers who were consciously aware of the association acquired the trace conditioning.

So in humans, some forms of classical conditioning (including delay conditioning) are acquired whether or not the subject is consciously aware of the association, but at least one form (trace conditioning) is acquired only if the subject is consciously aware of the association. Thus trace conditioning, although Pavlovian, seems to require consciousness.[3]

---

[3] To the extent that the performance of animals on delay and trace conditioning are analogous to that of humans, this suggests that those animals are consciously aware of the association that they are acquiring when they acquire trace conditioning. In this way it is significant that Thompson et al. (1996) found that the performance of rabbits on trace conditioning varies in the same ways that it does in humans. On delay conditioning, rabbits acquired the association equally well, regardless of age group. But with trace conditioning, there was variability both within and across age groups. Within age groups, individuals varied in terms of how well they acquired the association, and the length of the delay made the association harder to acquire. Across age groups, they found increasing variability among individuals with advancing age. These same variations are found in human subjects, suggesting that rabbits too only acquire trace conditioning if they are consciously aware of the relationship between the unconditioned and conditioned stimuli.

(ii) Instrumental Conditioning in the Spinal Cord

The second reason for denying that while instrumental conditioning requires consciousness, classical conditioning does not, is that the spinal cords of rats have been shown to acquire instrumental conditioning, independent of the brain. Assuming that the spinal cord is not conscious, then, instrumental conditioning does not always require consciousness.

This surprising result was obtained in psychologist Jim Grau's lab at Texas A&M. The lab prepares what they call "spinalized" rats, by completely severing the spinal cord at the second thoracic vertebra. This leaves the front limbs unaffected, but produces paralysis in the hind limbs and ensures that no information is exchanged between the brain and the part of the spinal cord that innervates the hind limbs. Grau's lab showed that spinalized rats still learn to hold their rear legs up away from a saline solution if contact triggers a shock to the leg. Long ago it had been shown that the severed legs of cockroaches do this, too (for a description, see Varner 1998, p. 31), but the mechanism there involved "a tetanuslike effect that produces a cumulative increase in muscle tension" (Grau 2002, p. 82). In another set of experiments, Grau's lab has shown that this could not explain the spinalized rats' performance, however.

These experiments simultaneously established that "learned helplessness," which inhibits instrumental learning, can develop in the spinal cord. Learned helplessness occurs when prior exposure to a "non-contingent" (i.e., random) shock undermines the animal's later ability to learn to avoid similar shocks. The classic experiment, performed by Seligman and Maier (1967), involved pairings of "master" and "yoked" dogs. In the first phase of the experiment, the dogs were suspended in slings. While the master dogs were administered electrical shocks that they could terminate by pressing on a panel, the yoked dogs received inescapable shocks. Both groups of dogs were subsequently placed in a "shuttle box," where they could escape a shock from the floor of one chamber by jumping over a wall into the adjacent chamber. Whereas master dogs quickly learned to avoid the shocks (which were signaled a few seconds in advance by dimming lights), yoked dogs failed to learn this.

Grau expected that learned helplessness would not develop in the spinal cord, because it had been thought that the phenomenon depended on "the organism's *perception* of whether the environmental event is controllable or uncontrollable" (Grau 2002, 83—italics in original), something that presumably required the brain's involvement. The Grau lab's experiment that showed otherwise involved spinalized "master" and "yoked" rats. While the master rats' paralyzed hind legs were taught to avoid shocks by flexing, the yoked rats' hind legs received a shock whenever the master rat's legs did, regardless of the position of the yoked rats' legs. If spinalized rats' "learning" to hold their legs up had been produced by "a tetanuslike effect that produces a cumulative increase in muscle tension," then the yoked rats should have "learned" to hold their legs up, but they did not. This showed that the spinal cord of these rats was really capable of instrumental learning, independent of the brain.

Moreover, when these yoked rats were later exposed to the instrumental learning situation, in which they could escape shock by flexing, they did not learn to. In a related experiment, the yoked rats were given the anti-anxiety/"opiod antagonist" drug naltrexone, which is known to overcome learned helplessness in intact animals. Administration of naltrexone reduced the learned helplessness effect in the Grau lab's spinalized rats. These findings proved, quite surprisingly, that learned helplessness can also occur in the spinal cord.

These and other findings[4] led Grau to formalize a distinction between "instrumental" conditioning/learning and "operant" conditioning/learning. The two terms are often used interchangeably, but Grau suggests that we reserve "operant conditioning" for instrumental conditioning that has certain special features. He suggests the following minimal criteria for instrumental learning:

1. Instituting a relationship between the response and an outcome produces a change in behavior (performance).
2. The effect is neurally mediated.
3. The modification outlasts (extends beyond) the environmental contingencies used to induce it.
4. The behavioral modification depends on the temporal relationship between the response and the outcome (Grau 2001, p. 768).

His lab's research on spinalized rats has shown conclusively that the spinal cord is capable of instrumental learning characterized in terms of conditions #1–4; but Grau characterizes operant learning as meeting these two additional criteria:

5. The nature of the behavioral change is not constrained (e.g., either an increase or decrease in the response can be established).
6. The nature of the reinforcer is not constrained (a variety of outcomes can be used to produce the behavioral effect).

He still believes that a brain is required for operant conditioning that meets his conditions #5 and #6.

§5.5(B). EXAMPLES OF *OPERANT* INSTRUMENTAL LEARNING

In this subsection, I describe several kinds of learning that count as *operant* rather than merely instrumental by Grau's criteria. Each of these is a plausible candidate for a kind of learning that requires consciousness. Each of them

---

[4] Other phenomena that Grau and associates have demonstrated in the spinal cord are learning-related anti-nociception, meaning when exposure to an aversive event weakens the aversive effect of subsequent, similar events; and "overshadowing," which occurs when previous exposure to a stimulus undermines its acquisition as a conditioned stimulus (Grau 2002, pp. 78–80).

would also make the behavior of an animal dramatically more flexible than that of an animal capable only of classical conditioning, and significantly more flexible than the behavior of an animal capable only of instrumental conditioning as described by Grau's criteria #1–4. So each of these forms of learning would also provide an animal with an adaptive advantage, thus making the effects of conscious learning "visible" to evolution.

In chapter 2 of *In Nature's Interests?* (Varner 1998), I argued that two forms of learning that have been studied in a variety of animals are suggestive of conscious problem solving, because humans report that they are consciously formulating hypotheses and testing them when they exhibit these two kinds of learning. The first was multiple reversal trials. Here the subject is rewarded for responding in one of two possible ways (e.g., pressing the bar or touching the disk on the left) until it responds at criterion (e.g., 80 percent correct responses) across a block of trials. The reward pattern is then reversed (so that the subject is rewarded for pressing the bar or touching the disk on the right) until it again responds at criterion. The pattern is reversed multiple times and the speed with which the subject adjusts to reversals is compared across time. In all animals, it takes longer to relearn the reward pattern the first time it is reversed, but some animals relearn the reward pattern faster after multiple reversals. This "progressive adjustment" is found in mammals, birds, and herps, but not in fish (Varner 1998, p. 32). Among invertebrates, it has been found only in octopi (Mackintosh & Mackintosh 1964). The second kind was probability learning. Here the subject is rewarded a fixed percentage of the time for responding one way (e.g., it is rewarded 30 percent of the time for responding to its left) and the rest of the time it is rewarded for responding the other way (70 percent of the time it is rewarded for responding to its right). Animals exhibit a range of "strategies" on such tests. Fish and herps exhibit a random response pattern that roughly matches the reward ratio (30 percent left, 70 percent right). Rats and monkeys typically "maximize," that is, they respond almost 100 percent of the time to the more frequently rewarded alternative. Birds behave like fish on certain kinds of problems (those in which the shape or color of the response target is salient) but match on others (those in which the location of the response target is salient). Only mammals exhibit "systematic" matching, by systematically choosing the alternative that was rewarded on the preceding trial, and only primates exhibit the strategy of systematically choosing the alternative that was not rewarded on the previous trial (Varner 1998, p. 33). To the extent that humans are consciously thinking about how to solve the problem when they exhibit the maximizing and systematic matching strategies, the fact that birds and mammals employ these strategies suggests that consciousness is also involved when they learn how to better solve probability learning problems.

Another kind of learning that counts as operant by Grau's criteria involves "the formation of learning sets," and this too has been described as involving hypothesis formation and testing: "The learning that is of primary importance to the primates, at least, is the formation of learning sets; it is the *learning how to learn efficiently* in the situations the animal frequently encounters. This learning to learn transforms the organism from a creature that adapts to a changing environment by trial and error to one that adapts by seeming hypothesis and insight" (Harlow 1949, p. 51). In the classic learning set paradigm, a subject is presented with pairs of objects across "blocks" of trials. Within each block of trials, the problem is the same: the same two objects are presented, and the subject is rewarded for selecting the same object; but with each new block of trials, two new objects are used. Subjects' performance is measured in terms of how often they give the correct answer on the *second trial* across numerous blocks. The idea is that, whatever response the animal gives on the first trial in a block, if it has formed the appropriate learning set, then it will have learned what the correct answer is by the second trial in every block (the same response if it was rewarded on the first trial, and the other response if it was not rewarded on the first trial).

J. Warren's figure (1965, p. 262, figure 4) summarizing some of the early data on learning set acquisition has been widely reproduced. It showed cats outperforming rodents and monkeys outperforming cats. Harry Harlow and followers assumed that performance on the classic learning set paradigm could be treated as a trans-species IQ test, and when data on chimpanzees and young children are added to Warren's figure, this suggests the following hierarchy of intelligence (Watanabe 2006, p. 624):

1. Humans,
2. Other great apes (gorillas, chimpanzees and bonobos, and orangutans),
3. Old-world monkeys (e.g., rhesus monkeys),
4. New-world monkeys (e.g., squirrel monkeys and marmosets),
5. Carnivores (e.g., cats and dogs), and
6. Rodents (e.g., rats and squirrels).

For chimpanzees have been shown to choose correctly on the second trial in nearly every block after only about a dozen iterations (Schusterman 1964, p. 154, figure 1), and 6- to 7-year-old children acquire learning sets even faster (Delius & Delius 2006, p. 557).

This neat hierarchy has some commonsense appeal, but it was upset by subsequent research. First, rats' poor performance as reported in Warren's figure may have been due in part to their heavy reliance on the sense of smell. On similar kinds of problems, rats were subsequently shown to do much better when the cues are olfactory rather than visual (Eichenbaum et al. 1986; Zeldin & Olton 1986). Second, subsequent research on a wider variety of carnivores showed that they vary widely, with the performance of minks and ferrets being closer to that of rhesus monkeys than cats (Doty et al. 1967, p. 1580, figure 2). Finally, the numbers of animals used in many of the earlier studies cited in support of the above hierarchy was small, and the statistical significance of some of the findings has not been demonstrated (Macphail 1982, p. 279).

Among birds, one study compared corvids (e.g., crows and ravens) to pigeons. Although pigeons are a stock laboratory subject, and they are capable of making very subtle visual discriminations, the corvids are commonly thought to be among the most intelligent birds (along with parrots). While the corvids in this study performed at something like the level of monkeys, reaching 60 percent to 70 percent accuracy after about 300 blocks of trials, the pigeons continued to choose correctly on the second trial only 50 percent of the time even after 800 blocks of trials (Macintosh 1988, p. 514, figure 2, citing B. J. Wilson's unpublished 1978 dissertation). In one early study, young chickens were shown to rise from 35 percent to 70 percent correct choices on the second trial after only 50 blocks (Plotnik & Tallarico 1966, p. 195). Chickens are commonly thought to be relatively unintelligent birds, so this study threatens to upset common sense thinking about the intelligence of various birds. Euan Macphail (1982, pp. 224–225) criticizes the chicken study, however, for probably reusing various objects across blocks of trials, and because it turns out that while the chickens' performance on the second trial in each block was an impressive 70 percent, their performance on the third trial in each block was only 58 percent, suggesting that they had not formed the relevant learning set after all.

Each of the kinds of learning discussed in this section meets Grau's criteria #1–6, and each would make the behavior of an animal significantly more flexible than that of an animal capable only of types of instrumental conditioning that meet criteria ##1–4 without also meeting #5 and #6. It also seems plausible to say that each of these kinds of learning requires consciousness, insofar as each involves hypothesis formation and testing, and human subjects report that they do this consciously. Note, however, that each of these kinds of learning has to date been studied only in vertebrates.

So the study of the kinds of learning surveyed in this section does not, at present, suggest that any animals are sentient that are not also conscious of physical pain. In the following chapter, other kinds of learning that would seem to require consciousness and that count as operant by Grau's criteria will be discussed, including tool making and the use of a "theory of mind." These too have been studied only among vertebrates, but I will argue that some of these are indicative of what I call "near-personhood," which gives certain animals' lives greater moral significance than the lives of those that are "merely sentient."

§5.6. Conclusion

To sum up, the main conclusions of this chapter are the following. First, "sentience" refers to the capacity for phenomenally conscious suffering and/or enjoyment.

Second, while consciousness of pain is not, strictly speaking, a necessary condition for sentience, the ability to feel pain will figure prominently in good ILS rules regarding our treatment of animals. This is because pain effectively marks the outer boundary of sentience among animals on earth. With the extremely rare exception of humans with congenital insensitivity to pain, all animals that we know are sentient in other ways are also capable of feeling pain.

Third, the standard argument by analogy for consciousness of pain suggests that, with the exception of cephalopods, all and only (normal) vertebrates are capable of feeling pain, and we should rely on this argument in formulating ILS rules regarding our treatment of animals. For, although future research may strengthen the case for thinking that some other invertebrates can feel pain, the standard argument (as updated in table 5.3) employs the presently available evidence on which ethical and public policy decisions should be based (this is "the Rumsfeld response" to Allen's first objection).

Finally, although various kinds of learning might require phenomenal consciousness, to date there is no evidence that there are any non-human animals that are capable of these kinds of learning that are not also capable of feeling pain. In chapter 8, I will argue that we are justified in developing different, and more demanding ILS rules governing our treatment of animals that are capable of certain of these kinds of learning. The arguments of the present chapter show, however, that lacking these kinds of learning cannot be used to justify excluding animals from moral consideration. Any animal that is capable of feeling pain is capable of suffering, and this means that a good set of ILS rules will include at least some rules governing our treatment of any animals that are capable of feeling pain.

{6}

Personhood and Biography

§6.1. What Makes Humans Special?

In the Western moral tradition, it is commonly thought that *something* about human beings makes us special, morally speaking. Some believe either that only humans have souls, or that humans are special because we are chosen by or "made in the image of" a god. Others have claimed that merely being a member of our species is what makes humans special. In these secular times, the previous rationale has lost adherents, and in the animal ethics literature the latter is dismissed as "speciesist." Nevertheless, even the most prominent "pro-animal" philosophers tend to think that something about human beings (at least normal ones) gives our lives special value in comparison to many, most, or all non-human animals. The veritable Moses of the animal rights movement, Peter Singer, defines "persons" as "rational, self-conscious beings"[1] and he holds that it is not speciesist to judge that it is worse to kill a person than it is to kill animals that are not persons (1990, pp. 18–19 [= 1975, p. 20]). And even Tom Regan, who claims that his "rights view" supports the abolitionist stance of activist groups better than Singer's utilitarianism, holds that the death of a normal human being is a worse thing than the death of any non-human animal (Regan 1983, sections 8.10, 8.13).

Clearly, then, the intuition that something about normal human beings gives their lives special moral significance is deeply rooted in our psyches. From a Harean perspective, I think that there are good reasons for this: good utilitarian reasons can be given for embodying in the ILS rules of common morality, law, and professional ethics some heightened form of respect for humans. But those reasons have little to do with the standard candidates for a uniquely human ability: rationality, tool use (or tool making), and language. As we will see, rigorous scientific studies have shown that various non-human

[1] While this is Singer's standard definition of "person" in *Practical Ethics* ([1979] 1993), he actually characterizes persons in a variety of ways, as detailed in §9.4(A).

animals are capable of various kinds of reasoning, the use and manufacture of simple tools, and mastery of simple syntax. I think that what makes humans special has something to do with language, but it is not *having* language per se. Rather, it is something that all normal, adult humans *do* with language that makes us special. Specifically, it is telling stories about our lives and our cultures.

In this chapter I describe how storytelling gives us a biographical sense of self, and why there is no good evidence that any non-human animals have a biographical sense of self. In chapter 7 I will defend the claim that having this biographical sense of self gives the lives of persons special moral significance in comparison to the lives of non-persons. That will complete my general defense of my favored stipulative definition of "persons" as individuals whose lives have special moral significance because they have a biographical sense of self (§1.2). Although we have no good evidence that any non-human animals are persons in this sense, in chapter 8 I will discuss research that suggests that at least some animals are nevertheless "near-persons." This is because they have a type of forward- and backward-looking consciousness that is relevantly similar to normal humans' biographical sense of self. Thus I will defend a moral hierarchy among:

1. persons,
2. near-persons, and
3. the merely sentient.

By "the merely sentient" I mean animals identified in the previous chapter as capable of feeling pain, but for which we have no good evidence that they have any form of consciousness that is relevantly similar to normal humans' biographical sense of self. The above list represents the claim that while the lives of persons have special moral significance in comparison to those of both near-persons and the merely sentient, the lives of both persons and near-persons have special moral significance in comparison to those of the merely sentient.

Defending this moral hierarchy in this and the following two chapters will set the stage for arguing, throughout part III of this book, that we should adopt different ILS rules to govern our treatment of human persons, any non-human animals that we have good reason to believe qualify as near-persons, and all other sentient animals, that we presume to be "merely sentient."

§6.2. Human Beings as Storytellers

In my first book (1998, chapter 4), I used the terms "ground project" and "categorical desire" in an effort to describe why humans' lives have special moral significance. I borrowed these terms from Bernard Williams, who defines them as follows: a ground project is "a nexus of projects . . . which are closely related to [one's] existence and which to a significant degree give a meaning to [one's] life" (1981, pp. 12, 13), and a categorical desire is one that answers the question, "Why is life worth living?" (1973, pp. 85–86), so a ground project is a nexus of categorical desires. Since then I have become convinced that *storytelling* is both the best candidate for a capacity that sets humans apart from animals, and a capacity that gives our lives special moral significance. For this capacity explains both the centrality of ground projects to humans' lives and a range of other features commonly attributed to normal, adult human beings.

It certainly seems that our identities are strongly tied to our biographies, which are stories we can tell about ourselves. Famously, Aristotle described a complete story as having three parts (*Poetics* 1450b), and these are identifiable in the stories of our lives: the beginning tells where we came from, the middle explains where we are now, and the end includes both aspirations for the future and an understanding of our mortality. Human communities also have biographies, and part of our cultural identity is being able to understand and repeat those stories. These stories have the same three parts: they begin with an account of how our community was founded, the middle describes how we understand the nature and structure of our community today, and the end describes what we aspire to as a community. We also think of our species as having a story, whether it is embodied in a creation myth coupled with an idea of an afterlife, or in scientific cosmology and evolution coupled with political ideals. The emergence of burial practices and art in Neanderthals and Cro-Magnons in Europe around 35,000–40,000 years ago is sometimes cited as evidence of the advent of religion, but we might as well say that it is evidence of the advent of storytelling. To better understand how central storytelling is to our identities as humans, consider the following three examples.

First, consider a man who watches the destruction of his uninsured business in a fire and exclaims, "That business was my life!" We might say that this man had only one "ground project" in Williams's sense. But unpacking his statement in terms of the "story of his life" reveals a richer texture. He means not just that all of his goals for the future revolved around the business, or that running the business was what made life worth living to him, but that his "life story," from beginning to end, can be understood by reference to it; where he came from and how he became the single-minded businessman that he is today can also be understood by reference to it. Fully expressing "who he is" and what "his life" is requires telling such a full story about how his past and his aspirations for the future both shaped his present.

Second, suppose that we notice a woman playing a cello on a city street corner. I might say, "I wonder what her story is?" Providing her story would involve more than giving her name, or explaining that she recently became unemployed when the symphony orchestra was cut from the city budget. Those are important details of her story, to be sure, but the full story would include, at a minimum: where she was born and raised, how she became committed to being a professional musician, and how playing on the street corner fits into her plans for her future.

Finally, suppose that I encounter a three-legged cat in a new friend's home. I might ask the same question, "What's her story?" But here we might be satisfied with a much sketchier account, such as: "I found her along a road outside town last year. She had apparently been hit by a car. I took her to the vet, who had to amputate her leg, but she's doing pretty well on three legs now. I just can't let her roam outside like she must have been used to doing." There is obviously more to the cat's story than this. We might notice that opening a can draws her attention, and conclude that she remembers being fed canned food. And even if we can never come to know them, there are facts about when and where she was born, how she came to be in the road and got hit, and so on.

There are profound differences, however, between the roles that their stories play in the humans' lives and the role that the cat's story plays in her life. Most fundamentally, the cat presumably is not conscious of her story in the same way the humans are. Telling stories about ourselves and others requires far more than the simple syntax that has been taught to some great apes, dolphins, and parrots, and without something very much like a full-blown human language, it may be impossible to represent one's life as a story, even to oneself (considerations supporting this conclusion are given below in §6.4 and §6.5). So while the cat may consciously remember certain events, she presumably does not think about her life *as a story*.

Relatedly, notice that the cat's story above is told from the human's perspective, and it is dominated by the impact of events on the cat. Unlike the cellist's story, there is nothing in the cat's story about striving to become a certain kind of individual or aspirations for the future. This suggests that autonomy, in the sense of choosing both to acquire certain character traits and to achieve certain things, requires a storied perspective on one's life.

Thinking about one's life as a story can also profoundly affect the way that one experiences the present. Nietzsche emphasized the backward-looking aspect of this phenomenon in his short monograph, *On the Advantage and Disadvantage of History for Life*. Nietzsche offers no explanation of why animals lack a sense of history, he simply assumes that they have none, describing the effects thus:

> Consider the herd grazing before you. These animals do not know what yesterday and today are but leap about, eat, rest, digest and leap again; and so from morning to night and from day to day, only briefly concerned with their pleasure and displeasure, enthralled by the moment and for that reason neither melancholy nor bored. It is hard for man to see this, for he is proud of being human and not an animal and yet regards its happiness with envy because he wants nothing other than to live like the animal, neither bored nor in pain, yet wants it in vain because he does not want it like the animal. Man may well ask the animal: why do you not speak to me of your happiness but only look at me?

> The animal does want to answer and say: because I always immediately forget what I wanted to say—but then it already forgot this answer and remained silent: so that man could only wonder. . . .In this way the animal lives *unhistorically*: for it goes into the present like a number without leaving a curious fraction; it does not know how to dissimulate, hides nothing, appears at every moment fully as what it is and so cannot but be honest. ([1874] 1980, pp. 8–9)

For Nietzsche, this suggests that happiness is impossible for humans, because we can never forget our histories and lose ourselves in the moment:

> With the smallest happiness as with the greatest happiness, however, there is always one thing which makes it happiness: being able to forget or, to express it in a more learned fashion, the capacity to live *unhistorically* while it endures. Whoever cannot settle on the threshold of the moment forgetful of the whole past, whoever is incapable of standing on a point like a goddess of victory without vertigo or fear, will never know what happiness is, and worse yet, will never do anything to make others happy. (p. 9)

I would say just the opposite, however. The animals that Nietzsche describes may be *content*, but only individuals who think of their lives as stories can be said to be truly *happy* (or *unhappy*). To be happy in the fullest sense of the term, an individual must live a good life story. The story can begin anywhere under any circumstances, but it must have a good trajectory, which involves the individual aiming to do certain things with his or her life or acquire certain character traits, and succeeding at it. Animals that are incapable of striving for such goals can be described accurately as content, but not happy.[2]

---

[2] Relatedly, Aristotle entertains the idea that events after one's death can affect one's happiness, saying that it would be both odd to say this and odd to deny it (*Nicomachean Ethics* 1100a15ff). On the view described here, it makes some sense to say that this can happen. Suppose, for instance, that a man has devoted his life to turning his father's failing business around, works at this well beyond normal retirement age and then, just after passing the now thriving business into his son's hands, dies. At that point, his is a happy story, but that story is adversely affected if the business immediately goes bankrupt due to his son's incompetence.

I am inclined to think that a desire can only accurately be described as "frustrated" while the desiring individual is both alive and conscious that the desire is not being satisfied. On the other hand, so-called "external" desires sometimes play an important role in persons' understanding of how well their lives go, and for this reason (as Aristotle observed) persons might meaningfully be said to be harmed by events that occur after they are dead. An external desire is a desire *for something to be the case, whether or not one is capable of experiencing its satisfaction*. So, for instance, parents want their children to thrive long after they are themselves dead, and revolutionaries want the republics that they have dedicated their lives to founding to flourish long after they are gone. Whether or not external desires should be counted in the utilitarian calculus is an important, but controversial issue. Since the focus of this book is on animals, I will not pursue it further here, since we have no evidence of external desires in near-persons. But this underlines another, potentially morally significant difference between persons and near-persons: for if persons *can* be harmed when external desires go unsatisfied, then that is another way in which persons' special cognitive capacities allow them to be harmed in ways that individuals lacking those capacities cannot be harmed.

What I have just suggested is that happiness is a backward-looking phenomenon and one that applies properly only to beings who think of themselves as having a history or biography, because individuals can be described accurately as happy rather than merely content only if they can understand their lives as stories that have turned out well so far. In a typically cryptic passage from the *Philosophical Investigations*, Wittgenstein seems to make an analogous claim about the forward-looking phenomenon of hope:

> One can image an animal angry, frightened, unhappy, happy, startled. But hopeful? And why not?
>
> A dog believes his master is at the door. But can he also believe his master will come the day after to-morrow?—And *what* can he not do here?—How do I do it?—How am I supposed to answer this?
>
> Can only those hope who can talk? Only those who have mastered the use of a language. That is to say, the phenomena of hope are modes of this complicated form of life. (If a concept refers to a character of human handwriting, it has no application to beings that do not write.) ([1953] 1958, p. 174)

Wittgenstein appears to believe that non-linguistic beings cannot think about the future at all. That would certainly explain why hopes are not attributable to them, but as we will see in §8.4(F), there are good reasons for thinking that at least some animals can entertain conscious thoughts about their futures, at least out to the next day.

How else, then, could language be involved in the phenomenon of hope? Elsewhere, Wittgenstein emphasizes the way in which attributions of hope only make sense in their temporal "surroundings": "Could someone have a feeling of ardent love or hope for the space of one second—*no matter what* preceded or followed this second?—What is happening now has significance—in these surroundings. The surroundings give it its importance. And the word 'hope' refers to a phenomenon of human life" ([1953] 1958, section 583).[3] I do not pretend to understand Wittgenstein well, but this passage is at least consistent with the view that it makes sense to attribute hope, in a strong sense, only to beings that think of their lives as stories. What I mean by "hope, in a strong sense" is best explained by contrasting it with the minimalist sense of hope captured in the *Oxford English Dictionary*'s first definition of the word: "To entertain expectation of something desired; to look (mentally) with expectation." Any animal—human or non-human—that is

---

[3] The other interesting passage about "hope" in the *PI* is this: "When someone says 'I hope he'll come'—is this a *report* about his state of mind, or a *manifestation* of his hope?—I can, for example, say it to myself. And surely I am not giving myself a report" (1953, section 585). There are also references to "hope" in sections 545 and 574, but these add nothing significant to Wittgenstein's treatment of the subject.

capable of forming expectations about the likelihood of their desires for the future being fulfilled can be said to hope in this minimalist sense. In contrast, consider this ascription of hope to a group of humans: "After Kennedy's May 1961 speech to Congress, NASA engineers hoped to reach the moon by 1969." This ascription of hope involves a complex story. The engineers understood that the president's speech changed the context, and this gave them hope. What they then hoped for was to be able, after solving a series of very complex problems, to achieve a long-term and very ambitious goal. When we describe such a hope as "fulfilled," this involves a success story. In this way, a story can provide the "surroundings" that turn a bare expectation into a bona fide hope.[4]

Finally, the centrality of storytelling to our identities also helps to explain how the death of a human being can be tragic in a sense that an animal's death cannot. Obviously, tragedy is a style of storytelling, so presumably only human beings can understand their own lives as tragedies, but when we say that a person's *death* was tragic, the subject of the story is no longer around. I would say, instead, that what makes a person's death tragic is the involvement with their life story that the person had while alive. Autonomy can be understood as achieved when one sets out to achieve certain things and to acquire certain character traits and then succeeds; that is, when one tells a story about the future, and then lives it. Just as storytelling beings understand that who they are in the present is a result of what they have experienced and done in the past, a storytelling being can resolve to become (or remain) a certain kind of individual in the future. I suggest that a person's death is tragic, in the fullest sense of the term, when a good prospect of achieving autonomy (understood as living a success story) is interrupted by unforeseen events—that is, when an individual who has been telling a story about his or her own future—and resolutely acting in ways that make that future likely to become a reality—dies unexpectedly. Even without dying, a person's life can become a tragedy when, in similar circumstances, one suffers an unexpected reversal of fortune that dooms one's chances of living a success story.

In these ways, the lives of storytelling beings can be richer and more complex than those of beings that lack the ability to tell stories. This complexity is richer than is revealed by simply noting that we have complex aspirations for the future, as I argued in chapter 4 of my first book.

---

[4] Perhaps something similar can be said of regret: a story can provide the particular kind of "surroundings" that turn a bare complaint into bona fide regret. You can't regret not doing something else with your life unless you understand it as unfolding according to the logic of narrative. I can say both "I regret that Bush went to war in Iraq" and "I regret that I didn't continue to study Spanish after junior high school," but there's a difference. If I had made mastering Spanish a goal, I'd be able to speak it today, but nothing I could have done (by myself, anyway) would have stopped Bush from invading Iraq.

§6.3. Schechtman on Narrative Self-Constitution

The preceding section describes where my thinking on the subject of storytelling was when, in late 2006, I ran across Marya Schechtman's book *The Constitution of Selves*. Other, more prominent philosophers—most notably Alasdair MacIntyre (1981, chapter 15; for a long list of examples, see Schechtman 1996, p. 93)—have claimed that persons' identities are constructed through the adoption of a narrative, but reading Schechtman's book crystalized my thinking about personhood and storytelling in several ways that will play into my defense of persons' special moral status in the following chapter. Schechtman's focus is entirely on humans, mainly on metaphysics, and she does not claim that persons' lives have special moral significance (1996, pp. 99–100). Nevertheless, her "narrative self-constitution" view will help me explain, in the following chapter, why having a biographical sense of self gives the lives of persons special moral significance.

Almost half of her book is devoted to a detailed critique of other philosophical accounts of personal identity. It would be going too far afield to review that entire critique here, but a brief, partial description will illustrate how Schechtman's account is supposed to improve on the dominant, psychological continuity account of personal identity. Contemporary discussions of personal identity are filled with thought experiments involving brain transplants, body-switches, and *Star Trek*–style "transporter" machines. A number of authors have appealed to our intuitions about such cases to defend variations on the psychological continuity account. What these variations have in common is the view that personal identity is preserved only if there is a certain kind of psychological continuity, which gets cashed out in terms of similarity of memories, beliefs, desires, and character or personality traits. They differ on the degree of continuity required, on the transitivity of the identity relation, and on what sort of basis the continuity must have. Derek Parfit, for instance, arbitrarily stipulates that so long as at least 50 percent of the psychological connections that normal persons usually have across days of their lives is preserved each new day, personal identity is preserved (Parfit 1984, p. 206). This commits him to saying that the same person can have radically different, or contradictory psychological traits, so long as the change occurs gradually enough. David Lewis, by contrast, held that sufficient dissimilarity across periods of time could make it literally true that "in later life I will be a different person" (although he claimed that "For us short-lived creatures, such remarks are an extravagance"; Lewis 1983, p. 30). And while Parfit is willing to say that when Captain Kirk "re-materializes" on the surface of a planet, he is the same person who entered the transporter aboard *The Enterprise*, Jeffrey McMahan holds that the requisite psychological continuity must be caused by continued functioning of the same brain (McMahan 2002, pp. 57, 67).

The main argument against all psychological continuity views, as Schechtman reports, is that they "are unable to explain the importance we attach to identity." More specifically, she claims that identity in a strong sense is required to support what she calls "the four features": (1) personal survival, (2) moral responsibility, (3) self-interested concern, and (4) compensatory fairness.

> Survival involves the continuation of the same experiencing subject; moral responsibility requires that the experiencing subject who commits a crime be the one to experience the punishment; self-interested concern requires that the person having an experience in the future be the one who anticipates it, and compensation demands that the same experiencing subject who suffers a sacrifice enjoy the later benefits. In order to capture the connection between personal identity and the four features, then, we need a view according to which personal identity implies sameness of experiencing subject. (p. 149)

Schechtman argues that psychological continuity views fail to provide *sameness* of experiencing subject; they provide only *similarity*, and this does not suffice to ground the four features.

She illustrates this objection with respect to the first feature, personal survival, using a thought experiment modeled on those that psychological continuity theorists use to defend their theories. Those theorists appeal to such things as "body switches," in which my body dies (taking my brain with it), and a neighbor suddenly acquires all of my memories, desires, and personality traits, thus achieving a perfect psychological continuation of me. Schechtman counters: "I would rather wake up tomorrow with partial amnesia than be smothered in my sleep by the evil genius who has also brainwashed my next-door neighbor to exhibit my psychological makeup. That my neighbor may believe herself to be me and acts as I would adds an eerie dimension to the whole ordeal, but it does not bring me any closer to surviving it." She argues that psychological continuity alone cannot ground the other three features either, giving in each case a more real-world example to illustrate why (pp. 52–53). Regarding moral responsibility: "It is only fair to punish me for my actions, and not for the actions of anyone else, no matter how similar to me they are." Regarding self-interested concern: "It may be the case that I care *more* about the future pain of a kindred spirit than about the pain of someone very alien, but I do not *anticipate* the pain of either." And regarding compensatory fairness: "Making Sally work after school so that she can go to college is one thing; making her twin sister work after school so that Sally can go to college is quite another."

Schechtman's critique is much more detailed than this, and includes reasons for rejecting the requirement that the requisite continuity must be caused by continued functioning of the same brain (pp. 22–25), which would make McMahan's view immune to all of the above examples. For present purposes, however, what I want to emphasize is the alternative that she proposes and how, when coupled with what she calls "the articulation constraint" and "the reality constraint," it supports "the four features."

§6.3(A). THE ARTICULATION AND REALITY CONSTRAINTS

Schechtman calls her view "the narrative self-constitution" account of personal identity.[5] On it, persons are self-creating and their lives are understood as narratives: "a person creates his identity by forming an autobiographical narrative—a story of his life . . . On this view a person's *identity* . . . is constituted by the content of her self-narrative, and the traits, actions, and experiences included in it are, by virtue of that inclusion, hers" (pp. 93–94). On Schechtman's view, then, it is only minimally metaphorical to say that persons are the authors of their own lives.

The narrative need not be consciously formulated, however, let alone written out. For persons, Schechtman proposes, "The sense of one's life as unfolding according to the logic of narrative is not just an *idea* we have, it is an organizing principle of our lives. It is the lens through which we filter our experience and plan for actions, not [just] a way we think about ourselves in reflective hours" (p. 113). In this way, an individual can have an "*implicit* self-narrative" (p. 115), but a person must still be capable of responding to questions and challenges about one's implicit life story. Schechtman calls this "the articulation constraint" (p. 114):

> To be a person one must realize that one's beliefs, desires, values, and traits are *supposed* to support one another. This involves acknowledging the legitimacy of certain kinds of questions and challenges . . . to make intelligible a trait that seems anomalous or out of character. A person may not always be able to comply with such a request, but to have the right kind of self-conception one must at least recognize that it is legitimate, thereby showing a basic understanding of what the life story of a person *should* look like. (p. 98)

If one did not recognize the legitimacy of such questions and challenges and at least generally try to answer them, one would not be thinking of oneself as a person in the sense of an individual whose life unfolds according to the logic of narrative. Schechtman acknowledges that sometimes a person is *incapable* of articulating some elements of his narrative, but on the narrative self-constitution view these elements are "attributable to him to a lesser degree than those aspects of the narrative he can articulate," and "if the inability to articulate one's narrative

[5] Readers familiar with the literature on personal identity may find it interesting that, according to Schechtman, Locke is best interpreted as endorsing a narrative self-constitution view rather than the psychological continuity view, as proponents of the latter view almost always interpret him (pp. 105ff).

is sufficiently widespread and severe, the overall degree of personhood can also be compromised" (p. 117).

Thus persons are the authors of their own, at least "implicit" stories, but there is also a constraint on the content of those stories. This is "the reality constraint," the requirement that "an identity-constituting self-narrative [must] fundamentally cohere with reality" (p. 119). Here is how Schechtman explains the need for this constraint:

> Take, for instance, the case of someone who believes himself to be Napoleon, insists that he led the troops at the Battle of Waterloo, and experiences deep shame and remorse about the outcome of that battle. Such a person might be said to have a self-narrative that includes Napoleon's actions at Waterloo, but we certainly cannot endorse a view of identity that attributes these actions to him. The narrative self-constitution view is saved from this unwelcome implication by the reality constraint. (p. 121)

The reality constraint prevents persons from claiming or disowning actions and accomplishments by simply incorporating them into or leaving them out of their self-narratives. This corresponds to common sense and relates to one of the four features in the following way. When categorizing such people as insane, we judge them to be less than fully persons insofar as we deny that they are responsible for their actions in the same way that others are.

Of course, a person's self-narrative can fail to match the public record in less dramatic ways than full-blown delusions. Schechtman discusses this in some detail, emphasizing that a person's narrative can get wrong the facts "in the gross and literal sense" or the "facts about which characteristics are *truly* his," with the former "merely appearing in his history" and the latter "defining who he is" (pp. 76–77). In normal individuals, errors about the former are "not recalcitrant" and are readily revised in the face of contrary evidence (p. 123), unlike the delusions of the insane. Schechtman calls the second kind of errors "errors of interpretation," that is, erroneous judgments about the significance of events. Some of these are fairly easy for others to detect. She gives the example of a paranoiac who believes that the CIA is after her:

> Assuming that the individual in question suffers no hallucinations (as paranoiacs sometimes do), her narrative contains no obvious errors about observable facts—presumably people are looking over at her, or taking notes, or wearing blue suits. The conclusions she draws from those facts seem so absurdly unwarranted, however, that we can easily recognize that the part of her narrative which involves the persecution is mistaken. (p. 126)

Other times, errors of interpretation can be difficult to detect and to correct; it is sometimes difficult or even impossible to decide between two narratives' interpretations of the same events.

I think that a comparison to dreams may be instructive here. On one hand, the events depicted in dreams often resist the imposition of narrative. In terms of Schechtman's account of personal identity, this means that the articulation constraint is not imposed on dreams. But then we don't think that dreams are constitutive of personal identity in the same way that real world actions and experiences are. That's why we don't think that an individual's personhood is compromised by their inability to tell the stories of their dreams in the way that we think that personhood is compromised by the inability to tell a story about one's life. On the other hand, when someone tells the story of a dream, we either don't impose the reality constraint at all or we impose it very liberally. Remember that the reality constraint has two foci: it requires that a self-narrative both conform to the publicly accessible facts of what happened in the past, and it requires that the person's chosen self-narrative impose a plausible interpretation on the flow of events. We don't impose the reality constraint at all, insofar as we don't expect the story a person tells about her dream to correspond to the way real events unfold; and we impose it in a very liberal way insofar as we accept, as a correct interpretation of the events depicted in a dream, whatever story line the subject imposes on it. In the interpretation of both dreams and our lives, observation creates the phenomenon observed.[6]

I think it fair to characterize Schechtman's attitude toward the choice of narrative in the following way. Like a person writing a memoir—which is generally more "personal" than an autobiography—the reality constraint allows an individual to interpret the significance of events in any plausible and comprehensible way, and—like an author deciding which events to mention in a memoir—this constraint allows the individual to focus on the most significant events and leave out the insignificant ones. This description mirrors empirical research on autobiographical memory in humans that shows it to be rife with the first kind of error and that subjects selectively remember events which convey the general flavor of periods of time or were pivotal in determining the future course of events. A famous example is John Dean's testimony before the

[6] Relatedly, we can see what may be misleading about saying that scientists will one day be able to monitor persons' dreams while they sleep. People seem, to a large extent, to interpret their dreams after waking, and for that reason there seems to be no fact of the matter about what story was told in a person's dream until they interpret it. One possible explanation of why people differentially remember dreams might relate to storytelling. Everyone dreams during REM sleep, insofar as brain scans reveal that everyone's brain is active in similar ways. Indeed, "The physiological correlates of REM sleep have been found to exist in nearly all mammals" (Siegel 2001, p. 1058), so probably all mammals experience the replay of images, events, emotions, and so on, that people experience during REM sleep. We only remember some dreams, however, and some people rarely remember their dreams. So it could be that the images and other information that are "presented" during REM sleep are only remembered afterward if the individual imposes a narrative structure on them. This mirrors the implication that there is no fact of the matter about what the story of a person's life is until the person chooses (at least implicitly) a self-narrative. And this could mean that only humans *remember* their dreams.

Watergate hearings, in which he confidently asserted various particular details about conversations and specific wordings used in them. Tape recordings subsequently proved these details to be inaccurate, even though his account of the conversations conveyed accurately the intentions of and wrongdoings of the principals in the affair (Schechtman 1994, section II).

§6.3(B). THE FOUR FEATURES

Relatedly, this explains why Schechtman believes that the attribution of characteristics to persons "can be a matter of degrees" (p. 79). On her analysis, actions are attributable to a moral agent to the extent that they "flow naturally from the rest of her life story" (p. 159). That attribution is a matter of degree in this way is a major reason that the narrative self-constitution view supports the four features. For just as characteristics are in this way attributable to persons to varying degrees, the four features are also matters of degree.

Regarding moral responsibility, Schechtman claims that this explains why young children (who have no life narrative at all) and the insane (whose life stories are profoundly incoherent and/or fail the reality constraint) are exempt from moral responsibility, and how responsibility is attenuated when a person "acts out of character" in an isolated case: "we determine that the hired killer is more culpable than the woman who kills to protect her infant or the person who kills accidentally when his car has a blowout and he loses control" (p. 81). Presumably she means that attribution, and therefore moral responsibility, strengthens across these three cases in the following order: the woman's action is positively misleading about "who she truly is," and the second man's action is an incidental characteristic of his life, but the hired killer's action is right in character.

As I understand it, the reality constraint also supports moral responsibility in the following way, although I am not certain that Schechtman would agree. *The reality constraint ensures that the very same, identical person is present across periods of profound personality change.* Without the reality constraint, persons could claim or disown past actions simply by including them in or excluding them from their self-narratives, but the reality constraint ensures that, even if the person changes dramatically over time, she must include in her narrative what might now seem out of character. Understanding one's life as a narrative requires one to include the publicly known facts about oneself in one's narrative, and this ensures that the aged Nazi war criminal can be held to account, even after he is no longer a racist.

Schechtman argues that the second feature, self-interested concern, also varies in a way that is related to self-narratives:

> First of all there is a certain primitive concern we have for the character of any state that will be ours at all—we care whether it will be pleasurable or

> painful. Furthermore, we have a special concern for our desires and goals and the beliefs they express. The goals and desires that occur in a person's history are not, however, all equally his, and so a person should not be equally concerned with the fulfillment of every desire or goal that occurs in his history. For a person to have *self-interested* concern for the fulfillment of a desire, that desire must, to be sure, at least be part of his history. Among those goals which fulfill this minimal requirement, however, the degree to which any particular goal is attributable to a person is at least part of what determines the degree to which he should be concerned about it . . . (p. 85)

In this way, a person's chosen self-narrative determines, to a very significant extent, what is best for him on the whole; it is the basis for making judgments about what is in a person's interests, "all things considered" (for more on this point, see §7.3).

Relatedly, when compensation is called for, it matters how the compensation relates to one's self-narrative. Giving him a new four-wheel-drive pickup truck will not compensate a man whose self-narrative is all about renouncing modern technologies in the way that it will compensate a man who fancies himself a sort of George Hayduke.[7]

On Schechtman's account, the final feature, personal survival, also comes in degrees. She describes persons with advanced Alzheimer's thus: "To imagine this happening is to imagine one's self—the self one wishes to continue—ebbing away. Although this is not as complete an example of psychological demise as irreversible coma, it does not seem entirely hyperbolic to consider it a type of personal death" (p. 87). She emphasizes, however, that there are even less extreme cases: she describes a chronically depressed person as "in a very real sense, *more* alive" after treatment, and says that "what the addict, prisoner of war, or abused spouse is being robbed of is, in a very real sense, his or her *life*" (pp. 88–89).

---

[7] Atonement is a richer process than compensation, involving reparation but also apologizing, feeling a sense of guilt, resolving to act morally in the future, and imposing penance or self-punishment (Radzik 2004, pp. 148–149); but the extent of atonement that is appropriate for past wrongdoing similarly varies with strength of attribution. To see how, compare the following two cases. In the first, a mature man regrets that as a teenager he was effectively brainwashed into acting on the racist attitudes of his family and community. He says that as an adult, he doesn't recognize himself in the youth who did those things; what he did then is misleading about "who he really was"; he was "acting out of character." In the second, an elderly man regrets that he acted on racist attitudes that he had in his forties. He has changed dramatically, but actively overcoming the racism that characterized him during that period is an integral part of his life story; a racist is "who he really was" back then, and the transformation that he has achieved is an essential part of who he is now. While the first man need hardly feel guilt and owes nothing more than an apology, it might be appropriate for the second man to make atonement an active project for the rest of his life, including feeling a deep sense of guilty, self-imposed penance, and ongoing reparations.

§6.4. A Working Hypothesis about Personhood and Language

Schechtman's account is plausible for several reasons. First, as just explained, it accounts for how "the four features" (moral responsibility, self-interested concern, compensation, and personal survival) are all matters of degree, because "attribution" of both traits and actions varies with the degree to which actions and events are woven into one's chosen self-narrative. At the same time, the articulation and reality constraints ensure that while persons are in a very real sense the authors of their own life stories, personhood is compromised to the extent that one is unable to answer questions about the coherence of one's chosen self-narrative and to correct factual errors. Finally, there is a whole body of empirical research, mainly by psychologists, on how individuals construct stories of their lives, and how important this process is to human psychology (see, e.g., Rubin 1986; Conway 1990; Fivush and Haden 2003).

If a "person" is defined as an individual with a biographical sense of self, however, then it seems very unlikely that any non-linguistic organisms could be persons, or at least it seems very unlikely that we would ever be able to know that they are. For a person must be capable of "using language" in much more than the minimal sense of understanding some basic syntax. Having a biographical sense of self requires having a suite of sophisticated concepts, including:

1. self,
2. birth (or at least some kind of "origin"),
3. death (or at least some kind of end to a phase or type of existence), and
4. personality- or character-shaping experiences.

We have at least *some* evidence for most of the above concepts in some non-human animals: in various great apes, dolphins, and elephants, mirror self-recognition suggests a concept of self (§8.3), and there is at least some evidence for second-order desires, which is a kind of personality-shaping (§7.3[B]).

As detailed in the following section, however, animal language studies conducted to date provide no good evidence that any non-human animal has the biographical sense of self characteristic of all normal, adult humans. To appreciate how difficult it would be to represent one's life as having a narrative structure—either to oneself or in communication with others—note that most natural languages employ a suite of tenses. English, for instance, uses not only the simple past, present, and future, but also progressive and perfect variations on these tenses. The importance of these tenses can be illustrated with the following biographical statement: "Because I had struggled with my sexuality as a teenager, in my twenties I was trying to find alternative role models beyond the traditional nuclear family." There is evidence of self-awareness and autonomy here, insofar as the individual describes himself as having had a second-order desire to acquire new first-order desires (the new sexual role model). The sentence also illustrates the way in which the perfect and progressive tenses are used to express the role that such second-order desires played in one's past. The speaker, reflecting back on a period of his life, uses the past progressive ("I was trying to find . . .") to represent to himself and his listeners the fact that his younger self had a settled, second-order desire during one significant period of his life. He uses the past perfect ("I had struggled . . .") to refer to a still earlier period of his life, and he expresses the belief that it was characteristics of that earlier period that caused him to form the second-order desire in question ("*because* I had struggled . . . I was trying to find . . .").[8]

Without a complex representational system much like humans' natural languages, individuals could not be held subject to what Schechtman calls the articulation and reality constraints. For these require recognizing and responding to challenges to the coherence and content of one's chosen self-narrative, and that can only be done via a very complex system of public communication. It is not *inconceivable* that some animals have such a complex system of public communication. Science fiction is populated with species that communicate telepathically, and, as detailed in §6.5(B), it *could* be that the complex vocalizations of real-world cetaceans and elephants actually have a structure complex enough to support a biographical sense of self. At present, however, there is no good evidence that this is so, as detailed in the following section.

For these reasons, I adopt, as a working hypothesis, the view that competence in a natural language is a necessary (but obviously not sufficient) condition on personhood. I characterize this as "a working hypothesis," because future, innovative research on animals (or on abnormal or pre-linguistic humans) could prove it false. For instance, innovative scientists might someday show that some non-human animals do indeed take a narrative perspective on their lives. I can imagine, for instance, somehow using movies of past events to pose relevant questions (and get reasonably clear answers) about the coherence of an animal's chosen self-narrative, and staged scenes about the future to query animals about their plans for old age. The following section provides a fairly comprehensive review of animal language studies emphasizing, however, that what has been accomplished to date comes nowhere close to making possible this kind of complex, interspecies discussion.

---

[8] What of an individual who used only ASL (which lacks the tenses listed) to communicate? First, notice that almost no one uses *only* ASL. Otherwise normal individuals might rely heavily on ASL, but these individuals also read and write in all these tenses, and they understand and respond to questions posed in all of these tenses. Second, even if an individual used only ASL, so long as they were capable of understanding and responding to questions posed in all of the above tenses, they would comprehend the necessary elements of language, even if they were incapable of producing grammatical sentences using them.

§6.5. Why Animal Language Studies Provide No Good Evidence of Personhood

Studies of language in non-humans have focused on three categories of animals: great apes, cetaceans, and parrots. In this section, I provide an overview of this research. In my judgment, it clearly shows that some non-human animals can master syntactic rules and that they can use these rules in spontaneous and creative communication. I will emphasize, however, that it provides no good evidence of any non-human animal having a biographical sense of self.

§6.5(A). GREAT APES

From the 1930s through the 1950s, four reported attempts were made to teach chimpanzees spoken language. In these "enculturation" studies, very young chimps were raised in homes, treated much like very small children (including potty training), and sometimes raised with the researchers' age-matched children as controls (see Kellogg 1968 for a summary). The most successful of these early studies was by Hayes and Hayes (1951). They raised a female chimpanzee from three days after birth until six and one half years of age, but they succeeded in teaching her to say only three words. Prior to experiments like this one, it was thought that perhaps chimpanzees would learn spoken language without explicit tutoring, the way human children do. But the Hayes only succeeded to the very limited extent that they did by hand-molding the chimp's vocalizations. They taught the chimp "mama" first, by manipulating her lips as she made an "ah" sound, and then succeeded somewhat more easily with "papa" and "cup."

These early projects now seem doomed to fail because non-human apes lack several adaptations essential to spoken language, including the ability to completely close off the nasal passages from the mouth, which is essential to producing consonant sounds. Consonants are easier to discriminate than vowels and make spoken language easier for listeners to parse into words. Thus non-human apes cannot make the sounds necessary for speech, but mammals generally can discriminate consonants, so they might be able to understand spoken language even if they are congenitally incapable of producing it.

For this reason, subsequent research programs have employed sign language or keyboards with "lexigrams" on them, although in most studies the trainers have continued to speak to their subjects in English, a technique that is sometimes called "simultaneous or total communication." The first sign language study began in 1966, when Allen and Beatrice Gardner took into their home a one-year-old female chimp whom they named Washoe. At first, they too had to use a molding technique, actively shaping Washoe's hands into the signs they were trying to teach her. But then observational learning took over, and they claimed that Washoe eventually acquired a vocabulary of 132 signs. Around the same time that the Gardiners began teaching American Sign Language (ASL) to Washoe, David Premack, a University of California, Santa Barbara, psychologist, began teaching four chimpanzees to use a simple, artificial language by manipulating plastic chips on a "language board" between the chimp and the trainer.

One of the strongest criticisms of ape language studies grew out of Herbert Terrace's study of ASL use by a chimp he named Nim Chimpsky. The name that he gave the chimp was a play on the name of Noam Chomsky, evincing Terrace's skepticism about Chomsky's claim that language competence is uniquely human. Nim learned a total of 200 signs (Terrace 1979a, pp. 166–167) through molding or imitation, and after a time, Nim "began to give his hands to his teacher to mold" when he did not know the word he needed for something (p. 139). However, in his book *Nim* and in an article published the same year (1979a, 1979b), Terrace provided a sophisticated deflationary analysis of his own data and claimed that no study to date had shown that an ape had acquired the same mastery of syntax that a young child normally has.

Terrace analyzed over 19,000 of Nim's videotaped sign combinations. They ranged in length from 2 to 16 signs, but a majority of them were limited to 2 or 3 (1979a, p. 179), and the one combination of 16 signs was the highly repetitious "give orange me give eat orange me eat orange give me eat orange give me you" (p. 210), which hardly expanded on the meaning of a much shorter combination that might have sufficed, such as "give orange me eat." Nevertheless, Terrace argued that, given the number of distinct combinations recorded, Nim's generation of combinations could not be explained in terms of conditioning. For instance, Nim frequently used the combination "give X" to request things long before his trainers began routinely asking him to give them things (pp. 180–181). Also, statistically speaking, Nim clearly preferred certain word orders that happen to match the most natural uses in spoken English, for example placing the name of a desired item or action after, rather than before, the word "more" (p. 180).

However, Nim's mean length of utterance did not increase in the way that those of children (both speaking and signing) do (pp. 210–211), and when Nim's most frequent 3- and 4- sign combinations were compared to his most frequent 2-sign combinations, almost all of the longer combinations were redundant in the same way that the 16-sign combination quoted above was redundant (pp. 212–213). Also, when videotapes of Terrace and the other teachers interacting with Nim were reviewed, it turned out that the vast majority of Nim's utterances employed signs that had just been used by his teacher. Similar studies of young children show that, although it is often an adult who initiates an exchange, less than 20 percent of the words in the child's contributions mirror the adult's recent speech. By contrast, about 40 percent of Nim's utterances were composed entirely of signs that his teacher had just employed (pp. 215, 218). In the videotapes, Terrace also noticed that Nim did not, like a child, show a good sense of taking turns during conversation (p. 219), although other authors have emphasized that unlike users of spoken language, users of ASL commonly sign simultaneously and do not regard this as interrupting.

Terrace concludes that nothing in the available research "suggests anything other than a consistent tendency for the teacher to initiate the signing and for the signing of the ape to interrupt and mirror the teacher" (p. 220), and that he "must therefore conclude—though reluctantly—that until it is possible to defeat *all* plausible explanations short of the intellectual capacity to arrange words according to a grammatical rule, it would be premature to conclude that a chimpanzee's combinations show the same structure evident in the sentences of a child" (p. 221). Terrace ends the book with some optimistic observations, however, noting that before the Premacks' and the Gardners' work, it was not even thought that chimps could learn a significant vocabulary of words. He also notes that Nim was reared under less than optimal conditions, having had over 60 different teachers and several significant changes in housing arrangements and primary caretakers over his approximately four years with the project (p. 221). And he suggests that a return to the "enculturation" approach might yield more impressive results by starting the work at an earlier age and making communicating with human handlers a significant part of life within a stable and secure environment, and in a social situation where the rewards for signing were not narrowly food- and drink-dependent (pp. 223–226).

Francine ("Penny") Patterson's work with her gorilla Koko has done precisely these things. She has been working with Koko in ASL since July 1972, when Koko was just barely a year old. Patterson has lived in close proximity with Koko ever since, working with her in a trailer home. She uses the "simultaneous communication" approach, signing to Koko while speaking in English sentences during a daily routine of language training, play and exercise, housecleaning, unstructured interactions, looking at books, and so on.

Patterson claims that Koko learned signs faster than Washoe. After three years of training, Koko had learned 127 to Washoe's 85, and after 51 months (4.25 years), Koko had learned 161 to Washoe's 132 (Patterson and Linden 1981, pp. 87–89). Her web site now makes the following claims: "Koko has a working vocabulary of over 1,000 signs. Koko understands approximately 2,000 words of spoken English. Koko initiates the majority of conversations with her human companions and typically constructs statements averaging three to six words" (http://www.koko.org/world/, accessed November 12, 2010). Unfortunately, as detailed by Eugene Linden, a fluent ASL signer and her coauthor of the popular book *The Education of Koko* (1981), Patterson has actively shielded Koko from contact with other researchers and has not made systematic data available to the scientific community, while making frequent claims about Koko's abilities in the popular media. He describes Patterson as adopting a "bunker mentality," not coding and analyzing data thoroughly, "excommunicating" people whom she perceives as slighting her in some way, and blocking research by interested assistants (Linden 1986, pp. 115–129). This makes it impossible to objectively assess Patterson's claims about Koko's accomplishments.

Sue Savage-Rumbaugh's work with bonobos has also embodied the virtues of the "enculturation" model, but her group has produced some carefully controlled experiments, showing that one of their bonobos compares well with a two-year-old child in terms of comprehension of spoken English. Since 1975, Savage-Rumbaugh and colleagues have worked with several bonobos ("pygmy" chimpanzees), first at the George State University Language Research Center in Atlanta (Savage-Rumbaugh et al. 1998) and now at a facility outside Des Moines (http://www.greatapetrust.org/). They use a board of lexigrams and the simultaneous communication technique. That is, the trainers speak to the bonobos in English sentences while simultaneously touching related lexigrams. By 2005 the board included lexigrams for various

> objects: including "orange juice," "raisin," and "hamburger";
> activities: including "wait," "go," "chase," "tickle," and "dig";
> locations: including "bedroom," "outdoors," "sandpile," and "Sue's gate";
> times: including "now," "before," "later," "today," "yesterday" and "tomorrow"; and
> emotional states: including "sorry," "mad," "want" and "happy." (Segerdahl et al. 2005, pp. 215–218)

Starting with very young bonobos, the researchers have not so much actively trained them to use the board of lexigrams as modeled its use during normal daily activities. They do sometimes reward the bonobos for matching lexigrams to pictures, but mostly they just engage the bonobos in various activities, such as preparing meals, playing, and traveling around their facilities, while conversing as naturally as possible with the bonobos about what they are doing.

Savage-Rumbaugh's star pupil is Kanzi, who was brought to the Georgia State facility as an infant. There he was fostered by an adoptive mother, Matata, who was being trained to communicate using the board of lexigrams. No systematic attempt was made to teach Kanzi to use the board, and although he touched lexigrams occasionally, he was not very accurate (e.g., touching "banana" after being given an orange). When Matata was taken away (for breeding) when he was about two years old, however, Kanzi's keyboard use quickly outstripped that of his foster mother. Male bonobos are "unusually passive" for apes, and this has allowed the team to continue working very closely with Kanzi long past the age at which the chimpanzees in the studies described earlier were all retired (for some sad details on their "retirement," see Linden 1986). Kanzi is now claimed to understand over 100 spoken words and to use about 250 lexigrams (Segerdahl et al. 2005, p. 215).

Despite being open to the scientific community in a way that Patterson is not, Savage-Rumbaugh's work has faced persistent skepticism. Some of this might be caused by her highly nuanced interpretations of the bonobos' thoughts and intentions. For instance, in order to discourage Kanzi from climbing onto the roof of a building where a lot of heating and air conditioning equipment was, she told Kanzi that there were "monsters" up there. Later, as they observed a man working on the top of the building, she describes the bonobo as "fascinated" by the fact that this man was casually walking in the area and, she adds, "it must have puzzled Kanzi" that no one warned the man about the monsters and that no monsters jumped out at him (Savage-Rumbaugh et al. 1998, p. 32).

In response, Savage-Rumbaugh and colleagues designed a tightly controlled test of language comprehension that could be applied to both Kanzi and a normal two-year-old child. In the experiment, the sentences to which the subjects were to respond were randomly generated, the person speaking was not visible to the subject, and the people that were in the room to evaluate the subject's response wore headphones that prevented them from knowing what the correct response would be. The experiment is described in detail in Savage-Rumbaugh et al. 1993, and, in my estimation, the results clearly indicate that Savage-Rumbaugh is correct about her more circumspect claim that Kanzi understood spoken English about as well as a two-year-old child.

The child in this study, named Alia, was reared in importantly similar respects to Kanzi. Alia's mother was one of Kanzi's caretakers, and she spent each weekday afternoon with her child, for about a year prior to the study, in the trailer that would be used as the test site. To a large extent, the child's training paralleled Kanzi's, with Alia's mother intentionally making her "language input and caretaking behavior . . . similar for both subjects" (p. 44). Moreover, the child was also exposed, from three months of age, to a lexigram board just like that used with Kanzi. At seven and a half months, Alia began spontaneously employing lexigram symbols on the board, more than three months prior to her first intelligible speech. (Alia was a normal child, slightly advanced for her age, but one of the goals of the project was to develop lexigram boards for use with autistic children; Savage-Rumbaugh 1998, pp. 190, 208.)

The team presented both Kanzi and Alia with several types of sentences. Some were instructions, for example to put one object on another, or to move an object from one place to another, to give a specified person in the room an object or pair of objects, or to perform an action on an object, for example "Knife the sweet potato." Others simply announced information, either about the location of an interesting object, for example "The surprise is hiding in the dishwasher," or about the imminent action of a person in the room, for example "Liz is going to tickle Kanzi." Still others made use of toys in "pretend actions," for example "Make the doggie bite the snake."

The subjects were initially tested to be sure that they were familiar with the words used in the sentences, and both were made accustomed to responding after the speaker disappeared behind a one-way mirror. The speaker in Alia's case was her mother, and in Kanzi's, it was Savage-Rumbaugh. The non-blind, training phase involved over 200 trials. Then, in the blind phase, about 400 more trials occurred, with the objects, places, persons, and actions used in the sentences randomly generated from among those with which each subject was familiar. The probability of a random response being correct was extremely low. For example, if there were seven objects and six locations from which to choose in response to a sentence of the form "Take object X to location Y," then the probability of a random pairing of object and location being correct was only 2.4 percent. The trials were videotaped and then scored separately by two observers as correct, partially correct, or incorrect. The percentage of inter-observer agreement on these scores was 98 percent for Kanzi and 89 percent for Alia. Kanzi's responses were scored as correct 74 percent of the time, Alia's 65 percent. Savage-Rumbaugh and colleagues conclude that "The overall high performance level of both subjects provided strong evidence of their ability to comprehend most sentence types" presented (Savage-Rumbaugh et al. 1993, p. 76).

Ape language studies tend to be a sort of Rorschach blot, with those who already think that apes are persons seeing clear confirmation in the work and those who already think that apes are not persons seeing the work as clearly flawed. My own response is mixed. In my judgment, Savage-Rumbaugh's work conclusively demonstrates that apes can learn simple syntactic rules and can comprehend simple, present-tense English sentences. It also seems clear that the apes engage in spontaneous and creative communication.

On the other hand, I see in ape language studies nothing approximating the complex suite of tenses and concepts that are required for personhood under my working assumption. In particular, there is no good evidence that apes understand or use language to express thoughts about the non-immediate future. Various subjects in the above studies use what they have been taught to request things or interactions. Kanzi, for instance, commonly uses sign combinations to request objects or interactions, for example to be given something, to be chased, to be carried, and even to be carried to a particular place (1998, p. 63). On the first day that his mother was gone, Savage-Rumbaugh and colleagues report that Kanzi began to tell her "not only what he wanted me to do for him but what he *planned* to do next" (1998, p. 25, italics in original). And while traveling around the forest at the Georgia State facility, she reports that Kanzi would repeatedly display a photograph of their destination and direct her appropriately at trail junctions (1998, pp. 40–41). Kellogg (1968, p. 425) reports that chimps in early language studies similarly used gestures to indicate short-term intentions such as to go to sleep or to use the bathroom. But nowhere have I seen evidence that they understand references to the long-term future and the distant past that persons include in their self-narratives.

Regarding thoughts about the past, Savage-Rumbaugh does report that Kanzi sometimes "combine[d] symbols to tell us . . . something that had happened earlier in the day, when we were not present" (1998, p. 63) but other researchers could (presumably) confirm that the reported events occurred. Similarly, Patterson reports an incident in which Koko referred to an event that occurred three days earlier (in such transcriptions, all-caps words indicate signs made by the apes):

> P: What did you do to Penny?
>
> K: BITE.
>
> P: You admit it? (Koko had earlier called the bite a SCRATCH.)
>
> K: SORRY BIT SCRATCH.
>
> (Penny shows the mark on her hand; it does resemble a SCRATCH.)
>
> K: WRONG BITE.
>
> P: Why bite?
>
> K: BECAUSE MAD.
>
> P: Why mad?
>
> K: DON'T KNOW.

One general problem facing apes in these studies is that neither American Sign Language nor the boards of lexigrams used include tenses; temporal references must generally be inferred from the context, and in these studies, that context is provided by the English sentences uttered by the human trainers.

The one would-be exception to the claim that there is no good evidence of apes understanding or using language to express thoughts about the non-immediate past relies on a large amount of such assumed context. Patterson has made remarkable claims about one gorilla's communication of early memories. Michael was a gorilla who trained along with Koko from 1976 to 2000. As an infant, Michael was captured when poachers shot and killed his mother. Patterson claims that Michael repeatedly told her about the incident. Here is Patterson's description of a related video posted on the Project Koko web site: "In this video, Michael tells a story in sign language about his mother, who was killed by poachers when he was just a baby in Africa. He retold this story several times when asked by Penny about his mother." The captions interpreting Michael's signing in the video read, in their entirety: "SQUASH MEAT GORILLA. MOUTH TOOTH. CRY SHARP-NOISE LOUD. BAD THINK-TROUBLE LOOK-FACE. CUT/NECK LIP(GIRL) HOLE" (http://www.koko.org/world/michael_story_vid.html, accessed November 12, 2010). Since Patterson claims that Michael has "told this story" on several occasions, it could be that less ambiguous versions were not caught on tape. There are clear cases in which a handler's knowledge of context allows him to make sense of what would otherwise be a confusing series of signs. For instance, in response to the question "When do you feel hungry?" Koko once responded, "FEEL TIME." That puzzling reply makes perfect sense, though, when we learn that "Koko's most frequent and emphatic use of the sign TIME is to tell her human companion that it is time to deliver the next meal" (Patterson and Cohn 1994, p. 283).

However, even Patterson's sympathetic coauthor Eugene Linden doubts her claim that Michael was telling stories about his mother's death:

> On several occasions, Penny has cited some of Michael's statements, which she interprets as referring to his capture in the wild. I did not think that we should mention these in our book [*The Education of Koko*] because she could offer little evidence to support her conclusions. However, Penny has related this story and her interpretation of it to a number of people, as well as in the newsletter she publishes. (Linden 1986, p. 126)

So I conclude that Patterson's remarkable interpretation of Michael's "story" provides no good evidence that apes understand or use language to express thoughts about the non-immediate past.

Even if we accepted Patterson's interpretation, however, Michael's memory hardly constitutes a *story*. A full-blown biographical story consists of more than an ordered series of events; to understand one's life as unfolding according to the logic of narrative, one must understand events as shaping one's character and aspirations across various periods of it. Again, how could one tell such a story, even to oneself, without something like the suite of concepts and tenses described in the preceding section?

Relatedly, remember that Savage-Rumbaugh and colleagues' impressive experiment with Kanzi and Alia involved language *comprehension* rather than *production*. They begin their report by saying, "The underrated importance of comprehension is heavily emphasized here" (Savage-Rumbaugh et al. 1993, p. 12). Later they admit that their bonobos "have not progressed in the development of language skills at the extremely rapid pace shown by normal human children" (Kanzi was 8 years old when he was tested along with Alia; Savage-Rumbaugh et al. 1998, p. 45), and the bonobos "have not gone as far as normal human beings" in language development (1998, p. 207). They say that "None of these deficits, however, cause the abilities of Kanzi and Panbanisha to stand outside the realm of what we usually think of when we apply the term 'language' to behavior" (p. 208). Fair enough, but on my view it is not language comprehension that is necessary for personhood, it is *storytelling*, which refers to a particularly complex form of language *production*. To understand events as shaping one's character and aspirations across various periods of one's life, one must be capable of telling the story, at least to oneself. Surely this is one of the ways in which Kanzi and other apes "have not gone as far as normal human beings."

§6.5(B). ELEPHANTS AND CETACEANS

So current research on language acquisition in apes gives us no good reason for thinking that any non-human apes can think about their future, or their past, in the way that individuals with a biographical sense of self can. Some scientists who study elephants and dolphins have suggested that they are better candidates for personhood than the great apes, in part because they both use—*in the wild*—complex vocalizations that *could be* a functional analogue of human language.

No language comprehension studies have yet been done with elephants, but some carefully designed studies have demonstrated that dolphins can learn the same kind of simple syntax that apes can learn. This research involves training dolphins to manipulate objects in their pool in response to syntactically structured commands like "Put the frisbee on top of the air mattress!" and "Put the air mattress on top of the frisbee!" In various studies, the elements of the commands have been delivered either as symbols on cards arranged in a row or via a sequence of computer-generated whistles. In such studies, dolphins respond correctly about two-thirds of the time (see, e.g., Herman 1986, p. 232), which compares favorably with Kanzi's performance in response to similar commands delivered in spoken English.

About such research, Denise Herzing and Thomas White write:

> That dolphins demonstrate the capacity to operate within an artificial representational and rule-governed system is highly suggestive of the fact that in their own lives they employ a capacity (the details of which have yet to be identified) that is equally as cognitively complex as the human capacity for language without being analogous in structure and form. (Herzing and White 1998, p. 75)

The same could be said of bonobos, chimpanzees, and gorillas, of course, whose performance in artificial language studies has been just as impressive as that of dolphins. The vocalizations of wild cetaceans, however, seem to be more complex than the vocalizations of any non-human ape, or even the combination of apes' vocalizations with facial expressions, arm gestures, and so on. A significant amount of cetaceans' vocalizations are inaudible to humans. Dolphins, for instance, can hear at much higher frequencies than humans (up to 150,000 Hz compared to 20,000 Hz for normal humans) and in addition to the very high frequency "clicks" that they use for "sonar," they "whistle" above the range of human hearing. One study of recorded vocalizations of dolphins does suggest, based on a complicated application of information theory, that dolphins' vocal repertoire compares favorably to that of two-year-old humans and adults (McCowan et al. 1999, p. 413).

Similar things have been suggested about elephants (see, e.g. Poole 1998, pp. 93–94; Poole & Moss 2008, pp. 83–85), whose hearing extends beyond humans' in the opposite direction from that of dolphins. Elephants are known to use a wide variety of calls, many of them infrasonic (below the level of human hearing), in coordinating group behavior and searching for mates. The infrasonic calls usually have a humanly audible component, but these higher frequency aspects of the calls attenuate very quickly with distance (up to 23dB/100m), so that in the wild, human observers were often astonished by "sudden coordinated movements of large groups of elephants when no signal was apparent to the observer" (Payne et al. 1986, p. 300). The low-frequency components do not attenuate with distance to anything like the same degree, so that elephants can hear the infrasonic calls of conspecifics from as much as several kilometers away. Since the late 1980s, researchers have used special equipment to systematically study elephant communication in the wild (Poole et al. 1988) and in captivity (Payne et al. 1986), and they have discovered that researchers were missing about one-third of all the calls made while they were observing (Poole et al. 1988, p. 387).

So in the case of both cetaceans and elephants, we have reasons to say that they *could be* using a communication system that is complex enough in relevant ways to support a biographical sense of self, even though at present we have no evidence that this is so. In the case of non-human apes, however, there is no reason to believe that wild individuals use any such complicated communication system, and the "enculturation" studies described earlier have yielded no good evidence that non-human apes can acquire an artificial language with the relevant complexity of human languages.

So cetaceans and elephants are special, vis-à-vis apes, in the limited sense that they vocalize in apparently complex ways that we do not yet understand, and therefore it *could be* that they are using a communication system complex enough to support a biographical sense of self. That's a "*could be*" that we don't have in the case of apes. I emphasize that it is only a "*could be*," however, and that at present we have no good evidence supporting the claim that any non-human animal uses a system of communication complex enough to support a biographical sense of self.

§6.5(c). PARROTS

Irene Pepperberg's research on the celebrated African grey parrot Alex suggests that parrots have an impressive ability to understand questions about categories and numbers of objects, and to distinguish "same/different" objects. For instance, after initially learning to identify a variety of objects of various colors (e.g., blue and red) and kinds (e.g., keys v. toy cars), and to say how many items (up to 6), when a mixture of types and colors of items were presented in a random arrangement on a platter, Alex was able to answer questions like "How many blue key?" As in Savage-Rumbaugh's language-comprehension study with Kanzi, Pepperberg gathered data on Alex's responses to questions that he had never heard before. Alex responded correctly on 45 out of 54 (83%) of these probe trials (Pepperberg 1994, p. 40). According to Pepperberg, this "replicates that of humans in a comparable study" (2002, p. 250).

Although the origins of the word are obscure (according to the *Oxford English Dictionary*, "parrot" may either have been derived from the diminutive of the given name "Pierre," or from a shortening of *perroquet*, French for "parakeet"), "parrot" now refers to both the birds and their celebrated ability to mimic a wide variety of sounds. Pepperberg's work has not employed tenses, however, and I am aware of no reported evidence of the birds referring to the non-immediate future or past. So although I believe that Pepperberg's work clearly shows that her parrots can master simple syntax, her work provides no good evidence of storytelling in them. Also, there is no reason to believe that parrots communicate using sounds that humans cannot hear. So, in comparison to both dolphins and elephants, there is less reason to think that parrots *might be* using a communication system complex enough to support a biographical sense of self.

§6.6. Conclusion

As explained at the beginning of §6.4, Schechtman's "narrative self-constitution" account of personal identity is plausible for several reasons. Although she does not claim that the lives of persons so characterized have any special moral significance (1996, pp. 99–100), in the following chapter I will argue precisely that.

For the reasons given in §6.4 and §6.5, however, if "persons" are defined as individuals with a biographical sense of self, then we have no good evidence that any non-human animal is a person. So if some animals' lives are to be attributed special moral significance in good ILS rules, there must be some other morally relevant differences among animals. In chapter 8, I will present evidence that some animals indeed have a form of forward- and backward-looking consciousness that is relevantly similar to normal humans' biographical sense of self. Having this "autonoetic consciousness" of the past and future makes them what I call "near-persons," and gives their lives special moral significance vis-à-vis those of "the merely sentient."

{7}

Moral Significance and Autonoetic Consciousness (AC)

§7.1. The Autonoetic Consciousness Paradigm

According to what I call *the autonoetic consciousness paradigm*, the lives of individuals with a robust, conscious sense of their own past and future have special moral significance in comparison to those of the merely sentient. According to the *Oxford English Dictionary*, "autonoetic" means "self-perceiving," but the term "autonoetic consciousness" is used by some contemporary psychologists as a label for conscious awareness of one's own past, present, and future (see §8.2). Examples of philosophers who endorse the autonoetic consciousness paradigm include Singer (1987, 1993), Jeff McMahan (2002), Michael Tooley (1983), Dale Jamieson (1983), and myself (1998), among others.[1] These authors all agree that having a robust, conscious sense of one's own past and future gives one's life a special kind of value.

In this chapter, I show why Hare, who never directly addressed the moral significance of autonoetic consciousness, should have endorsed the autonoetic consciousness paradigm and with it the distinction between persons and the merely sentient. As I noted in §1.2, Hare never argued for any stipulative definition of "person," but he did at one point say that a stipulative definition of the term can serve to summarize an argument for the special moral significance of certain traits. In section 2 of this chapter, I give two utilitarian arguments to the conclusion that the lives of individuals with autonoetic consciousness have special moral significance vis-à-vis those of the merely sentient. In section 3, I then use these same arguments plus a third to show that the same is true of the

---

[1] Tom Regan would also qualify as a proponent of the autonoetic consciousness paradigm if he believed both (1) that what he calls "subjects of a life" (SOALs) have a robust, conscious sense of their own past and future, and (2) that there are also some merely sentient animals that don't. He seems to believe the former (Regan 1983, p. 243), but I suspect that he thinks that any individual with conscious desires for the future—even the very immediate future—counts as a SOAL. So what counts as a "robust," conscious sense of the future for Regan is probably something very much weaker than what counts as a "robust" one for McMahan, Singer, Tooley, or me. See also the related points in §1.7 and 8.5(C).

lives of individuals with a biographical sense of self vis-à-vis the lives of those that have autonoetic consciousness but no biographical sense of self. In the remaining sections of this chapter, I reply to several objections.

In the following chapter, I will survey various kinds of evidence for the claim that at least some non-human animals are near-persons—that is, although they lack a biographical sense of self for the reasons given in the preceding chapter, they nevertheless have a fairly robust, conscious sense of their own past and future. To the extent that there is good evidence for this, good ILS rules will distinguish between these animals and merely sentient animals, which we presume lack any robust, conscious sense of their past and future.

Before proceeding, I need to remind the reader of two qualifications that were discussed in the introduction. One concerns the notion of "moral significance." The arguments that I give in support of the autonoetic consciousness paradigm will all emphasize how having the capacity for autonoetic consciousness can make an individual's life go better *or worse* than it could otherwise. This is because it is possible to benefit *or harm* an individual with autonoetic consciousness in ways that it is impossible to benefit or harm an individual that lacks autonoetic consciousness.

Pop culture recognizes this. On the forward-looking side, this is why it makes sense for the character David Aames in the 2001 movie *Vanilla Sky* to be a "pleasure delayer"; as Daniel Gilbert puts it in *Stumbling on Happiness*, "Forestalling pleasure is an inventive technique for getting double the juice from half the fruit" (2006, p. 17). On the backward-looking side, the Jedediah Leland character in the 1941 movie *Citizen Kane* at one point says, "I can remember absolutely everything, young man, that's my curse, that's one of the greatest curses ever inflicted on the human race: memory." And in the 2006 film *Eternal Sunshine of the Spotless Mind*, a couple find themselves in a relationship that began happily but has deteriorated in predictable ways. As long as they are aware of the history of their relationship, they cannot enjoy themselves together. After they have their memories of each other erased, however, they are content to start again.

Those popular pieces convey the idea that having the abilities to consciously anticipate the future and to consciously remember the past dramatically affects the quality of humans' lives. As I emphasized at the end of §1.8, this does not mean that the lives of persons are always preferable to those of near-persons and the merely sentient, but it does mean that their lives are more (so to speak) *morally charged* than those of the merely sentient. By that I mean that we should take special care in our dealings with them, and this is what justifies endorsing different ILS rules to govern our treatment of individuals with autonoetic consciousness.

The other qualification concerns my notion of a "merely sentient" individual, by which I mean one that "lives entirely in the present." As I noted in §1.7, this may turn out to be a hypothetical construct. Since having desires means wanting things to change, it may be that any animal that has conscious desires has some conscious sense of at least the immediate future. Thus the animal kingdom may not divide neatly into persons, near-persons, and the merely sentient: it may be that all sentient animals have desires (at least the desire to avoid pain), and thus some minimal conscious sense of the future. So the animals for whom we have good evidence of autonoetic consciousness may just have a *more robust* sense of their future, and persons, in turn, an *even more robust* sense of their future. For the purpose of explaining my arguments in this chapter, however, I will assume that there are sentient animals that literally live *entirely in the present*, and this is what I mean when I refer to "merely sentient animals."

§7.2. The Special Moral Significance of Autonoetic Consciousness

According to my stipulative definition, persons are individuals with a biographical sense of self. As described in §6.3(A), persons are capable of responding to questions about the coherence of their self-narratives, and I presume that these responses are made consciously. So persons are paradigm examples of individuals with autonoetic consciousness. The arguments advanced in this section do not depend on individuals having a biographical sense of self, however. These arguments apply to any individual that has a robust, conscious sense of its own past and future in comparison to any merely sentient individual. Thus these arguments support the conclusion that the lives of both persons and near-persons have greater moral significance than those of the merely sentient.

§7.2(A). THE ARGUMENT FROM CONSCIOUS MEMORY AND CONSCIOUS ANTICIPATION

From a utilitarian perspective, the most basic reason for thinking that autonoetic consciousness adds value to the life of an individual turns on the fact that the abilities to consciously remember the past and to consciously anticipate the future allow the individual to reexperience good (and bad) states of consciousness and to anticipate (and dread) future experiences. (Jamieson 1983 gives a version of this argument; p. 145.)

The merely sentient may experience a sense of ease based on what psychologists call "implicit memory" or anxiety based on what we might call "implicit anticipation." An implicit memory is one that affects one's choices, but without being available for conscious recall. For example, in some of the research described in the next chapter, human subjects are asked which one of a pair of words was on a list that was shown to them some time ago. On such problems, subjects can often choose the correct word even though they are unable to consciously recall being presented with it. A merely sentient animal is, by definition, incapable of consciously recalling the past, so all of its memories are implicit. Nevertheless, a merely sentient animal might, as a result of having relevant implicit memories, be put at ease when placed in a cage where it has always been left undisturbed.

Similarly, we might describe as "implicit anticipation" the anxiety that a merely sentient animal presently experiences when placed in a location from whence it was previously transferred into a painful experimental procedure. The implicit anticipation that a merely sentient animal experiences has no forward-looking content that the animal can consciously attend to. In this respect, the resulting psychological state is analogous to anxiety neurosis, in which the patient presents with extreme anxiety without being able to say what, specifically, he or she perceives as threatening. Nevertheless, a merely sentient animal experiencing implicit anticipation might react in adaptive ways that are based on its previous experiences—attempting to flee in response to the event that immediately preceded its previous aversive experience, for instance. But the future event that is—functionally speaking—the focus of its anxiety cannot be played out in conscious anticipation. The merely sentient animal that flees in implicit anticipation of an aversive event flees in response to "it-knows-not-what."

So implicit memories and implicit anticipations can make a merely sentient animal's life go better or worse for it, but they do not add as much value or *dis*-value as consciously reexperiencing a satisfying event or consciously dreading an anticipated event. To see why, compare the experience of relaxing in the "afterglow" of a sexual encounter to the experience of relaxing in the same "afterglow" while consciously replaying memories of the encounter. The two experiences contain the same pleasant "glow" of present relaxation, but the second experience also contains the pleasure of recalling—"reliving"—the encounter itself. So, however valuable the first experience is, the second is that valuable and more. And the same would be true of the experiences of "implicit anticipation" and conscious dread of a future event: whatever *dis*-value the first contains, the second contains that *dis*-value and more.

From a utilitarian perspective, then, their ability to consciously remember the past and to consciously anticipate the future gives the lives of individuals with autonoetic consciousness special moral significance, in contrast to the lives of animals that "live entirely in the present." Notice that this argument does not depend on our being able to quantify or in any way measure the value of the experiences compared. It depends only on the principle that if we know that an experience *A* contains some value, and that experience *B* contains all of that value and more, then we know that experience *B* contains more value than experience *A*. This is what Ralph Barton Perry called "the principle of inclusiveness" (1926, p. 647). For a detailed discussion of this principle, including responses to some objections to it, see Varner 1998, chapter 4.

§7.2(B). THE ARGUMENT FROM MORE INCLUSIVE DESIRES

Another reason for thinking that the lives of individuals with autonoetic consciousness have special moral significance involves their ability to plan for the future. I will first state this argument by comparing persons with the merely sentient. As I will go on to explain, however, it also shows that the lives of animals with more limited forms of autonoetic consciousness have greater moral significance than those of the merely sentient.

The utilitarian holds that the positive conscious states accompanying desire satisfaction have intrinsic value, and that negative conscious states accompanying the frustration of desires have intrinsic *dis*-value. Both persons and merely sentient animals are capable of experiencing the satisfaction of immediate desires for food, drink, and so on, but merely sentient animals are capable only of this. The person is capable of these serial satisfactions, but also has long-term, more complicated desires, projects, and plans. So whatever the value of the momentary serial satisfactions of which each is capable, a person's life includes, superimposed upon these, whatever additional value comes from the satisfaction of such long-term projects. (A version of this argument is given in Varner 1998, chapter 4.)

In saying this, I do not mean to deny that the merely sentient are capable of "planning" in an unconscious sense. Consider, for instance, a squirrel hoarding acorns. For the sake of making it clear how different are the conscious lives of the merely sentient and those with an ability to consciously plan for the future, let us assume both (1) that the squirrel consciously desires to get each acorn into its stash, and (2) that the squirrel is completely unconscious of the purpose of its hoarding behavior. Accomplishing the goal of hoarding enough acorns will obviously improve the squirrel's future well-being, insofar as its future experiences will be more pleasant because it has enough to eat. If the squirrel is not conscious of the long-term goal, however, it can achieve no sense of satisfaction when it has stashed enough acorns. It cannot, in effect, say "There, I've accomplished *that*!" since *that* (the goal of laying up enough acorns for the winter) is something of which it is not conscious.[2] Thus, however much achieving its goal improves the conscious life of the squirrel, an animal that is conscious of the goal enjoys both that improvement *and* the sense of satisfaction that comes with being conscious that the goal has been achieved.

---

[2] Could the content of "*that*" instead be what Joel Feinberg once described as a mere brute longing for I know not what? Feinberg denied that such a "brute longing" would constitute a desire at all (1974, pp. 52–53), and with good reason. For it seems that the only function of such a conscious desire would be to keep the acorn stashing going until some number, sufficient to the cause of sustaining the squirrel through the winter, had been stashed. But then, what purpose would the "mere brute longing" be playing? The squirrel would behave just the same if it just kept having desires to stash acorns, one after another, until it stops having them (or if the desires became progressively weaker as more acorns were stashed).

Here again, the argument does not depend on our being able to quantify or in any way measure the value of the experiences compared. It depends only on Perry's principle of inclusiveness: the life of the merely sentient individual includes whatever pleasant experiences the achievement of the goal makes possible, but the life of the conscious planner contains all of those pleasant experiences *plus* whatever sense of satisfaction comes with being conscious of the achievement. At the same time, the ability to consciously plan for the future increases the possibilities for *dis*-satisfaction, but this just means that, rather than ensuring that an individual's life goes better from a utilitarian perspective, the ability to plan for the non-immediate future gives the lives of individuals with autonoetic consciousness "special moral significance," as specified in §1.8: their lives are more "morally charged" as a result of having autonoetic consciousness, and we should exercise some kind of special care in our dealings with them.

Although I initially stated this argument in terms of persons' relatively complex ability to plan for the future, this argument also applies to animals with more limited forms of autonoetic consciousness. Between human persons and merely sentient animals there may lie a continuum of animals with varying degrees of conscious awareness of the future. For instance, the evidence surveyed in §8.4(F) suggests that both great apes and some birds can consciously plan at least a day in advance. This is nothing in comparison to persons' ability to consciously plan the rest of their lives, but it is enough to augment the value of their lives.

From a utilitarian perspective, then, the ability to plan for the non-immediate future gives the lives of individuals with autonoetic consciousness special moral significance vis-à-vis those of the merely sentient.

§7.3. The Special Moral Significance of a Biographical Sense of Self

For the reasons given in the preceding chapter, however, there is no good reason to believe that any non-human animal has a biographical sense of self, and careful attention to the nature of autobiographical memories and to the kind of conscious anticipation of future events of which persons are capable shows why the lives of persons have greater moral significance than those of any animals that have a more limited form of autonoetic consciousness.

§7.3(A). The Argument from Conscious Memory and Conscious Anticipation (Again)

To the extent that telling and understanding stories is a conscious process, being in the habit of telling stories augments the value of a life in two obvious ways. First, to the extent that I am in the habit of telling stories about my own past and future (to others or just to myself), I will in a way be reexperiencing past events and anticipating future ones. I say "in a way," because telling a story about my past does not necessarily involve having an episodic memory. Similarly, telling a story about my own future does not necessarily involve consciously anticipating events. Nevertheless, to the extent that I consciously enjoy telling and understanding stories about my life, doing so augments the value of my life.

This points to the second obvious way that consciously engaging in telling and understanding stories augments the value of persons' lives: even if the stories are not about us—in fact, even if they are fictional—to the extent that we enjoy (or suffer from) the experience, the practice of storytelling augments the value of our lives. Obviously, some of this augmented value comes from empathizing with the characters in the story, but I should emphasize that this is not the only way in which storytelling augments the value of our lives. Humans seem to simply enjoy storytelling. The amount of time and money spent on novels, movies and television, songs, poetry, theater, and so on testifies to how habitual and satisfying an activity storytelling is.

What I have just been saying was premised on the assumption that the telling and understanding of stories is a *conscious* process. As Schechtman emphasizes, however, a self-narrative is often only *implicit*. Having a narrative self-identity still alters the way in which an individual experiences things when the self-narrative is implicit. Schechtman says, for instance, that a person who was in the past

> convinced by beatings and parental disapproval that the only thing that can make her worthwhile is to excel in athletic competition may well suffer much more intensely in losing, and be willing to go to far greater lengths to win, than the person who has come to believe that his athletic talent is just one of the many wonderful and lovable things about him, and that his life has many other rewards and adventures in store. (1996, p. 112)

Similarly, she describes how a graduate student preparing for an academic career and a father who expects to always have trouble supporting his family will experience a period of economic hardships very differently because their expectations for the future are so different. These persons' experience of the present is shaped by both their understanding of their past and their expectations of how their present is preparing them for the future, whether or not their chosen self-narratives are implicit, whether or not they have ever consciously told anyone the story of their lives.

So, just like autonoetic consciousness in general, storytelling in particular is a two-edged sword. Obviously, humans enjoy storytelling as an activity, but it can also augment our lives in negative ways. We can enjoy a frightening story as long as we know that we are really safe and in control, but stories can also take hold of our psyches involuntarily and with damaging results. The female athlete in Schechtman's examples above is so much the worse off for having internalized a self-narrative that makes athletics all-consuming, and cultures can use stories to inculcate false and pernicious beliefs. And my racy example of a sexual encounter in §7.1(A) provides a striking illustration of how having a biographical sense of self can spoil one's experience of an otherwise pleasant moment. For suppose that, rather than reliving the encounter while you bask in its pleasant afterglow, you remember that you've slept with this person several times before, that each encounter was followed by rejection and humiliation, and that you have repeatedly vowed never to let it happen again.[3] Under those circumstances, thinking of one's life as a narrative would spoil an otherwise pleasant moment.

As I emphasized when I introduced the notion in §1.8, however, this is just what it means for one kind of life to "have greater moral significance" than another. The ability to tell and understand stories can augment the value of a person's life in both positive and negative ways, and it is possible to harm or benefit persons through kinds of treatment that would be indifferent in terms of their effects on non-persons' lives.

§7.3(B). THE ARGUMENT FROM MORE INCLUSIVE DESIRES (AGAIN)

Having a narrative conception of their lives allows persons to form a particular kind of complex desire, and this too gives their lives greater moral significance than those of near-persons. I have in mind *the desire to become a certain kind of person.*

Let me first illustrate this claim with a backward-looking example. Schechtman describes a scene from the 1942 film *Now, Voyager* in which Bette Davis's character (Charlotte) undergoes a "miraculous treatment" that transforms her from a "stocky, unattractive and highly neurotic spinster" to "a delightful, slim, and attractive woman" (p. 112). In the scene in question, the woman has gone on a cruise and has begun an affair with an unhappily married man (Jerry). Trying to help her lover understand who she is, she shows him an old picture, saying "I'm the fat lady with the heavy brows and all the hair." "That response," Schechtman writes, "speaks volumes," because

> a particular significance has been bestowed on her cruise, and on her relationship with Jerry, that would not have existed if she were not the fat lady with the hair and the brows. She feels exhilarated, frightened, proud, and somewhat fraudulent in the way she does because she knows where

[3] I owe this variation on the example to Marya Schechtman, who gave it while commenting on my presentation of some material from this chapter at the Central Division meeting of the American Philosophical Association in April 2011.

> she has come from and how she got there. Her story is not the story of the always-beloved, but the story of the sorry caterpillar who has suddenly become a beautiful butterfly; these are two different stories, and it is something quite different to live the one rather than the other. Charlotte undoubtedly remembers her past all too well, but she does more than just remember it, she lives it—experiencing the present and planning the future in its light. (p. 113)

That is, the cruise and the affair *feel different to her* precisely because she understands her life as the story of a sorry caterpillar transformed into a beautiful butterfly, rather than the story of the always-beloved. The use of the present tense—"I *am* that woman in the picture"—imports the significance of the past into the present; it has the effect of underlining how the past affects the present.

Although this is a backward-looking example, it is best understood in terms of Charlotte's desire, earlier in her life, to become a different kind of person: she is exhilarated in the present because she understands that this desire has been fulfilled. A near-person can consciously reexperience past events, and this gives their lives special moral significance vis-à-vis those of the merely sentient. Only persons can understand their lives as unfolding according to the logic of narrative, however, and only within such a narrative conception of one's life can one appreciate that such a complex desire has been fulfilled.

Relatedly, I said earlier (in §6.4) that "we have at least *some* evidence" for second-order desires in some animals. I can now explain what I meant by comparing and contrasting the kind of second-order desires of which persons and near-persons are capable. A second-order representation is a representation of a mental state, either one's own or that of another. By a "second-order *desire*" is meant, specifically, a desire about what desires one has, or at least whether or not one will act on a desire that one has.

Examples of non-human animals forming second-order desires are very rare (at least plausible examples), but here is one involving a group of wild chimpanzees for whom Jane Goodall and colleagues had been putting out food. Notice how Goodall interprets the behavior of the subordinate male, whom she calls Figan:

> One day, some time after the group had been fed, Figan suddenly spotted a banana that had been overlooked—but Goliath was resting directly underneath it. After no more than a quick glace from the fruit to Goliath, Figan moved away and sat on the other side of the tent so that he could no longer see the fruit. Fifteen minutes later, when Goliath got up and left, Figan, without a moment's hesitation, went over and collected the banana. Quite obviously he had sized up the whole situation: if he had climbed for the fruit earlier Goliath, almost certainly, would have snatched it away. If he had remained close to the banana he would probably have looked at it from time

> to time: chimps are very quick to notice and interpret the eye movement of their fellows, and Goliath would possibly, therefore, have seen the fruit himself. And so Figan had not only refrained from instantly gratifying his desire, but had also gone away so that he could not give the game away by looking at the banana. (quoted in Whiten and Byrne 1988, p. 219)

Obviously the skeptic can find reasons for doubting Goodall's interpretation. In the extreme case, it could be that Figan had given up and was just leaving the area when he went and sat behind the tent, but when he saw Goliath go by he remembered the banana and ran for it. The interpretation that Whiten and Byrne give is less deflationary: they interpret Figan's behavior as evidence of a second-order representation; they interpret Figan as representing *Goliath's* attentional state (Whiten and Byrne 1988, p. 219). If we read Goodall's interpretation literally, however, Figan's second-order representation is actually a second-order *desire*: she interprets him as going to sit out of sight of the banana *because he wants not to act on his desire to look at it.*

To the extent that Goodall's interpretation of Figan's behavior is correct, then there is *some* evidence of second-order desires in near-persons. In adopting a complete self-narrative, however (even if only implicitly), a person operates at an entirely different level. A person does not just have desires about which desires she has or which desires she will act on; a person understands that her present personality or character as a whole has been shaped by previous experiences, and she aims at becoming or remaining a certain kind of person, with a certain personality or constellation of character traits. These understandings and intentions are expressed using the complex suite of tenses found in natural language (as explained in §6.4). How this is so comes out more clearly when we ask how Figan's second-order desire stacks up against Henry Frankfurt's account of "second-order volitions" and Charles Taylor's notion of being "a strong evaluator."

In "Freedom of the Will and the Concept of a Person" (1971), Frankfurt adopts a Hobbesian definition of "will"[4] as "an *effective* desire," one that actually moves an agent or would actually move him if and when he acted (p. 8), and he defines a "second-order volition" as a second-order desire about what one's will or effective desires should be (p. 11). It is, he says, "in securing the conformity of his will to his second-order volitions, then, that a person exercises freedom of the will" (p. 15). At first blush, it looks like Figan has a second-order volition in Frankfurt's sense, insofar as Figan (according to Goodall's interpretation) wants not to act on his desire to look at the banana. But Figan satisfies his second-order desire not to act on his desire to look at the banana by positioning himself where he *cannot* look at the banana, rather than by

[4] Hobbes defined "will" as "the last appetite or aversion immediately adhering to the action or to the omission thereof," that is "*the last appetite in deliberating*" ([1651] 1958, p. 56).

making it the case that this first-order desire is not his effective desire, that is, by changing his will. This is the difference between Odysseus having himself strapped to the mast and really being able to weather the sirens' song.

Taylor's emphasis can also be said to be on changing one's will, but in "Responsibility for the Self" (1976), he identifies being what he calls "a strong evaluator" with making judgments about *the desirability* of having various desires:

> Someone who evaluates non-qualitatively, that is, makes decisions like that of eating now or later, taking a holiday in the north or in the south, might be called a simple weigher of alternatives. And the other, who deploys a language of evaluative contrasts ranging over desires we might call a strong evaluator.
>
> Now we have seen that a simple weigher is already reflective in a minimal sense, that he evaluates courses of action, and sometimes is capable of acting out of that evaluation as against under the impress of immediate desire. And this is a necessary feature of what we call a self or a person. He has reflection, evaluation and will. But in contrast to the strong evaluator he lacks something else which we often speak of with the metaphor of "depth."
>
> The strong evaluator envisages his alternatives through a richer language. The desirable is not only defined for him by what he desires, or what he desires plus a calculation of consequences; it is also defined by a qualitative characterization of desires as higher and lower, noble and base, and so on. Where it is not a calculation of consequences, reflection is not just a matter of registering the conclusion that alternative A is more attractive to me, or draws me more than B. Rather the desirability of A over B is something I can articulate if I am a strong evaluator. I have a vocabulary of worth. (Taylor 1976, p. 287)

That Taylor does not identify "desirable" with "*morally* desirable" is clear from a related criticism that he makes of utilitarianism. He says that "The complete utilitarian would be an impossibly shallow character, and we can gauge how much self-declared utilitarians really live their ideology by what importance they attribute to depth" (p. 288). That is, according to Taylor, a strong evaluator must judge the desirability of his desires on non-utilitarian grounds.

The Harean rejoinder to this disparaging remark about utilitarians should be obvious from §1.5 and chapter 3, *passim*: real-world humans cannot be "complete utilitarians," because they need ILS rules, and ILS rules have a "deontological flavor." It is the internalization of such ILS rules that provides what Taylor calls the "deep," non-utilitarian aspects of strong evaluators' thinking, although these rules are "ultimately" utilitarian because their adoption is justified in utilitarian terms. So I would modify Taylor's conception to think of "a strong evaluator" as one who strives to change what desires she has and which desires she acts on, in light of specifically moral considerations, whether these are articulated in non-utilitarian ILS rules or in explicitly utilitarian critical thinking—there is nothing "shallow" about the latter.

We now see that there are at least three kinds of second-order desires. Figan's case seems to meet the minimal criterion of having a desire that is—in some way or other—about what one's desires are, but in Figan's case we do not have a clear instance of what Frankfurt calls a second-order volition, and we clearly do not have a case of what Taylor calls "strong evaluation." In the lives of normal persons, however, we have clear examples of the latter two. To the extent that the ability to form and satisfy such complex desires augments the value of their lives, persons' lives have greater moral significance than those of near-persons.

§7.3(C). ONLY PERSONS HAVE "LIVES-AS-A-WHOLE"

The preceding subsections have shown how the arguments originally used to establish that the lives of individuals with autonoetic consciousness have greater moral significance than those of the merely sentient can be used to show that persons' lives also have greater moral significance than those of near-persons. In this subsection, I present a new argument, one that is related to a metaphysical or ontological claim about personhood. Specifically, I argue that the development of a narrative conception of self creates a new and radically different type of interest that only persons have.

In making this argument, I intend to be distancing myself from something that I said in my first book. There I said that "*almost no* non-human animals have ground projects" (1998, p. 93, emphasis added), allowing that "Perhaps some of the great apes (gorillas, chimpanzees, and orangutans) or some cetaceans have ground projects, but none of the animals the average person comes into contact with on a regular basis would appear to" (p. 92). I now think it incorrect to say that *any* known non-human has ground projects. A "ground project" is a nexus of "categorical desires," and Bernard Williams (from whom I've borrowed those two terms) describes a categorical desire as one that "makes life worth living." But when persons think about the value of their *lives*, they are thinking of their lives as *stories*, and this is something that (I am assuming—§6.4) only linguistically competent individuals can do.

As noted in §6.3(B), however, self-interested concern for future events varies as a function of how central they are to the person's chosen self-narrative. As Schechtman puts it:

> First of all there is a certain primitive concern we have for the character of any state that will be ours at all—we care whether it will be pleasurable or painful. Furthermore, we have a special concern for our desires and goals and the beliefs they express. The goals and desires that occur in a person's history are not, however, all equally his, and so a person should not be equally concerned with the fulfillment of every desire or goal that occurs in his history. For a person to have *self-interested* concern for the fulfillment of

> a desire, that desire must, to be sure, at least be part of his history. Among those goals which fulfill this minimal requirement, however, the degree to which any particular goal is attributable to a person is at least part of what determines the degree to which he should be concerned about it. . . . (1996, p. 85)

When I quoted this same passage in §6.3(B), I said that this means that a person's chosen self-narrative determines, to a very significant extent, what is best for them on the whole; it is the basis for making judgments about what is in a person's interests, "all things considered." This is because in adopting a self-narrative, an individual expresses commitments to various things, and without this kind of commitment there is no way to say what is best for an individual on the whole, what is worth sacrificing or risking for what.

For both near-persons and the merely sentient, then, the concept of individual well-being is completely aggregative—how well or badly their lives go is purely a function of how much positive and negative conscious states they contain. We might say, using a term introduced by Steve Campbell (2006), that their "phenomenal well-being" is all that matters. A person, on the other hand, thinks of his life as unfolding according to the logic of narrative, and this means that how well his life goes is not completely addressed by asking how good it felt, on the whole, to live that life. In this way, a person's chosen self-narrative determines, to a very significant extent, what is best for him on the whole; it is the basis for making judgments about what is in a person's interests, "all things considered."

To drive home the significance of this point, remember that, on the narrative self-constitution account of personal identity, it is only minimally metaphorical to say that persons are the authors of their own lives. This is because "a person creates his identity by forming an autobiographical narrative . . . a person's *identity* . . . is constituted by the content of her self-narrative, and the traits, actions, and experiences included in it are, by virtue of that inclusion, hers" (Schechtman 1996, pp. 93–94). Thus there is no fact of the matter about what is most important to a person before she or he chooses (at least implicitly) a self-narrative, and thus there is no answer to the question "What is in this person's best interests?" independent of that narrative. For a near-person, as with a merely sentient animal, what is in the individual's best interests is always a function of how pleasant the life in question is; in the case of a person, there is something more: to live a good story, to be a certain kind of person, to achieve certain things. This is the major difference between persons and non-persons, and the difference arises because persons are storytellers.

Related to this point about the moral significance of persons' lives is the metaphysical or ontological point to which I referred earlier: persons are inherently diachronic in a way that a merely sentient animal is not. As Schechtman puts it, "the organization of experience into a narrative self-conception actually changes the nature of subjectivity in a way that is reasonably described as the creation of a temporally extended experiencing subject who is conceptually prior to its experience" (pp. 144, 145). Schechtman's account contrasts with all the available alternatives, which she divides into "substance" and "reductionist" views:

> A common way of describing the relation between the different temporal parts of a person's life as understood by a sameness of substance view (a view that personal identity consists in either sameness of body or of soul) is by analogy with beads on a string. The string represents the continuing person, and the individual beads the person's distinct experiences, which are all his by virtue of being on the same string. Reductionists, on the other hand, use the metaphor of a river or stream to express the kind of unity persons have. Sameness of river does not, they point out, imply sameness of water, but rather continuity of flow. Similarly, reductionists claim, persons consist in the steady stream of different experiences.
>
> In contrast to both of these conceptions, the narrative self-constitution view sees the composition of a person as more like that of a complicated soup or stew. A soup is, of course, made of different ingredients, and these must exist prior to the soup itself. Once they are mixed together, however, they interact to produce something that is not best understood as a mere collection of ingredients laid out in some particular arrangement. Each ingredient contributes to the flavor of the whole and is itself altered by being simmered together with the others. A soup can, of course, be divided into portions, but the character of each portion is determined by the soup from which it came.
>
> In a like manner the experiences woven together into a person's narrative interact and alter one another in such a way that the narrative itself becomes the primary unit. The narrative is like the soup into which experiences are thrown, seasoning and altering one another—the past is reinterpreted and experienced in a new light in virtue of the present; the expectation of the future gives a different taste to current experience; and future experiences will have their character within the context of the whole. The experience of a person is thus had by an extended narrative subject, and not by a time-slice. (pp. 143–144)

In this way, a person is *a fundamentally different kind of thing* from a merely sentient animal.

The significance of this, as I see it, is that only persons have what I call "lives-as-a-whole." As Schechtman puts it, "the formation of a narrative brings into being a temporally extended subject who has this concern for her whole self" (p. 156):

> Formation of a narrative creates a persisting subject who is conceptually prior to its temporal parts. This subject clearly has an interest not only in the quality of the present moment, but in that of the narrative as a whole. At each moment the possessor of an autobiographical narrative is claiming the whole story to be *her* story, and a story with a happy ending is generally more desirable than a story with a sad ending. (p. 157)

To lead a good life is one thing; to lead a good life-as-a-whole is another. Only an individual with a biographical narrative can have a life-as-a-whole, so this is a kind of good that is only available to and at issue regarding the well-being of persons.

It may make my point more clear if I compare it to Michael Tooley's famous treatment of abortion and infanticide (1983). Tooley endorses what he calls the "interest principle," according to which an individual *A* can have a right to/an interest in *P* if and only if *A* has a concept of *P*. Tooley uses this principle to argue that only individuals with a concept of continued life can have an interest in continued life, and thus a right to life. Tooley's interest principle applies in the present context as follows. Only individuals who can conceive of their lives as wholes can have an interest in how their lives-as-wholes go, and only an individual with a biographical sense of self can conceive of its life as a life-as-a-whole. This explains a crucial difference between a *person*, understood as an individual who conceives of his or her life as a biographical whole, and a *near-person*, who has a robust sense of his or her own future and past but lacks this biographical conception of his or her life-as-a-whole.

So it certainly makes sense to say that my cat Nanci "lived a good life" in the sense that her experiences were, on balance, pleasant. And insofar as she enjoyed anticipating pleasant experiences and enjoyed recalling pleasant experiences, this awareness of her future and past added value to her life. Since she had no biographical sense of self, however, it makes no sense to say that Nanci "lived a good life-as-a-whole," for she in fact *had no life-as-a-whole*.

In summary, because persons have a biographical sense of self, two things are true of them: persons are capable of autonomy and they have an interest in how their "life-as-a-whole" goes. Autonomy is achieved when they succeed in living out their chosen self-narratives, and their chosen self-narratives determine, to a significant degree, what is in their best interests "all-things-considered." These two facts about persons give their lives special moral significance, and this is reflected in a third fact about persons: that they can be described as "happy" in the fullest sense, rather than merely "contented" (see §6.2).

These two facts also contribute to what a Harean utilitarian can say in response to Williams's argument that utilitarianism is a threat to human "integrity" (§4.12). If preserving one's integrity means sticking to one's chosen self-narrative, then yes, doing what Harean critical thinking requires will sometimes mean sacrificing one's integrity. It should now be more clear why the same kind of integrity ought to be valued very highly by a utilitarian, however, and why good ILS rules will embody a special form of respect for individuals who are capable of achieving it. At the same time, critical thinking is explicitly utilitarian and strictly impartial, so it can call for individuals to sacrifice their autonomy/integrity for the sake of aggregate utility. This is indeed a disquieting feature of human life (§4.11), but thankfully such critical thinking is called for only in unusual cases, and special respect for individuals with a biographical sense of self will be called for by the ILS rules that should govern in other cases.

Regarding the specific case of George the conscientious job seeker (§4.7, case #5, which was discussed in §4.11), this may very well be a case that calls for explicitly utilitarian critical thinking. But if, as I have argued in this section, a person's choice of self-narrative determines, to a significant extent, what is best for him on the whole, then it is not obvious that critical thinking will lead him to take the job at the chemical and biological weapons plant. To the extent that George's opposition to such weapons is central to his self-narrative, his preference for declining the job is not a garden-variety preference. If it is really central to "who he is," then however weakly he feels it at the moment, it is an especially important kind of preference that should be counted as such, and it is not so obvious that he should take the job. On the other hand, to the extent that his opposition to such weapons is a peripheral interest, then however vehemently he feels it at the moment, he should recognize that his self-narrative will not be significantly affected by the decision and that he should "suck it up," for the sake of his family, and take the job.

§7.4. The "Intelligence Is a Species-Specific Trait" Objection

In his fine book on dolphins, Thomas White argues that "'intelligence' is best understood as a *species-specific* trait" (2007, p. 167) and that associating linguistic competence with intelligence is a form of anthropomorphism. In this section, I want to make it clear why, in defining persons as individuals with a biographical sense of self, I am not making this mistake.

White motivates his point with a thought experiment:

> Imagine that all of the discussion about global warming has, in fact, colossally underestimated the problem. In reality, the temperature of the planet is going to climb so high that over the next few centuries the polar ice caps will completely melt and flood the planet. Also assume that, for one reason or another, it's going to be impossible for us to create floating communities that would let us preserve our land based technologies. However, life in the sea will persist. The bottom line is that ultimately it's going to be just us and the oceans. (p. 120)

In response to a flood of such Biblical proportions, White argues, the best evolutionary strategy would be to become like dolphins. We wouldn't need hands because we wouldn't be building those "land based technologies," instead, we would need to be very fast swimmers, with sleek, hydrodynamic bodies to avoid predators, and we would have no use for storing vast amounts of information in things like books (pp. 121–122). In short, "the problems of living in the water faced by dolphins are dramatically different from those faced by humans trying to survive on land. 'Intelligence' is ultimately just about the brain's ability to process information in ways that solve problems and enhance the survival of both individuals and species. So in dramatically different environments, 'intelligence' will look different" (pp. 153–154). In particular, he goes on to argue, language is only indicative of intelligence *in organisms with hands*.

In support of this claim, he discusses Frank Wilson's book, *The Hand: How Its Use Shapes the Brain, Language and Human Culture*. Wilson states that "It is a virtual certainty that complex social structure—and language—developed gradually in association with the spread of more highly elaborated tool design, manufacture and use" (1998, p. 30). This is reflected, Wilson claims, in the fact that children's manipulations of objects and their use of words mature together in a developmentally stereotyped way. "These two specific skills (manipulating objects and manipulating words), and the developmental chronology associated with the child's mastery of those skills, proceed in such transparently parallel fashion that the brain must be: (a) applying the same logic or procedural rules to both; and (b) using the same anatomic structures as it does so" (1998, p. 165). White concludes that:

> Wilson's theory gives us a paleontological explanation for the fact that philosophical discussions of the possibility of intelligence in nonhumans are dominated by a preoccupation with language and the linguistic abilities of nonhumans. Wilson argues that "The partnership of language and culture is so deeply woven into human history, and so compelling a force in our own personal development and acculturation, that we quite naturally come to regard language as the trait that both explains and defines our intelligence." (White 2007, p. 173, quoting Wilson 1998, p. 37)

That is, treating language as a sign of superior intelligence is anthropomorphic.

I agree; but my view does not depend on treating language as a sign of superior *intelligence* per se. To explain why, let me begin with an anecdote. For 20 years I have been attending a reading group on cognitive ethology at Texas A&M. For me, the experience has been valuable, because some of the readings (including some referenced in this book) have had obvious relevance to questions about the moral significance of animals' lives. Recently, we discussed some research showing that baboons perceive the so-called corridor illusion.

That is, when presented with figures identical in height, placed against a background image of a corridor so that one figure appears to be farther down the hallway, baboons perceive one figure as being taller than the other (Fagot and Barbet 2006, pp. 22–23). At the end of the discussion, I said: "I see that this concerns various cognitive processes, but none of the cognitive processing discussed in this chapter is accessible to consciousness. What I'm interested in is how cognition shapes consciousness, because it's consciousness that's important from the moral point of view."

That my view does not rely on an anthropomorphic conception of intelligence should be obvious when "intelligence" is understood, as White understands it, as "just about the brain's ability to process information in ways that solve problems and enhance the survival of both individuals and species." The brain can—and does—do that in all sorts of ways without the results ever being accessible to consciousness, because all sorts of cognitive processing occur that are not accessible to consciousness. For instance, the visual cortex does an enormous amount of processing to produce the three-dimensional images we are conscious of, but none of that cognitive processing is consciously accessible. In normal humans, the results are consciously accessible (we can enjoy the resulting scenery), but the adaptive results of cognition need not be conscious in any way, shape, or form. On the utilitarian view adopted in this book, a form of intelligence adds moral significance to a life only if *it enhances the conscious experiences* of the organism that possesses it. For the reasons given in this chapter, a biographical sense of self does just that.

§7.5. The "Super-Aliens" Objection

A common objection to views like mine usually comes in the form of a question: "If we someday encounter space aliens who are as cognitively sophisticated relative to us as we are to the animals you call near-persons, or even those you call 'the merely sentient,' would you say that their lives have special moral significance vis-à-vis ours?" Let us call such hypothetical beings "super-aliens."

In assessing this objection, a crucial thing to remember is that, as I emphasized in the preceding section, my argument is not that persons' lives have special moral significance because they have special cognitive abilities; my argument is that persons' lives have special moral significance because these special abilities allow their lives to contain more conscious states of value (and *dis*-value) than those of individuals that lack those abilities. To give the super-aliens objection traction, then, the objector must be able to explain how this would be so for the aliens vis-à-vis us, but neither contemporary science nor science fiction gives us any real guidance on how this might be so.

Astronomers who speculate about it generally think it very, very likely that intelligent life exists elsewhere in the universe. But here "intelligent life" does not mean super-aliens, it means something drastically more modest. Presumably it means species that use various technologies, including communications technologies that might allow us to detect transmissions from across vast tracts of interstellar space. Beyond the fact that they would be able to communicate electronically, which we're quite capable of ourselves, the search for extraterrestrial intelligence currently gives us no idea what space aliens would be like.

Science fiction isn't much help either. The aliens' spaceships travel faster, they have fancier weapons, they can read our thoughts, or they can change their appearance at will, but they are portrayed as looking, thinking, and acting pretty much like us. What reason do we have for thinking that their lives would contain some kind of valuable experiences that ours lack?

Perhaps this is asking too much of the objector. For just as I am assuming (§6.4) that individuals lacking language cannot imagine what it is like to have a biographical sense of self, we would not be able to understand what the experiences of the super-aliens are like, and we could not represent them adequately in fiction. In considering the super-aliens objection, then, we just have to stipulate that their lives have greater moral significance than ours in the same way that persons' lives have greater moral significance than those of near-persons or merely sentient animals. This makes the super-aliens what I called in §2.3(A) and §4.8 a "strictly hypothetical" case, but of an especially peculiar kind.

In a strictly hypothetical case, we assume that, unless uncertainty about some things is stipulated in the description of it, we know everything there is to know about the case and with absolute certainty. In the super-aliens case, we are supposed to assume that we know that their lives have greater moral significance than ours, and yet we are incapable of understanding how their lives differ from ours, which is the only way we could justify that claim. On a standard view in epistemology, however, having a justification for what one believes is what turns true belief into *knowledge*. In the super-aliens case, we must assume that we know, with absolute certainty, something that we are incapable of giving any reasons for believing—the only justification we can give for the belief is that making this assumption is essential to stating this objection to views that, like mine, distinguish persons from near-persons and the merely sentient on the basis of cognitive capacities.

Apart from how strained that makes the very statement of the case, what would a Harean say about the arrival of super-aliens? The assumption effectively stipulates that this is a case in which critical thinking would reach the conclusion that the super aliens' lives have greater moral significance than ours, and the objection is that this is counterintuitive. But according to Hare's theory, this conclusion is *supposed* to sound counterintuitive, because our ILS rules shape our intuitions and our ILS rules are not designed to cover such a situation. Humans' ILS rules commonly attribute special moral significance to humans' lives, because the only persons we encounter are human beings and all normal, adult human beings are persons. So the super-aliens objection is on all fours with the Bloggs case that was introduced in §1.5. As explained in §4.8, Bloggs-type "test cases" for utilitarianism do not succeed in showing how Hare's theory would "get the wrong answer" in any real-world case, and at the same time, his theory explains why the conclusions that critical thinking would reach in such strictly hypothetical cases should seem counterintuitive.[5]

Aside from explaining why the conclusion that critical thinking must reach in this hypothetical case is bound to sound counterintuitive, I would want to ask if the objector is willing to abandon the conclusion that the lives of normal, adult humans have greater moral significance than those of what I call the merely sentient. For if we refuse to allow that cognitive differences could ground this conclusion in the case of the super-aliens, we must also deny this in our own case.[6]

§7.6. The Problem of "Marginal Cases"

This is an appropriate place for an initial discussion of the so-called "problem of marginal cases." The problem arises for any view that ascribes special moral significance to persons and defines personhood in terms of cognitive

---

[5] A related and particularly interesting direction that a super-alien might take is being a Harean archangel. If the super-alien were Hare's archangel, knowing her would be analogous to knowing a god, and just as problematic. For if we know that she is an archangel, then we know that she has perfect factual knowledge and perfect powers of critical thinking, and that she will presumably issue *precisely* the ILS rules that we should internalize. Having internalized them, however, we will still need to exercise critical thinking in unusual situations, when the ILS rules conflict, and when violating them will "clearly" maximize utility. And as individuals, we will still need to internalize meta-principles adapted to our own weaknesses, foibles, and so on. Moreover, if the archangel makes relatively few public appearances, then squabbles will develop as different people claim to have heard different things from her, and over time these squabbles will calcify into churches. Let one of these churches seize political power, and faster than you can say "Spanish inquisition" there will be oppression, torture, and war. That is why political liberalism has gotten hold of a very good ILS rule in requiring separation of church and state.

[6] While commenting on my presentation of some material from this chapter at the Central Division meeting of the American Philosophical Association in April 2011, Marya Schechtman raised an objection that deserves a brief reply. Suppose there were a species of animal that experiences physical pleasures in the here-and-now so intensely that the life of such an animal could embody greater value than that of a well-lived human's life. Such animals would be merely sentient, yet we would have good reason to attribute equal or greater moral significance to their lives vis-à-vis the lives of normal human persons.

Considered as a strictly hypothetical case (§2.3[A]), a Harean utilitarian must acknowledge that the lives of these merely sentient animals could have more value than those of normal human persons. However, this case provides no challenge to the arguments that I have given in this chapter for embodying special moral respect for persons vis-à-vis merely sentient animals in the ILS rules of real-world human societies. For, although we can imagine there being such a species of merely sentient animals, we have no evidence that such animals exist, and Harean ILS rules are not designed to handle such purely hypothetical cases.

capacities that not all human beings have. If competence in a natural language is a necessary condition for personhood, as I am assuming (§6.4), then:

1. Young children are not (yet) persons;
2. The permanently comatose have ceased to be persons; and
3. As Schechtman points out, adults can lose their personal identities to varying degrees through dementia and mental illness.

So on the view defended here, a significant number of human beings do not qualify as persons, and this strikes many people as highly counterintuitive.

A full response to this problem will have to wait until after I have discussed candidates for near-personhood in the following chapter and, throughout part III, the diverse ways in which ILS rules for persons, near-persons, and the merely sentient are sensitive to background ecological, technological, economic, and cultural conditions. But the general outlines of a Harean response can be gotten on the table now. Like the Harean responses to standard "test cases" for utilitarianism discussed in §§4.7ff, a Harean response to the problem of marginal cases will be complex, involving the distinction between critical and intuitive-level thinking, and emphasizing the importance of various facts and the ways in which our knowledge of relevant facts is often incomplete.

The basic point is this: although critical thinking would appear to lead us to the conclusion that the lives of "marginal" humans are no more valuable than those of certain animals, there are good reasons for building special protections for them into the ILS rules of common morality, law, and professional ethics.

Regarding the first specific category mentioned above, various studies have consistently concluded that children acquire autonoetic consciousness around three to four years of age (Suddendorf and Busby 2005, pp. 114–115 and 122). But people who study autobiographical memory (the backward-looking part of having a biographical sense of self) do not think that children develop full-blown autobiographies until adolescence. Prior to that, "temporal coherence applies mainly to single autobiographical events rather than to connections between different events" (McAdams 2003, p. 192). However, young children will, in the normal course of events, become persons, and this gives us good reasons for treating them more like persons than we would similarly cognitively sophisticated animals. Although there are obviously reasons for treating them differently from adults (as emphasized in §9.7[B]), children should be treated as "persons in training." Regarding the comatose, our knowledge of relevant facts is always imperfect. In many cases, there is genuine uncertainty about when to declare a patient "permanently" comatose, and some people who "come back" from comas claim to have been aware of things when doctors claimed that they couldn't be. So given that codes of medical ethics should respect patient autonomy, there are good reasons for allowing patients to issue advance directives or designate others to make treatment decisions. Finally, cases of dementia and mental illness lie on a continuum, interested parties can have reasons for exaggerating their severity, and all of us are at risk of slipping into dementia and mental illness. So we have good reason for "erring on the side of caution" in related legal protections.

I will return to the problem of marginal cases in §9.7(B), where I will consider a broader range of "marginals," I will have more to say about how common morality, law, and professional ethics currently treat them, and I will discuss in detail how my Harean response to the problem of marginal cases compares to that of Singer.

§7.7. Summary: My Stipulative Definition of "Person"

For the reasons given in this chapter, I believe that the lives of individuals with a biographical sense of self have special moral significance in the sense described in §1.8, and that this justifies calling for a special kind of treatment or respect for persons in the ILS rules of a society. I therefore favor, as a stipulative definition of the term: "person" =df an individual who deserves special treatment or respect because he, she, or it has a biographical sense of self.

It is now clear why I said, in §1.2, that the first three descriptive features in the family two (descriptive) definitions of "person" that I chose to include in table 1.1—being rational and self-conscious, autonomous in the sense of having second-order desires, and a moral agent—are included in my conception of persons as individuals with a biographical sense of self. For being rational and self-conscious and having second-order desires are part and parcel of adopting a narrative conception of one's self as described in §6.3 and §7.3(B); and, since I am assuming that only human beings are persons (§6.4 and §6.5) and all normal adult humans are moral agents, it follows that individuals with a biographical sense of self normally are also moral agents.

It is also clear why I said, in §1.2, that the special respect that persons are due in virtue of having a biographical sense of self can aptly be described, within a Harean framework, in terms of having legal and moral rights. For as I noted in §1.5 and §3.1, rights language is a useful way of expressing the "deontological flavor" that ILS rules need to have, and more will be said of this later, in §11.2. The third example that I chose to include in table 1.1 of a way to express the special respect that persons are due in virtue of having a biographical sense of self—being "not replaceable"—will be discussed in detail in chapter 9 (§§9.3–9.6).

For the reasons given in the second section of this chapter, I believe that the lives of individuals with autonoetic consciousness (i.e. a robust, conscious sense of their own past and future) have special moral significance vis-à-vis those of the merely sentient, even if they lack the biographical sense of self-definitive personhood. It is for this reason that I introduced, in §1.7 and §6.1, the category of "near-persons," and the next chapter will address the question of which non-human animals (if any) are good candidates for near-personhood.

{ 8 }

Candidates for Near-Personhood

§8.1. Studying Autonoetic Consciousness in Animals

In this chapter, I discuss three areas of research relevant to answering the question of whether or not animals have autonoetic consciousness:

1. Episodic memory, which is the backward-looking portion of autonoetic consciousness;
2. Mirror self-recognition, which is suggestive of autonoetic consciousness in the present; and
3. Use of a theory of mind and certain kinds of planning, both of which seem relevant to the forward-looking element of autonoetic consciousness.

Having autonoetic consciousness doesn't make an individual a person, but it does give one's life special moral significance vis-à-vis the merely sentient. As defined in §1.7, "near-persons" have a robust, conscious sense of their own past, present, and future, and, for the reasons given in §7.2, this gives their lives special moral significance. So while having autonoetic consciousness doesn't give one a biographical sense of self and make one a *person*, good ILS rules will incorporate some kind of special respect for *near-persons*.

The strongest candidates for near-personhood will be those for whom we have the strongest evidence across the most categories. Animals for whom we have less strong and/or less comprehensive evidence will be weaker candidates. As with my discussion of pain in chapter 5, I need to emphasize several qualifications.

First, new findings in some of these areas are appearing all the time, which ensures that some of what I say in this chapter will be out of date by the time you read this.

Second, a cautionary note about generalizing within taxa is again in order. In the above three areas, relatively few species within a family or higher order taxon have been studied, and sometimes only a few individuals within one species have been studied. From such paltry data we can legitimately draw only very tentative conclusions, but as I emphasized in §5.4(A), ethicists and legislators must make decisions based on the best evidence currently available.

Relatedly, convergent evolution could produce, in a single or several species, a capacity that exists nowhere else in the family or order in question. The evidence surveyed in this chapter suggests that this may have happened with regard to autonoetic consciousness, because the strongest evidence for it is found among widely separated taxa: the great apes, elephants, dolphins, and corvids (the avian family consisting of ravens, crows, magpies, jays, and nutcrackers). Those three mammals have no common ancestor prior to 60 or 65 million years ago, and mammals and birds each diverged from reptiles long before that. As we will see, autonoetic consciousness is potentially highly adaptive, so we shouldn't be surprised if convergent evolution produced it in widely separated species.

On the other hand, it could turn out that autonoetic consciousness or at least some components of it are much more widespread than the small amount of currently available research suggests. I will discuss, for instance, one innovative piece of research that provides some evidence that rats have episodic memory.

The next three sections of this chapter are devoted to summarizing the available evidence in each of the three areas listed above. In §8.5, I summarize what the currently available evidence suggests, dividing candidates for near-personhood into "the usual suspects" and "some other contenders." And in the final section, I give my reasons for referring to individuals with autonoetic consciousness as "near-persons" rather than "borderline" or "quasi-persons."

This chapter is the longest in this book, but it will inevitably seem incomplete to some readers, and some of the information will be out-of-date before it is in print. For research on autonoetic consciousness is still in its infancy, despite the explosion of relevant research in the last decade. The survey of research in this chapter is intended to illustrate the ways in which autonoetic consciousness can be studied in animals and thus how we can systematically address the question of which animals we should treat as near-persons. As illustrated in chapter 10 and §3.3(A), good ILS rules for a society will vary with changes in background ecological, technological, and economic conditions. So, too, with the changing knowledge base relevant to answering this question. As I emphasized in §5.4(A), you have to make ethical decisions—including about what changes to make in the ILS rules of common morality and law in one's society—with the evidence you have, not the evidence might want or wish to have. As more research is done on autonoetic consciousness in humans, it may significantly alter the ways that we think about various animals. This chapter at best provides a snapshot of where things stood in the first decade of the twenty-first century.

§8.2. Episodic Memory

Memory is the backward-looking third of autonoetic consciousness, and some animals are renowned for their memories. There is striking anecdotal evidence that elephants, for instance, have really good memories. Field researchers (e.g. Moss 1988, pp. 270–271, and caption to photo of Agatha, pp. 128–129) have reported that elephants linger among and closely examine the bones of long deceased kin, and captive elephants have been reported to remember both keepers (Lewis and Fish 1978, p. 96) and conspecifics (Poole & Moss 2008, pp. 86–87) that they have not seen in dozens of years. Some experimental evidence confirms that elephants have good memories. A classic study (Rensch 1957) showed that zoo elephants remembered which of up to 20 pairs of cards were the correct ones to choose for a reward as much as a year after initially learning them, even when the subjects had not been drilled in the interim. And when field researchers broadcast recordings of the calls of family members that had been absent from their groups for up to 12 years (either because they had died or emigrated to other groups), the remaining family members displayed "a strong affiliative response" (McComb et al. 2000, p. 1108).

However, psychologists studying humans distinguish at least three kinds of memory, only one of which definitely involves conscious awareness of the past. They draw a basic distinction between *procedural memory*, or "remembering how," and *propositional or declarative memory*, or "remembering that." The former is possible without the latter, as when a guitarist can play a complicated passage without being able to describe or in any other way answer questions about how to play it (besides just playing it, of course). The label "propositional or declarative memory" is misleading insofar as it does not presume mastery of a language, only the ability to somehow indicate answers to a problem or question, including, for instance, pointing out which of two images has been presented earlier in a lab trial, something that many animals can do. But it is possible to have propositional or declarative memory of events without consciously remembering them at all. Therefore, since the 1970s, a further distinction has been drawn within propositional or declarative memory, between *semantic memory*, or "symbolically representable knowledge that organisms possess about the world," and *episodic memory*, or the "remembering of personally experienced events" (Tulving 1985, p. 2, emphasis removed). For a prosaic illustration of the difference, think of a moviegoer who accurately recalls that she has seen a particular movie without having any conscious memory of when or where, or of what the plot or images in the movie were like. This person has no episodic memory of the event, but she has some semantic memory of it. Similarly, sometimes people can answer "forced recall" questions at far higher than chance accuracy about things of which they have no conscious memory. That is, they will "guess" right at a higher than chance rate, indicating that they do have relevant memories, but unconscious ones. This is another example of semantic memory without episodic memory. Such memories are also called "implicit" memories, in contrast to "explicit," episodic memories.

In 1985, psychologist Endel Tulving proposed a specialized account of episodic memory that he and colleagues have since expanded into a general account of what they call "autonoetic awareness," which they say

> occurs whenever one consciously recollects or re-experiences a happening from a specific time in the past, attends directly to one's present or on-line experience, or contemplates one's existence and conduct at a time in the future. Autonoetic awareness of the subjective past constitutes episodic retrieval. It represents the major defining difference between episodic and semantic memory. (Wheeler et al. 1997, p. 350)

Elsewhere, they variously describe autonoetic awareness as "the kind of consciousness that mediates from the personal past through the present to the personal future," as "providing the characteristic phenomenal flavor of the experience of remembering" (Tulving 1985, p. 1), and as making possible "mental time travel" (Tulving 1985, p. 5; Wheeler et al. 1997, p. 331).

For the celebrated cases of elephant memory described above to count in favor of the claim that elephants have autonoetic consciousness, we would have to be able to show that the elephants were having *episodic* memories. We would have to show that they were consciously recalling events rather than just "telling us," through their behavior, that those events had occurred. So, for instance, even if elephants clearly become emotionally aroused in the presence of a relative's carcass or an abusive trainer from long ago, they need not be consciously recalling events involving these former colleagues. For their present emotional responses could be triggered by implicit memories without them being able to experience, in addition, an episodic memory—that is, without being able to reexperience the past, as described in §7.2.

Tulving and associates have developed systematic ways of studying episodic memory, but none of their current research paradigms can be used on animals, because they all involve complex linguistic communication between the subjects and the experimenters. Primarily, they have used word list recall problems. When shown a list of words and asked to study them for later recall, subjects will respond with much higher than chance success when "cued" to recall words that they say they don't remember seeing. When given first a "free recall" test on the list, subjects will mention words that they say they consciously remember seeing, but then, if they are asked further "cuing" questions, like "Do you remember any words similar to 'snake'?" they mention that similar words, like "cake" or "snail," were on the list, and they are correct in the ones they suggest at a rate well above chance. In such cases, when asked "whether they actually 'remembered' its occurrence in the list or whether they simply 'knew' on some other basis that the item was a member of the study list," subjects will typically say that they "just knew it but don't actually remember it" (Tulving 1985, p. 8).

§8.2(A). "EPISODIC-LIKE" MEMORY IN SCRUB JAYS

Certainly it is hard to imagine how to distinguish episodic from semantic memory without relying on the subjects' introspection, and therefore on complex linguistic communication. But recent, innovative research illustrates how a fairly simple experiment can be used to show that some animals have at least "episodic-like" memory.

Beginning in the late 1990s, ethologist Nicola Clayton and colleagues developed a novel way of testing the hypothesis that scrub jays remember not only what happened and where, but also *when* (Clayton & Dickson 1998). Scrub jays cache food in the wild, and Clayton and colleagues used this natural behavior to test their hypothesis by allowing the birds to cache two kinds of food and retrieve them later. In one of the more sophisticated versions of the experiment, one of the foods was always peanuts, a stable, palatable, but less preferred food. The other was always an unstable but preferred food, either crickets or mealworms. The crickets and mealworms were treated to ensure that they would become unpalatable after specified periods of time, and the jays were allowed to learn this over a series of trials. Then the jays were tested to see which food they would try to retrieve after delays of various lengths, including some trials when the food items had been removed to ensure that they were not relying on visual or olfactory cues. Under two pairings:

1. peanuts paired with mealworms treated to be fresh through 4 hours but rotten by 28 hours, and
2. peanuts paired with crickets treated to be fresh through 28 hours but rotten by 100 hours,

the birds were allowed to cache both kinds of food in each pair and then retrieve food after three lengths of delay: 4, 28, and 100 hours. If the birds remembered not only what food was hidden where, but also *when*, then under pairing #1 above they should spend most of their time checking where they had cached mealworms after a 4-hour delay, but most of their time looking for peanuts after both 28- and 100-hour delays. Under pairing #2, they should look for mealworms after both 4- and 28-hour delays, but switch to peanuts after 100 hours. The birds did exactly as predicted, indicating that they indeed had memory not only of what and where, but *when* (Clayton et al. 2001, p. 1485).

However, Clayton and colleagues have always used, in the titles of their papers, the expression "episodic-*like* memory." In his first paper on the subject, Tulving described episodic memory as memory that "receives and stores information about temporally dated episodes or events" (1972, p. 385), suggesting

TABLE 8.1. Experimental Conditions in Test of "Episodic-like" Memory

	Peanuts	Crickets	Mealworms
4 hours	Good	Good	Good
28 hours	Good	Good	Bad
100 hours	Good	Bad	Bad

Source: Clayton et al. (2001)

that a memory's being somehow "time-stamped" (my term)[1] suffices to make it an episodic memory. In subsequent papers, however, Tulving began including references to conscious reexperiencing of past events in his definitions of episodic memory. Clayton and colleagues' scrub jays clearly had time-stamped memories, but that still didn't show that in using those memories they were consciously recalling caching the foods. Indeed, as one commentary on their work observed, the time-stamp could even be incorporated in a perception of the present: "one could ask whether the birds remember the event of caching worms in a particular location, a memory that is oriented to the past, or simply know the current state of the world—that caches are in particular locations, and have been there for particular periods of time" (Hampton & Schwartz 2004, p. 3).

However, in normal human subjects, positron emission tomography (PET scanning) confirms that episodic memory is associated with intense activity in the prefrontal cortex (PFC). This suggests a way that scientists could strengthen the claim that some animals—at least some mammals—have episodic memories, and a novel analysis of memory in rats suggests another way of identifying episodic memory in animals.

§8.2(B). THE ROLE OF THE MAMMALIAN PREFRONTAL CORTEX

When human subjects are asked to solve episodic memory problems (such as the word list problems described above) during a PET scan, intense activity occurs in the left PFC during the encoding of episodic memories, while intense activity occurs in the right side of the PFC during their recall (Wheeling et al. 1997, pp. 335–343). The frontal cortex as a whole is the area

[1] Note, however, that a memory's failing to be "stamped" with *a specific date* does not disqualify it as an episodic memory. As Zentall and colleagues note: "many vivid episodes cannot be accurately placed in time. One may have a vivid memory for what happened at one's last visit to the dentist but be unable to remember the day or even the month of the event. In fact, we often use semantic memory as an aid in placing episodic events in time (e.g., 'It must have been a Monday because that is the only day I have time to go to the dentist')" (2001, p. 685). For his part, at one point Tulving characterizes episodic memory of an event in terms of being able to "more or less accurately specify its temporal relations to other events" (1972, p. 388).

of the cerebral cortex forward of the central gyrus. Generally speaking, the frontal cortex as a whole is action-oriented, whereas the rest of the cortex is sensory-oriented (Fuster 1997, p. 3). The *pre*frontal cortex lies at the front tip of the frontal lobe, but it is defined in two different ways: in terms of how its cytoarchitecture (cellular architecture) differs from the rest of the frontal lobe, or in terms of its connections to other parts of the brain. Defined as "the part of the cerebral cortex that receives projections from the mediodorsal nucleus of the thalamus," all and only mammals have a PFC (Fuster 1997, p. 2).

If non-human mammals could be presented with problems which (like those faced by the Clayton lab's scrub jays) could only be solved using knowledge of *when* events occurred, and if, while solving those problems successfully, similar information on brain activity could be obtained, a strong argument by analogy could be made for the claim that those mammals have episodic (and not just episodic-like) memories. Such a study is impossible at present. Since subjects must lie still inside current PET scanning devices while being scanned, it is unclear how to give an animal subject a PET-scan while solving an episodic memory problem. In principle, however, other techniques for imaging brain activity could be developed and employed, and if the mammals in question both (a) solved the episodic-like memory problem and (b) displayed the same kind of lateralization of PFC activity while doing so, this would provide a relatively strong argument by analogy for the claim that they do have episodic memory.

§8.2(c). RECEIVER OPERATING CHARACTERISTICS (ROCS) OF EPISODIC MEMORY

A novel approach has led Howard Eichenbaum and colleagues to claim that rats probably use episodic memory to solve certain kinds of problems. Their novel argument (Eichenbaum et al. 2005, pp. 192–196) relies on a comparison of the performance of humans on word list recall tests when relying on "familiarity" and/or "recollection," to the performance of rats on analogous problems. Rats rely heavily on their sense of smell, so Eichenbaum and colleagues tested them on their ability to remember odors. The rats were presented with a set of odors and then, 30 minutes later, they were rewarded for correctly identifying which odors were and were not in the earlier set. Using some sophisticated mathematical analysis, Eichenbaum and associates showed that the rats' distribution of errors was similar in relevant ways to the distribution of humans' errors when solving problems using recollection rather than familiarity.

The terms "familiarity" and "recollection" are used in the literature on receiver operating characteristics (ROCs), where "recollection" is described in ways that strongly parallel descriptions of episodic memory:

> Recollection and familiarity are distinguished by their cognitive mechanisms, by their retrieval dynamics, and by their putative brain substrates. With regard to cognitive mechanisms, familiarity is determined by strength of a perceptual match to prior exposure and, consequently, is susceptible to variations in superficial sensory qualities of the stimuli. By contrast, recollection allows one to recover the prior episode in which the stimulus was experienced, and emphasizes conceptual properties—the meaning of the object to be recognized—as well as associations of the object, including the spatial and temporal context in which it was experienced. With regard to the dynamics of retrieval, familiarity grows incrementally depending on the amount of prior exposure and degree of perceptual match. By contrast, recollection occurs at a threshold, before which no information is recovered and after which the object plus its associations and context are re-experienced . . . (p. 193—reference removed)

So although the literature on ROCs does not use the term "episodic memory," Eichenbaum and colleagues argue that studying distributions of errors, as is done in the literature on ROCs, can provide evidence for episodic memory in non-humans.

When humans are asked to say whether or not certain words were on a list that was presented some time earlier, individuals who are more likely to sound "false alarms" (i.e., falsely report that a presented word appeared on the list they saw earlier) will have a higher percentage of "hits" (i.e., they will tend to correctly identify a higher percentage of the words presented later that in fact appeared on the list that they saw earlier), while those who are less likely to sound "false alarms" will have a lower percentage of "hits." For normal humans relying on a mixture of familiarity and recollection, the resulting curve begins well above zero probability of hits and arcs up with a decreasing slope to the point of 100 percent probability of hits and 100 percent probability of false alarms (something like curve A in figure 8.1).

Eichenbaum and colleagues state that in normal humans it is possible to "decompose" humans' performance on word list recall problems into two curves, however, with one representing performance based on familiarity alone, and the other representing performance based on recollection alone. The resulting curve for familiarity begins at 0 percent probability of hits and 0 percent probability of false alarms and arcs up in a fairly constant curve to the point of 100 percent probability of some hits and 100 percent probability of some false alarms (something like curve B in figure 8.1). On the other hand, the resulting curve for recollection starts from a point less far above 0 percent hits than the starting point of normal humans relying on a mixture of familiarity and recollection and runs in almost a straight line to the point of 100 percent probability of hits and 100 percent probability of false alarms (something like curve C in figure 8.1).

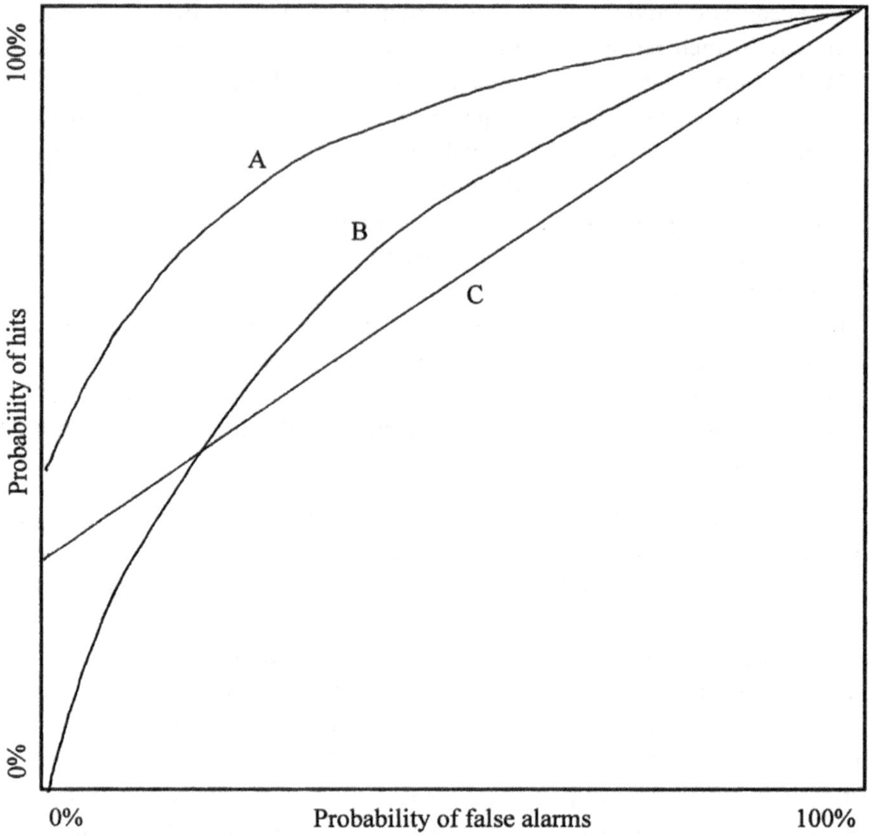

FIGURE 8.1. *Performance Curves for Humans Relying on a Combination of Familiarity and Recollection (A), Familiarity Alone (B), and Recollection Alone (C)*

Eichenbaum and colleagues' novel argument involves comparing these curves for humans to the performance curves of rats on the odor list test under various conditions. First, the performance of rats with hippocampal lesions was compared to that of preoperative rats and a control group that underwent a sham operation. This is significant, because in humans, hippocampal damage is known to impair recollection while leaving familiarity intact. Whereas the performance curves for both groups of rats with unlesioned hippocampi had the more irregularly curved shape, similar to that for normal humans who are relying on both familiarity and recollection (like curve A in figure 8.1), the performance curve for the lesioned rats had the more regularly curved shape, similar to that for humans relying solely on familiarity (like curve B in figure 8.1).

In humans, familiarity is known to fade more quickly than recollection, so next Eichenbaum and colleagues increased the delay that intact rats experienced between exposure to the odor list and testing from 30 minutes to 75. Under this condition, the intact rats' performance curve closely matched the very flat one for humans relying exclusively on recollection (like curve C in figure 8.1).

Eichenbaum and colleagues interpret these results as showing that the rats probably were relying on a mix of episodic memory and familiarity while the delay was limited to 30 minutes, and that the rats probably were relying exclusively on episodic memory when the delay was extended to 75 minutes. Their analysis is potentially very significant for the following reason. Most of the remaining research on episodic memory described in this section involves experiments in which there was a relatively short delay between the initial stimulus and the test of memory. As Schwartz and colleagues point out (2005, p. 230), if the delay is short enough, rehearsal in short-term memory rather than episodic memory could account for solving the problem. If Eichenbaum and colleagues' interpretation of the ROC curves is correct, however, then the distribution of errors that animals exhibit when solving problems after various delays could allow researchers to determine that episodic memory is involved, in addition to familiarity.

§8.2(D). THE ELEMENT OF SURPRISE

Another approach that has been used to study episodic memory in nonhuman animals involves asking them a "surprise" question about a past event. The rationale for this approach is summarized in this passage from one such study:

> To see more clearly how training may lead to a reliance on semantic or associative memory, rather than episodic memory, imagine the following scenario: We see a friend in the morning and ask him (unexpectedly) what he had for dinner last night. After a few seconds of reflection (presumably to search for the episode) he says, "Spaghetti." Imagine now that we ask him the same question every morning. Now, at dinner time, while eating chicken, he may say to himself, "When I am asked tomorrow what I had for dinner, I will say 'chicken.'" The next day when asked the question he says to himself, "I don't have to think back about what I had for dinner because I know to say 'chicken.'" This would be an example of semantic or associative memory rather than episodic memory, and it would be the basis for performance of any task with repeated trials. (Zentall et al. 2001, p. 686)

That is, if you're used to being asked a certain sort of question, you can form associations, based on which you can answer the predictable question without using episodic memory to recall the past event. But when you're asked a surprise question about a past event, you need to use episodic memory to answer it.

This experimental approach seems to build on Tulving's idea that episodic memory involves "mental time travel." Tulving acknowledges that forward-looking "mental time travel" is adaptive because it allows us to avoid making costly mistakes by playing out future scenarios in imagination (1985, p. 5). Analogously, backward-looking "mental time travel" would be adaptive if, when "reliving" an experience, we are able to extract information from it that we did not notice the first time. Retrieving an episodic memory would be in this way like replaying a videotape: there can be things in there that you didn't notice the first time,[2] and the element of surprise strategy forces subjects to recall information that they had no reason to notice the first time.

While a study of "Memory for Recent Actions in Dolphins" (Mercado et al. 1998) does not use the term "episodic memory," it has been cited as evidence for it in dolphins, because of the way in which it explains its subjects' ability to respond to surprise questions. The study involved training two dolphins a "repeat" command, which was rewarded if they performed whatever behavior they had last been instructed to perform. The dolphins were first trained to repeat several simple behaviors reliably. They were then tested on their ability to repeat 32 previously untested commands of varying complexity from their larger repertoire of known commands, including the command to do something novel (a behavior that had not recently been performed). Both dolphins were then trained to perform four new behaviors and were tested on repeating these. One of the two dolphins performed significantly better than the other: Elele successfully repeated behaviors from the familiar repertoire 90 percent of the time, Hiapo 57 percent; and Elele repeated the newly trained behaviors 79 percent of the time, Hiapo 50 percent (pp. 213–214). In the final phase of the study,

> Elele was tested on her ability to repeat behaviors multiple times on command. If the rule she had learned to follow when given a repeat command was *repeat the behavior corresponding to the previous gesture*, it would be difficult for her to respond correctly when the previous gesture was another repeat

---

[2] Here's a prosaic illustration from my own life. One Saturday morning, without consciously noticing anything about my morning coffee, and without consciously trying to recall anything about the night before, I suddenly experienced an episodic memory. Specifically I reexperienced (at least visually) taking a bag of unground coffee out of the freezer the night before. I do this when the coffee grinder is low and I'm about to grind coffee to set for automatic brewing the next morning. The significance of this episodic memory was this. I drink a mixture of 50 percent decaffeinated coffee. I store the beans in the freezer, sometimes pre-mixed and sometimes with the caffeinated and decaffeinated beans in separate bags. As soon as I "saw" myself taking the bag out of the freezer the night before, I realized both that the bag contained unmixed, 100 percent decaffeinated beans, and that I had poured only beans from that bag into the grinder the night before. At the time, I had not been aware that, because I chose that bag alone, I was preparing 100 percent decaffeinated coffee, but when I reexperienced the selection of the bag, I noticed something that I hadn't noticed the first time around. And I avoided getting a headache later in the day by brewing up a new pot of coffee that was only partially decaffeinated.

> command (because this command is not associated with a specific behavior). However, if the rule she had learned was *repeat the previous action(s)*, one would expect multiple repeat commands to be no more difficult than single repeat commvands. (p. 215)

Elele performed correctly on 100 percent of 16 double-repeat trials conducted. Emphasizing the dolphins' success on repeating self-selected (novel) behaviors and Elele's success on double-repeat trials, the authors conclude: "The simplest explanation that can account for all of the results in this study is that the dolphins can retain internal representations of their recent past actions in working memory that they can use to recreate those past actions. These representations must be flexibly accessible and detailed enough to allow for the repetition of fairly complex behaviors" (p. 216). Although Mercado and colleagues do not use the term "episodic memory," this language is reminiscent of Tulving's description of how "mental time-travel" can be used to retrieve details of past experiences that are not encoded in other memory systems, both when asking the dolphins to repeat "self-selected" behaviors and when they exposed Elele to double-repeat trials, Mercado and colleagues were employing the element of surprise.

Praising Mercado and colleagues' work as a "sophisticated approach to episodic memory in animals" (2001, p. 689), Zentall and colleagues claim to have obtained analogous results in pigeons. In a box with response keys that could be illuminated in various colors and a screen onto which patterns of lines were projected, they first trained the birds to peck or not peck the keys in response to presentation of vertical versus horizontal lines. If the birds responded to this problem appropriately, they were then presented with red and green lights, and they were rewarded if they pecked according to the rule "peck the key illuminated in red if pecking has just occurred and peck the key illuminated in green if pecking has not just occurred." The crucial part of the study occurred after the birds were responding correctly to this problem over 90 percent of the time. The birds were then presented with a novel stimulus: either an unilluminated response key or a response key illuminated with a color that they had not seen before. After a short interval, the keys were then illuminated in the familiar red and green colors. Without training, pigeons will tend to peck at an illuminated response key but not an unilluminated one, and indeed, the birds pecked the novel colored key 3.32 times per trial and the unilluminated key only 0.10 times per trial. The interesting question was whether or not the birds would peck the red key ("pecking has just occurred") after presentation of a key illuminated in a novel color and the green key ("pecking has not just occurred") after presentation of unilluminated keys. To maintain the element of surprise, each bird was presented with this novel question only once. On average, the birds pecked the red key after presentation of the novel color on a key 71.9 percent of the time (p. 689). While acknowledging that this result "may not provide definitive evidence for the capacity for episodic memory in pigeons" (p. 689), Zentall and colleagues conclude that "the pigeons demonstrated that they could remember specific details about their past experiences, a result consistent with the notion that they have the capacity for forming episodic-like memories" (p. 685).

The element of surprise approach was also used by Bennett Schwartz and colleagues in one of their experiments related to episodic memory in a captive lowland gorilla named King. King has had extensive interaction with handlers at Monkey Jungle in Miami across 20 years. He is able to answer some questions, posed in English, by selecting among pictures provided to him. To produce the element of surprise, they arranged for King to witness "unique events," involving people doing unusual things or new kinds of food and unfamiliar objects. Presented with three photographs 5 to 15 minutes after the events, King selected the photograph showing an aspect of the event 55 percent of the time. While "King's performance on this task was less than stellar . . . it was significantly higher than chance (33 percent)" (Schwartz et al. 2005, p. 234). Noting that presenting humans with "misinformation" about an event degrades their performance on episodic memory tasks, Schwartz and colleagues staged more unique events, but this time they presented him with a photograph that was either blank, carried correct information about the event, or carried incorrect information. In the first two conditions, King's performance remained about the same (53 percent), but it dropped to 40 percent correct in the misinformation condition (p. 235).[3]

§8.2(E). SUMMARY

In summary, the backward-looking portion of autonoetic consciousness is episodic memory. Although it is still unclear how to study episodic memory without relying on introspection and therefore language, innovative techniques for studying memory of dated events, and therefore at least episodic-like, or "time-stamped" memory, in non-humans are being developed. To date, studies have suggested that a wide range of animals have this kind of memory, including gorillas and dolphins, but also scrub jays and rats. In principle, similar studies of mammals could be coupled with PET scan imaging to

[3] In the same paper, Schwartz et al. report several other results. In one experiment, they showed that King could identify the order in which he had received three food treats five minutes earlier. He indicated the last item correctly 90 percent of the time, the second item 50 percent of the time, and the first item 60 percent of the time. Given that there were five treats to choose from and a food was used no more than once per trial, Schwartz et al. assume that chance level responses would have been 20 percent, 25 percent, and 33 percent, respectively (2005, p. 236). And in other experiments, they separately demonstrated that King has the ability to answer questions about who did what and where a short (five minutes) or long (24 hours) time ago.

develop a stronger argument by analogy to the conclusion that they have this backward-looking third of autonoetic consciousness, although this kind of research will not be possible in the foreseeable future.

§8.3. Mirror Self-Recognition

If "autonoetic consciousness" just meant "phenomenal consciousness," then it would be unnecessary to devote a separate section to autonoetic consciousness in the present. For, as discussed in chapter 5, the "standard argument by analogy" already shows that all vertebrates probably are conscious of pain, and that invertebrates (with the exception of cephalopods) probably are not.

Memory researchers usually describe autonoetic consciousness as awareness of a "personal" past, present, and future, however, and what "personalizes" phenomenal consciousness of the present is a sense of self. Assuming that animals who recognize themselves in mirrors are self-aware and that "awareness" here means *consciousness*, mirror self-recognition constitutes evidence of autonoetic consciousness in the present.[4] The best-known criterion for mirror-self recognition is the so-called "spot test," which was first described by psychologist Gordon Gallup in a 1970 paper (so it is also called "the Gallup test"). Gallup observed that when chimpanzees were first exposed to mirrors they reacted aggressively, as if confronting unfamiliar conspecifics; but they began, after further exposure, to act as if they recognized themselves in the mirrors.

> Such self-directed responding took the form of grooming parts of the body which would otherwise be visually inaccessible without the mirror, picking bits of food from between the teeth while watching the mirror image, visually guided manipulation of the anal-genital areas by means of the mirror, picking extraneous material from the nose by inspecting the reflected image, making faces at the mirror, blowing bubbles, and manipulating food wads with the lips by watching the reflection. (1970, p. 86)

[4] One might argue that mirror self-recognition should be treated as a necessary, rather than a sufficient condition for autonoetic consciousness in the present for the following reasons. Mirror self-recognition emerges in normal children between 18 and 22 months of age (Gallup et al. 2002, p. 328), while they are still affected by "infantile amnesia." This is amnesia with respect to episodic memories, an inability to consciously recall events from infancy. Most research indicates that infantile amnesia extends through at least two, and often three to four years of age (Howe & Courage 1993, pp. 305–306). If we think of autonoetic consciousness of the past, present, and future as developing together, then the fact that mirror self-recognition develops in human beings during the period covered by infantile amnesia suggests that mirror self-recognition is not sufficient for autonoetic consciousness in the present. But it could be that autonoetic consciousness of the present emerges sooner. Also, full-blown biographical consciousness, with its sophisticated understanding of the distant future, may not emerge until adolescence. The related "problem of marginal cases" is discussed in §7.6 and §9.7.

That the chimpanzees recognized their images in the mirror seemed obviously true, but Gallup proposed a simple, controlled test of the self-recognition hypothesis. He removed the mirrors and, under anesthesia, applied red colored marks above one eyebrow and at the top of one ear of the chimpanzees using a non-irritating, non-odorous dye that could not be detected by touch after drying. Following recovery from anesthesia, he first measured the frequency with which they touched the marked areas, then reintroduced the mirrors and measured again. Between them, the two animals tested touched the marked areas only once prior to reintroduction of the mirror, but over 20 times afterward (p. 87, fig. 4), and the amount of time they spent viewing themselves in the mirror quadrupled compared to the time they were spending in front of the mirror just prior to the markings being applied (p. 87, fig. 3). This, he argued, was clear proof that the chimpanzees recognized themselves. Gallup repeated the test on stumptailed, rhesus, and cynomolgus macaques and found both (1) no decline in "social responsiveness to the mirror image"—that is, behavior indicating that they thought they were seeing an unfamiliar conspecific in the mirror—and (2) no mark-directed behavior after marking under anesthesia (p. 87).

§8.3(A). CHIMPANZEES AND ORANGUTANS

On the assumption that "self-recognition of one's mirror image implies a concept of self," Gallup concluded that he "may have found a qualitative psychological difference among primates, and that the capacity for self-recognition may not extend below man and the great apes" (1970, p. 87). Thirty years later, Gallup and colleagues claimed that this was well confirmed (2002, pp. 325–326). By then, many individual chimpanzees had passed the spot test, as well as one orangutan (Suarez & Gallup 1981, p. 180), and one gorilla (Patterson & Cohn 1994, p. 275). In contrast, subjects from over a dozen species of other primates had failed to pass the spot test, despite the use of multiple variations designed to rule out possible causes of failure, including variations in the sizes, shapes, and locations of mirrors, the age-timing of exposure to mirrors, controlling for aversion to making eye contact, and training in the use of mirrors to solve problems in order to ensure that the animals are motivated to attend to details in the mirror images (Gallup et al. 2002, pp. 325–326).

A number of challenges have been raised to the Gallup mirror self-recognition test. Gallup and colleagues (2002) discuss some really bad objections that have been raised to their claim that self-consciousness is necessary to explain chimpanzees' spot-directed behavior (pp. 327ff). Two other kinds of challenges come from the opposite direction, giving reasons to think that self-aware animals might still fail the Gallup spot test.

One was raised by Menzel and colleagues (1985). Their study showed that while chimpanzees can use both mirrors and video monitors to guide their hands to foods that they cannot see, rhesus monkeys could not master the same task, and they cite two other studies finding that macaque species fail similar tasks (p. 215). They speculate that monkeys' failure on the Gallup test "might thus stem from differences in the organization of visual spatial perception and ability to adapt to spatially displaced images" (p. 216). So monkeys' failure on the spot test might result from some difference in their visual systems that makes them unable to guide their hands to the spots using mirrors, and, depending on how widespread is the inability to use mirror images to guide one's actions, many other animals might be self-aware without being able to pass the standard form of the spot test. (Noting that some fish have been shown to target prey by looking up at the mirror image available in calm surface water, Menzel and colleagues caution that "This does not, of course, imply that such adaptability is a sufficient mechanism for recognizing one's own image in a mirror"; p. 216.)

Other reasons for thinking that some self-aware animals might nevertheless fail the Gallup spot test have been suggested by researchers working with various animals. Parrots, for instance, can visually inspect almost every area of their bodies, so it is difficult to mark them where they would be unable to inspect it directly and could still see it using a single mirror; dolphins cannot touch parts of their bodies with their snouts or flippers; and some animals, like elephants, just may not be disposed to notice novel spots, because their bodies are constantly changing in appearance. Nevertheless, studies of mirror self-recognition have been published on each of these species.

§8.3(B). DOLPHINS

Reiss and Marino (2001) used an ingenious variation on the spot test to provide "the first convincing evidence that a nonprimate species, the bottlenosed dolphin, is capable of [mirror self recognition]" (p. 5942). Because they could not count the numbers of times the dolphins touched a spot, Reiss and Marion had to use a more complex experimental protocol than the original Gallup test. They used two dolphins who had been housed for several years in pools with various reflective surfaces. The dolphins were trained to go stay in a given area of the pool without reflective surfaces until given a release signal by their trainers, and while in that staging area they were either marked, sham marked, or not marked. One of the pools used had three degrees of reflective surfaces: the normal glass walls of the pool, a portion of glass wall with a black surface attached to the back to make it more highly reflective, and a portion with an actual mirror that was removed for certain trials. The other pool had no reflective surfaces except a single mirror. By comparing data on the dolphin's behavior in marked, sham marked, and unmarked trials, along with whether or not the actual mirror was present, Reiss and Marino were able to provide strong evidence for visual self-inspection without needing to see the animals touch themselves.

Pairs were randomly chosen from four trained observers to score videotaped behavior of the dolphins. Only if the two observers agreed in their ratings of the location, timing, and type of behavior in question was the rating added to the database, and the observers could not tell whether or not the dolphins had been marked. There was agreement regarding 80 percent to 85 percent of the videotaped behaviors, and the disagreements were due to visibility problems. The combination of trials run and statistical analyses performed are too complicated to summarize here, but their analysis shows, as they put it, "definitive evidence" that the dolphins "used the mirror (and other reflective surfaces) to investigate parts of their bodies that were marked" (p. 5942). In short:

1. Before the actual marking trials were first begun (in "early sham" trials), the dolphins did not tend to swim to the reflective surfaces when released.
2. Once actual marking trials had begun, however, the dolphins would swim directly to the best reflective surface available and inspect the parts of their bodies that had been marked or sham marked, and they would not swim to a reflective surface when they had not been marked.
3. When only sham marked, the dolphins would then spend less time examining themselves than when actually marked.

This is indeed an impressive result, and their methods could be adapted to other species that, like dolphins, cannot touch marked parts of their bodies that they cannot see without using a mirror.

§8.3(C). ELEPHANTS

Although several mirror self-recognition experiments have been done on elephants, the results are more mixed. Veteran elephant trainer Kari Johnson told me that one of their females "likes to look at herself" in the shiny sides of an elephant transport trailer (taped interview conducted October 21, 2002), and there is one unpublished[5] and one published report of an elephant passing the Gallup spot test. Generally, though, elephants exposed to the spot test have failed to show any interest in the marks.

The first two published experiments (Povinelli 1989 and Hyatt et al. 2003) failed to find any mark-directed behavior, even though one of them (Povinelli)

[5] Students in an animal behavior class at Sierra Nevada College claim that an elephant at a Nevada casino passed the Gallup spot test. The professor, Patricia Simonet, states that the elephant "went to the mirror and examined (gazed into the mirror and rubbed) marks only visible when looking into the mirror. She did not display such behavior when she was marked and no mirror available." No other details of the procedure are given, however, in her one-paragraph communication to *Elephants: The Publication of the Elephant Research Foundation* (Simonet 2000).

proved that elephants can use mirrors to retrieve food that they cannot see or smell. The third found significant mark-directed behavior, but only by one out of three elephants tested, and on only one of the four occasions that it was marked. The study used three female Asian elephants at the Bronx Zoo. After habituating them to the presence of a large (2.5 meter square) mirror, the elephants' handlers applied a mark on the right side of the elephants' heads, and a sham mark on the left side.

> One elephant, Happy, passed the mark test on the first day of marking. Caretakers did not notice her touching either the mark or sham-mark before being released into the elephant yard. After being released into the yard, she walked straight to the mirror where she spent 10 seconds, then walked away. Seven minutes later she returned to the mirror, and over the course of the next minute she moved in and out of view of the mirror a couple of times, until she moved away again. In the following 90 seconds, out of view of the mirror, she repeatedly touched the visible mark but not the sham-mark. She then returned to the mirror, and while standing directly in front of it, repeatedly touched and further investigated the visible mark with her trunk.

Happy was retested the next two days, and again two months later, but she never again showed mark-directed behavior. Neither of the other elephants (Maxine and Patty), who were tested twice and then again two months later, ever showed any interest in their marks. On the other hand, the authors emphasize that animals that have passed the spot test have generally progressed through four stages before facing the spot test:

1. In the first phase, animals react to their mirror image as if to an unfamiliar conspecific.
2. In the second phase, they inspect the mirror in various ways, including (usually) looking behind it.
3. In the third, they engage in "repetitive mirror-testing behavior," things like systematically moving back and forth while studying the image in the mirror.
4. In the fourth, they exhibit various kinds of "self-directed behavior," such as inspecting their teeth or other areas of their bodies that they could not see without using the mirror. (Plotnik et al. 2006, p. 17053)

The authors emphasize that:

> All three elephants displayed behavior consistent with mirror-testing and self-directed behavior . . . such as bringing food to and eating right in front of the mirror (a rare location for such activity), repetitive, nonstereotypic trunk and body movements (both vertically and horizontally) in front of the mirror, and rhythmic head movements in and out of mirror

> view; such behavior was not observed in the absence of the mirror. . . . On more than one occasion, the elephants stuck their trunks into their mouths in front of the mirror or slowly and methodically moved their trunks from the top of the mirror surface downward. In one instance, Maxine put her trunk tip-first into her mouth at the mirror, as if inspecting the interior of her oral cavity, and in another instance, she used her trunk to pull her ear slowly forward toward the mirror. Because these behaviors were never observed in . . . the initial, "no mirror" control conditions . . . they indicate the elephants' tendency to use the mirror as a tool to investigate their own bodies. (Plotnik et al. 2006, p. 17054)

The authors characterize the fourth phase of mirror acclimation as "the beginning of mirror understanding" (p. 17053), but it would be more accurate to say that animals who actively inspect otherwise invisible parts of their bodies using mirrors fully understand them, whether or not they pass the spot test. Various animals are known to acclimate to mirrors in the sense of progressing from stage one to two, when, after inspecting the mirror and looking behind it, they stop treating the mirror image as a conspecific. While this kind of acclimation probably shouldn't count as understanding, when an animal enters phase three it appears to be testing hypotheses about how the mirror works. That, I suggest, is "the beginning of mirror understanding," which is fully achieved as soon as the animal exhibits self-directed behavior, whether or not it shows any interest specifically in spots that are applied to it by experimenters.

Both of the earlier published studies of elephants mentioned similar examples of apparently self-directed behavior, while being somewhat more cautious than Plotnik and colleagues in their interpretations of them. If the elephants understand that the mirror images are their reflections, then why don't they more consistently pass the spot test? Some general facts about them suggest that the Gallup spot test might not reliably reveal self-recognition in elephants. First, although elephants are known to use branches to scratch parts of their bodies, elephants do not extensively groom themselves the way that primates do, and this may make elephants less likely to examine unusual spots on themselves and others. Also, elephants' eyesight is notoriously bad, they lack color vision, and their skin color changes noticeably as they spray themselves with water or mud, so they may simply disregard the kinds of marks traditionally used in the spot test (Nissani & Nissani 2007).

So while elephants' performance on the Gallup spot test has been mixed in the three studies published to date, the extent of self-directed behavior displayed by elephants exposed to mirrors still strongly suggests that elephants are capable of recognizing themselves in mirrors, and hence that they have the present-time element of autonoetic consciousness.

§8.3(D). BONOBOS

Neither of the two existing studies of bonobos exposed to mirrors administered the spot test, but both found significant self-directed behavior. Walraven and colleagues found that social behaviors toward the mirror almost completely disappeared after the first day, and four of their seven bonobos exhibited self-directed behaviors using the mirror (1995, p. 147). Hyatt and Hopkins obtained similar results, although when they exposed a group of chimpanzees under similar conditions for comparison, they found that "The reactions of some of the chimpanzees to the mirror were more immediately and obviously indicative of self-recognition than those of the bonobos" (1994, p. 250).

§8.3(E). ORANGUTANS

One of Gallup's earlier studies compared chimpanzees, orangutans, and gorillas on the spot test. Of the two orangutans tested, one both passed the spot test and exhibited significant self-directed behavior, while the other both failed the spot test and exhibited no self-directed behavior (Suarez & Gallup 1981, p. 180).

§8.3(F). GORILLAS

Among the great apes, gorillas are the most puzzling case. While one individual (Koko) has passed the spot test and exhibits significant self-directed behavior, the six other individuals who have been tested by Gallup and colleagues never displayed any self-directed behavior in front of the mirrors (Suarez & Gallup 1981, p. 177; Shillito et al. 1999, p. 1001).

Unlike her interpretation of Michael's "story" described in chapter 6, there is nothing suspect about Patterson's description of how the spot test was administered to Koko. Patterson and a colleague introduced a new mirror into a room where Koko's behavior was videotaped for ten minutes on each of six successive days. On the fourth and fifth days, a towel was used to wipe Koko's forehead. On the fourth day it was wet with water, on the fifth with paint. Two independent observers determined how many times Koko touched the area wiped with the towel on the six days of the test. On the fifth day, when the towel was wet with paint, she touched the area 47 times. In contrast, on the previous days, including the fourth day (when the towel was wet with water), she never touched the area more than once per session. On the sixth day, the towel was used again, but this time it was again wet with water. On that day, Koko touched the area three times. And they provide four photos that clearly show Koko exhibiting self-directed behaviors (Patterson & Cohn 1994, p. 277).

Patterson has noted that gorillas might fail the spot test in part because direct eye contact can be perceived as a threat by gorillas or because human observers are present in the room (Patterson & Cohn 1994, p. 286). Shillito and colleagues (1999) controlled for these concerns, however, by first exposing the gorillas to pairs of mirrors angled so that they could view their images without making eye contact (which is like using a small mirror to view the side of your face in the bathroom mirror), and then using videotaped recordings to gather data rather than observers in the room.

So why would Koko's behavior in front of mirrors be so different from all other gorillas tested to date? Two remaining factors are that Koko was exposed to mirrors from a young age and that mirror use may have been extensively modeled for her by her human handlers. By contrast, the six gorillas studied by Gallup and colleagues were all adults at the time of the studies and may have had no previous exposure to large mirrors across a significant amount of time. So it could be that early exposure and/or modeling is necessary for gorillas to understand mirrors. This clearly is not the case for chimpanzees, bonobos, and orangutans, however, since older individuals were always among those who exhibited self-directed behaviors in studies of those species,[6] and those individuals were zoo animals that received no modeling of mirror use.

§8.3(G). OTHER PRIMATES

No significant mirror-mediated, self-directed behavior has been observed in any monkey (Anderson & Gallup 1997, p. 1564), despite the fact that "dozens of species and scores of individual prosimians and monkeys" have been observed during exposure to mirrors (Gallup et al. 2002, p. 325). As noted earlier, these studies have used multiple variations designed to rule out possible causes of failure, including variations in the sizes, shapes, and locations of mirrors, the age-timing of exposure to mirrors, controlling for aversion to making eye contact, and training in the use of mirrors to solve problems in order to ensure that the animals are motivated to attend to details in the mirror images (Gallup et al. 2002, pp. 326).

§8.3(H). PARROTS

Using two birds from her vocal communication and cognition studies program, Irene Pepperberg and colleagues found limited and ambiguous evidence of mirror self-recognition in African grey parrots. They did not attempt the

[6] The gorillas' ages were 13, 17, 18, and 19 in Suarez and Gallup (1981), and 12 and 22 in Shalitto et al. (1999). The orangutan who showed self-directed behavior in the Suarez and Gallup study was 12 years old, and their chimpanzees (who both showed self-directed behavior) were 16 and 21. Finally, the bonobos who exhibited self-directed behaviors using the mirror in Walraven et al. (1995) were aged 2, 8, 15, and 23 (pp. 145 and 147).

spot test, because "Grey parrots, unlike apes and monkeys, do not often use their feet either to explore their bodies or to investigate items" (1995, p. 193), and second, parrots can "see all body areas except parts of their heads" (p. 186). So they simply exposed the birds to a mirror that was leaned against a wall (so that the birds could look behind it), and recorded the numbers of times the birds exhibited behaviors in various categories:

1. little or no interest,
2. social behavior toward the mirror (including aggressive acts),
3. searching behind the mirror, and
4. avoidance.

Although they did not include a category of self-directed behaviors, some of the birds' behaviors could have been interpreted that way. For instance:

> Both parrots often tapped the mirror with opened beaks. Such behavior resembles "beak wrestling" responses to a conspecific, during which birds approach one another with open beaks, then intertwine beaks and engage in pushing, pulling and sideways maneuvers. Opened-beak tapping might, however, be interpreted as a self-guided action. . . . Kyaaro engaged in what is called "peekaboo" in children: He stared at the mirror image, moved sideways until his head was just out of range of the mirror, then quickly moved back into view. In other sessions he said "you come," "you climb," or "tickle" to the mirror. (p. 184)

Their data indicated that one bird exhibited less aggressive behavior over time, but neither exhibited the increased lack of interest over time that is characteristic of monkeys exposed to mirrors, and both "exhibited search behavior" (i.e., looked behind the mirror). It is important to note that these birds' exposures to the mirrors was limited: 14.5 hours for one bird, and 5.75 for the other. This is significant, because animals tested in other studies commonly have much more exposure time. In the end, while planning further experiments, Pepperberg and colleagues concluded that their data from this experiment "suggest that both birds viewed the mirror image as a conspecific" (p. 186).

§8.3(I). SUMMARY

So to date, members of only five species have been shown to pass the spot test for mirror self-recognition: a number of chimpanzees, one orangutan, one gorilla, a couple of dolphins, and one elephant. If the criterion for mirror self-recognition is expanded to include exhibiting significant self-directed behavior, however, then most of the elephants and bonobos tested have exhibited mirror self-recognition. Although a study of mirror self-recognition in African grey parrots has been published, its results are inconclusive.

§8.4. Use of a "Theory of Mind" (ToM) and Conscious Planning for the Future

Finally, what about the forward-looking element of autonoetic consciousness? What evidence do we have or could we someday have that an animal "consciously . . . contemplates [its] existence and conduct at a time in the future" (Wheeler et al. 1997, p. 350)? In this section, I review the available evidence for theory of mind use in non-human animals, and I briefly consider some other kinds of planning for the future that seem to require conscious anticipation.

To use a theory of mind (ToM) is to interpret others' behaviors in light of their beliefs and desires. This constitutes using a theory insofar as the beliefs and desires are not observable; they are theoretical constructs about what is "inside the others' heads." All normal children acquire the human ToM without explicit training and in a developmentally stereotyped sequence, suggesting that humans are hard-wired to use ToM. (The following summary is based on Stone et al. 1998, pp. 640–641.) Around 18 months, children first become able to understand what others are looking at in general, and in particular that others are looking at the same objects. Between 18 and 24 months, they come to understand pretend play. By two they understand and use attributions of desires to others. Between three and four years, they come to understand that others can hold different, and sometimes false, beliefs about the world. By six or seven, they come to understand that others are attributing mental states to various people and they thus acquire an understanding of second-order false beliefs, that is, false beliefs about what others' beliefs or desires are. Finally, between 9 and 11 years, children come to understand the making of a faux pas, which requires a complex representation of multiple persons' beliefs and desires. To understand that "someone said something they shouldn't have said," one must understand both that the speaker did not know that he should not have said something and that the audience felt offended by what he said. (Interestingly, girls develop this ability by 9 years, while boys take 11 years.)

To suggest that some animals use ToM is to suggest that evolution has similarly programmed them to acquire at least some (but not necessarily all) of the abilities listed above. Lacking language, no animal displays an understanding of faux pas, but both anecdotal field reports and some experiments suggest that some primates use an understanding of others' desires and/or beliefs to predict and manipulate others' behaviors (Whiten 1997). If they do use this relatively sophisticated and later-emerging element of ToM, it seems plausible to say that they do so consciously for a couple of reasons.

First, it seems that when we use ToM to predict others' behaviors, we usually do so consciously. Introspectively, we describe our use of ToM in terms of "imagining future scenarios," and outside Freudian psychologists' offices, to describe someone as "imagining" something is to describe them as consciously thinking about it. This is confirmed insofar as the prefrontal cortex is involved in ToM use (Stone et al. 1998), since our prefrontal cortex is involved in conscious planning for the future. Relatedly, we often attribute beliefs and desires to others based on an understanding of what we ourselves would believe and desire under varying circumstances. In doing this, we seem to be consciously projecting ourselves into various circumstances, past, present, and future.

Finally, there is a general consensus that ToM use is closely related to what is called "executive control" or "executive function." "Executive function is a term for processes responsible for higher-level action control (e.g. planning, inhibition, coordination and control of action sequences) that are necessary in particular for maintaining a mentally specified goal and for bringing it to fruition against distracting alternatives" (Perner & Lang 1999, p. 337).

Exercising executive control involves "focusing," or consciously riveting one's attention on accomplishing something in the midst of salient distractions. Examples of tasks used to assess executive function in small children include the following. In "the day night Stroop task," the children are asked to say "day" when shown a picture of a nighttime scene and "night" when shown a picture of a daylight scene. In "the windows task," children first learn that when they point to a box that contains a reward, it will be given to another individual. They are then shown boxes with windows in them (so that they can see where the reward is) and asked to point. In both cases, success depends on suppressing what would otherwise be a salient cue. Introspective reports indicate that in humans, at least when initially presented with such a problem, this requires a conscious attempt to "focus." This is, in effect, a forward-looking version of the element of surprise that was discussed in regard to episodic memory in §8.2(D).

The tests commonly used to determine when ToM use emerges in small children similarly require suppression of attention to salient cues. In a first-order false belief task, a child and a collaborator watch the experimenter place something in a box. Then, while the collaborator is out of the room, the child observes the experimenter move the object from one box to another. The child is then asked where the collaborator will look when she returns and looks for the object. In an appearance versus reality task, the child is shown an object with unexpected qualities, such as a "rock" made of Styrofoam, and asked both: "What does this look like?" and "What is it really?" (These descriptions of the day/night, windows, and appearance/reality tests are based on Perner & Lang 1999, p. 339.)

As detailed in Perner and Lange (1999), although executive control and ToM are thought to be linked, some data suggest that executive function emerges prior to ToM, and other data suggest the opposite. For present purposes, suffice it to say that solving ToM problems (at least new or difficult ones) requires conscious attention, just like executive control problems, and, relatedly, both "are mediated by the same region of the prefrontal cortex" (p. 341).

§8.4(A). USE OF ToM IN PRIMATES

Unlike episodic memory, where the study of animals is in its infancy, research on ToM use in primates has been exploding for 20 years (Byrne & Whiten 1988; Whiten & Byrne 1997), and unlike mirror self-recognition, where the results are fairly uniform and clear-cut, there is great controversy about how to interpret the results of ToM research (Carruthers & Smith 1996).

Some of the most striking evidence of ToM use in primates is anecdotal. Field researchers studying a wide variety of primates have reported what seem to be sneaky behaviors and various kinds of deception. While many of the sneaky behaviors might require only the most rudimentary element of ToM (an understanding of others' gaze direction), others strongly suggest that apes and monkeys can represent and try to manipulate others' beliefs.

Subordinate monkeys and apes are both reported to engage in "sneaky copulation," out of sight of dominant males, but this could be explained in terms of simple association and first-order representations. A representation is first-order when it represents a fact about the physical world, for example that a dominant male is present. Using ToM requires at least second-order representations. These are representations of other representations, such as a belief about what someone else believes or sees or desires. Having associated aggression with situations in which he attempts to copulate with a female in the presence of the dominant male, a subordinate could learn to be sneaky about copulating without attributing any mental states to the dominant male. In other cases, however, the sneaky animal must understand, at a minimum, where another animal's gaze is directed, as in this example involving chimpanzees:

> Often a low-ranking male will sit with his upper arm resting on his knee and his hand loosely hanging down so that a female in front of him can see his erect penis, but apes on the side cannot see it. This inconspicuous form of concealment occurs together with quick glances at dominant males. Needless to say, the subordinate always uses the hand on the body side which is turned towards the dominant. (Frans de Waal, quoted in Whiten & Byrne 1988, pp. 215–216)

Here is an analogous case involving Hamadryas baboons, which are Old World monkeys. An adult female "spent 20 minutes in gradually shifting in a seated position over a distance of about 2 metres to a place behind a rock about 50 cm high." A young male interloper was crouched behind the rock, out of sight of the dominant male, and this allowed the female to groom him, something that the dominant male would not have permitted otherwise.

Other striking anecdotes suggest more than understanding where another animal's gaze is directed. A famous example involves vervet monkies. These Old World monkeys have three kinds of alarm calls, to which they respond differently. The call that a vervet normally issues when it sees a leopard causes the other vervets to climb high into the trees, whereas the calls they normally issue when they see an eagle or a snake cause vigilance directed toward the sky and the ground, respectively. Daniel Dennett described the vervets' use of these calls and the famous anecdote as follows:

> [V]ervets *seem* to be engaged in a practice that could involve learning, insincerity, trustworthiness, deception, divided loyalty. While it would be wildly romantic to suppose that a vervet could *tell a joke*, it is not so clear that there isn't room in their way of life for one to *tell a lie*.
>
> For instance, two bands or groups of vervets were once observed in a territorial skirmish; one group was losing ground and one of the losing-side monkeys, temporarily out of the fray, seemed to get a bright idea: it suddenly issued a leopard-alarm (in the absence of any leopards), leading all the vervets to head for the trees—creating a truce and regaining the ground his side has [*sic*] been losing. (Dennett 1998, p. 290).

Marc Hauser reports that a male vervet similarly used a leopard alarm call to stop other vervets' assault on a female who attacked him after she resisted mating with him (Hauser 2000, p. 156). To interpret these behaviors as deception involves attributing second-order representations to these monkeys. For the deception to be intentional, the vervets issuing the "fake" alarm calls would have to be representing to themselves what the others vervets' representations were. At a minimum, they would have to be attributing to their conspecifics simple beliefs with content something like "There is (not) a leopard nearby." For the deception to be fully deliberate, perhaps we must also attribute to them some beliefs about what will cause others' beliefs to change.

Some of the most complicated deceptions observed seem to involve third- and fourth-order representations. For instance, Diane Fossey describes the following incident that she observed as a group of gorillas was traveling between feeding sites along a narrow trail:

> S looks up into *Hypericum* tree and spies a nearly obscured clump of *Loranthus vine*. Without looking at those behind her, she sits down by the side of the trail and begins to intently self-groom until the others have passed her and all are out of sight some 15 foot ahead. Only then did S stop "self-grooming" to rapidly climb into the tree, break off the vine clump and descend with it to the trail to hastily feed on it before running to catch up with the group. (quoted by Whiten & Byrne 1988, p. 218)

Whiten and Byrne interpret this as "inhibition of attending," and they say that the gorilla S must have been representing to herself what the others would represent about her own representation of the espied food.

Here is an anecdote that Whiten and Byrne say might involve a *fourth-order* representation (I've added the *A*'s and *B*'s):

> One chimp [*A*] was alone in the feeding area and was going to be fed bananas. A metal box was opened from a distance. Just at the moment when the box was opened, another chimp [*B*] approached at the border of the clearing. The first chimp [*A*] quickly closed the metal box and walked away several metres, sat down and looked around as if nothing had happened. The second chimp [*B*] left the feeding area again, but as soon as he was out of sight [of *A*], he hid behind a tree and peered at the individual [*A*] in the feeding area. As soon as that individual [*A*] approached and opened the metal box again, the hiding individual [*B*] approached, replaced the other [*A*] and ate the bananas. (quoted in Whiten & Byrne 1988, p. 220)

A natural folk psychological explanation of this incident would be that *B* desires that *A* believe that *B* believes that *A* does not believe that there are bananas in the box.

Despite such striking anecdotal evidence of ToM use in both apes and monkeys, some controlled laboratory experiments have found that they perform poorly on some basic ToM tasks. In the 1990s, Daniel Povinelli and colleagues developed a simple ToM test that has subsequently been applied to other animals. Chimpanzees were given an opportunity to beg from one of two experimenters who were holding something just out of reach in front of their cages. Only one of the two experimenters could see the chimpanzee, either because only one was facing the cage, one had a bucket over her head, or one was blindfolded. On analogous tasks, children aged three and up consistently solicit the attention of the human who can see them, but Povinelli's chimps initially performed at chance levels. After further training, their best performance was still only 58 percent correct, in the bucket over the head condition (Povinelli & Eddy 1996, p. 65, table 3).[7]

This was a rather surprising result, given the striking anecdotal evidence of ToM use by chimpanzees described above. Reasoning that it was competition rather than cooperation that drove the evolution of ToM, Brian Hare and colleagues began studying the performance of both chimpanzees and monkeys when the gaze-monitoring problem involved competition rather than cooperation. In an experiment reported in Hare et al. (2000), they used adjacent areas with opaque walls to either side of an open door between them. Before the chimpanzees were introduced to this experimental area from their adjacent cages, an experimenter baited it to one of three conditions: Subordinate-Door (meaning that one piece of food was in the doorway between the two areas and the other was in the subordinate chimp's area up against the opaque wall so that the entering dominant could not see it), Dominant-Door (meaning that

[7] Moti Nissani (2004) applied the same paradigm to chimpanzees that were older than Povinelli's and, in some cases, wild-reared. His findings were inconsistent with Povinelli's. Of chimpanzees he tested, 6 out of 7 immediately performed "significantly above chance" (p. 248).

the situation was reversed), or Door-Door (meaning that both pieces of food were in the doorway, visible to both animals as they entered). During the experiment, 83 percent of all food obtained by the subordinate animals came from pieces on their side in the Subordinate-Door condition. The experimenters also note that "On a number of occasions in the Subordinate-Door condition, subordinate individuals behaved in especially strategic ways with regard to the piece of food that only they could see." On seven occasions, five different individuals approached the door and stood there until the dominant moved away, at which time they snatched the food that the dominant could not have seen, and on four occasions two different individuals moved to the side of the doorway, out of sight of the dominant, to take the food that the dominant could not see. They also note that in the Dominant-Door condition, the dominant retrieved the piece of food in the door first 74.8 percent of the time (obviously the best strategy if they understood that the subordinates could not see the other piece of food; pp. 775–776). In other experiments reported in the same article, they used three other protocols to rule out alternative, deflationary explanations of the subordinates' performance, but all of these confirmed the results of the first experiment just described.

Flombaum and Santos have subsequently obtained similar results with free-ranging Old World monkeys. Their subjects are rhesus macaques living in a Puerto Rico field station. They are accustomed to eating monkey chow provided by keepers, but they definitely treat their human observers as dominants. Flombaum and Santos have capitalized on this fact, devising several protocols in which the monkeys could see two grapes (a food prized relative to their familiar monkey chow), only one of which could be seen by an experimenter. In one of these experiments, they found that the monkeys selectively approached the experimenter who was not physically facing the grape (2005, p. 447). In a second study, the monkeys were faced with a single experimenter who held in front of her a device designed with two platforms separated by an incline, so that a grape could be caused to roll down the incline from one platform to the other. The platforms could be made visible or invisible to the experimenter by installing canopies, but the incline was always invisible to the experimenter. In front of the monkey, the experimenter would place two grapes on the upper platform and then trigger the descent of one to the lower platform. With the canopies in place, so that the experimenter could not see the one grape roll to the lower platform, the monkeys selectively approached the grape from the lower platform. With the canopies removed, however, so that the experimenter could see both grapes, the monkeys approached the two grapes randomly. This suggests that they understood what the experimenter was and was not seeing (unpublished study described in Santos et al. 2006, pp. 441–442).

Thus there is good evidence that both monkeys and apes use at least some aspects of ToM, because both can understand what others are seeing. Santos and colleagues (2006) have suggested a related physiological explanation of why some primates' brains might work this way. Studies of both human and non-human primates show that two areas of the brain are crucially involved in gaze tracking. Both the amygdala, a part of the basal ganglia in the forward part of the frontal lobe, and an area of the temporal lobe called the superior temporal sulcus are activated during attempts to track others' eye gaze, and damage to either area impairs this ability. The amygdala is also crucially involved in aggressive motivation, however, suggesting that "primates may fail in cooperative versions of ToM because these kinds of contexts fail to engage the amygdala emotionally, although competitive interactions obviously do" (Santos et al. 2006, p. 450). They note that future experimentation could evaluate this hypothesis by studying whether or not monkeys with lesions to the amygdala continue to follow eye gaze while being unable to use the information correctly in competitive situations.

§8.4(B). USE OF ToM IN ELEPHANTS

Elephants' social structure is similar to that of primates in ways that suggest that they too would benefit from using ToM, and both anecdotal field reports and the one ToM experiment published on elephants suggests that they do.

Elephants are long-lived, highly social organisms. They grow up in complex, changing social settings involving interactions with multiple generations of individuals, from variously related groups, over a life span of up to 65 years. Such animals would benefit both from understanding others' intentions and being able to predict behaviors in light of them, and from being able to actively manipulate others' behaviors by manipulating their beliefs. So elephants are at least good candidates for ToM.

In the only published study of ToM in elephants, Moti Nissani applied Povinelli's begging paradigm to two elephants, Wanda and Winky at the Detroit Zoo. He cautions that "the small sample, and the limited number of trials . . . do not allow us to carry out meaningful statistical tests," but the results at least suggested that the elephants understood who could and could not see them. For although the elephants begged correctly less than two-thirds of the time when a screen obscured one experimenter's gaze and when one experimenter had a bucket over her head, they begged correctly 81 percent of the time when one was lying face down and when one was standing sideways, looking away from the elephant. These were much better performances than Povinelli's chimps, whose best performance, in the bucket over the head condition, was still only 58 percent. (All of the figures in this paragraph are from Nissani 2004, p. 244, table 1.)

Joyce Poole, who has studied the elephants of Amboseli Park in Kenya since 1976, states that the available evidence, while predominantly anecdotal, is "very suggestive that elephants have a theory of mind" (1998, p. 107). She speculates that deception is less important in elephant society than among primates (p. 106), but she emphasizes field observations of helping or guarding sick or injured elephants, including unrelated ones (p. 103–104), and even one instance of what appears to be sympathy for a member of another species: "In India a working elephant named Chandrasekharan had been asked to place poles in pre-dug holes. He continued to follow his mahout's commands until he reached one particular hole when he simply refused to insert the pole. When the mahout examined the hole he found a sleeping dog at the bottom. Only after the dog had been removed would the elephant place the pole into the hole" (p. 95).

§8.4(C). USE OF ToM IN DOLPHINS

No controlled study of ToM in any cetacean has yet been published, but like elephants, cetaceans are long-lived, and some of them live in complex social groups where using a ToM could be adaptive. Dolphins are the best-studied cetaceans, and anecdotal reports suggest in various ways that they may use ToM. Here I will limit myself to three examples.

At the Dolphin Research Center in the Florida Keys, dolphins are housed in natural, but fenced lagoons. The dolphins can easily jump over the fence and one of them, Mr. Gipper, had been known to open a hole in the fence so that he could come and go without having to jump it. When an injured wild dolphin was brought to the facility for rehabilitation, the DRC staff mended Mr. Gipper's hole so that the weakened dolphin would not escape. Thomas White continues the story:

> Unknown to the staff, however, Gipper had re-opened the hole.
>
> Unfortunately, after the new dolphin was put into the pool, she found the hole and headed out to sea. Before the staff could do anything, however, Gipper and Little Bit went after her, got on either side of her, turned her around and escorted her back. Now, however, the new dolphin didn't want to come back in. Dolphins appear to have an instinctive aversion to going through restrictive openings, and teaching them to do so ("gating") is a difficult process. So it's not surprising that now that the new dolphin had apparently calmed down, she'd balk at going through the fence. Again, the dolphins took the initiative. While Gipper stayed with the sick dolphin, Little Bit swam back and forth through the hole—showing that it wasn't dangerous. The new dolphin followed Little Bit's example and swam through the hole. And then Little Bit and Mr. Gipper kept her away from the hole until the staff could fix it. Schuler [who was then a trainer at DRC] observed that the sick dolphin "was in such a weakened state that if she'd gotten lost, she would have died for sure." The former trainer believed that Mr. Gipper and

> Little Bit "kept her from dying that day; they knew this was the place she needed to stay." (White 2007, pp. 94–95)

As interpreted in the above passage, Little Bit and Mr. Gipper were representing the wild dolphin's mental states as they worked to coax her back into the enclosure, and their overall strategy expressed sympathy for it.

Denise Herzing and Thomas White state that "Cetaceans are the only animals known to actively seek out contact with humans in the wild" (1998, p. 71), and a number of these anecdotes involve dolphins interacting with humans in suggestive ways. One of the most dramatic is due to Wayne Grover, who claims that two adult dolphins solicited him to assist a baby dolphin who was bleeding around a large fish hook lodged in its flank.

> "Whether it was my imagination or a logical deduction," Grover writes, "I suddenly felt that I was being asked for help. . . . The large [dolphins] closed in on the baby from either side until they were touching it with their pectoral flippers. They settled the baby to the sea floor right in front of me, still holding it from each side." After discovering that the hook was in too deep to be removed by simply unwinding the fishing line and pulling it out, Grover steadied the baby on the ocean floor and used his diving knife to remove the hook. The two adults observed the entire procedure. After the hook came out, Grover reports,
>
>> The largest dolphin came to me, stopped at eye level, and looked into my eyes behind the mask. For a brief moment, we looked deeply into each other's eyes, and then the dolphin nudged me with its snout, pushing me slightly back.
>>
>> I had the distinct impression that we were communicating but, even as I thought it, my logical mind tried to dissuade me, saying it was imagined.
>>
>> Then the three dolphins were gone. Without a sound, they rapidly climbed upward toward the surface, leaving me alone again. (White 2007, p. 93)

As interpreted by Grover, the adult dolphins not only understood his intentions as he performed a painful procedure on the calf, they approached him in an attempt to manipulate his desires—they were trying to get him to want to try to help. (In another version of the story, the adults fended off a pair of sharks attracted by the blood in the water; Herzing and White 1998, p. 77.)

Finally, Herzing herself has managed a changing group of assistants studying wild dolphins at the Dolphin Research Center in the Florida Keys for over a dozen years, and she reports that a mother dolphin once expressed displeasure with her after a novice assistant disrupted the mother dolphin's training of her calf.

> What is striking about this action is that the mother dolphin tail-slapped in front of Herzing, not the offending swimmer. Given the context of this encounter and the history of Herzing's interactions with the [dolphin] community, is it possible that the mother targeted Herzing, and not the offending swimmer, because the dolphin recognized Herzing from years of encounters as the individual who was dominant in the hierarchy of humans and held her "responsible" for the actions of the other swimmers? (Herzing and White 1998, p. 72)

As interpreted by Herzing, the dolphin was trying to get her to control one of her assistants.

All three of these anecdotes have been highly interpreted by the people reporting them. But these are just three of many, many such stories told by a variety of dolphin researchers, not just by a few wild-eyed enthusiasts. This wealth of anecdotes at least suggests that dolphins use ToM in the wild. Given generally high intelligence, and their performance on syntax and episodic memory problems reported earlier in this chapter and chapter 5, that probably shouldn't surprise us.

§8.4(D). USE OF ToM IN SCRUB JAYS

More surprising is the result of a controlled experiment on scrub jays that certainly looks like the use of ToM. Nathan Emery and Nicola Clayton, whose work with scrub jays in the area of episodic memory was groundbreaking, designed the experiment inspired by the fact that scrub jays have been observed to re-cache their food, especially when a conspecific has observed their initial choice of cache location. In their study, Emery and Clayton controlled captive-born jays' experiences with pilfering to produce two groups of birds that could then be compared in their response to a novel situation. Although both groups of birds were allowed to watch other jays caching food, only one group was ever allowed to pilfer from other birds. They then compared the behavior of the two groups under the following conditions. Each group was allowed to cache foods, either with or without another bird watching them. They were then given an opportunity to re-cache the foods in private. Those birds who had never had the experience of pilfering food did not tend to re-cache their food, whether they had cached it in private or while another bird watched. In contrast, those who had themselves pilfered foods from conspecifics tended to re-cache their food if they had been observed caching it but not if they had done it in private. Emery and Clayton conclude: "Our results suggest that jays relate information about *their* previous experience as a pilferer to the possibility of future stealing by *another* bird, and modify their caching strategy accordingly" (2001, p. 443, emphasis added). They repeatedly hedge their bets this way, saying that the described result "suggests" that the jays were engaging in "tactical deception," "knowledge attribution [to] conspecifics," and "mental time travel" (p. 445), but it is unclear how else to explain this striking result than by saying that the jays were using ToM.

§8.4(E). UNDERSTANDING THE POINTING GESTURE INVOLVES ToM

An understanding of pointing is thought to involve using a ToM insofar as it requires—at a minimum—gaze following, and—perhaps—understanding others' intentions. Human infants come to understand pointing in a developmentally stereotyped way: beginning at about 9 months of age, they examine the pointing finger itself and objects near it, and by 12 months they are able to direct their attention to objects distant from the pointing finger.

Although great apes have not been observed using pointing gestures in the wild, the subjects of long-term "enculturation" projects (like Savage-Rumbaugh's bonobos) do begin to point with their arms, although they do not extend their index fingers the way humans in all cultures do, perhaps for anatomical reasons (Povinelli & Davis 1994). Controlled experiments have failed to show, however, that chimpanzees interpret humans' pointing gestures correctly when the objects are as little as 90 centimeters (about a yard) away from the finger (Herman et al. 1999, p. 348; citing Povinelli et al. 1997).

Experiments with dolphins indicate that at least some of them are able to interpret the human pointing gesture without training. Herman and colleagues (1999) found dramatically above chance performance of two dolphins when trainers, who had previously issued commands using hand-gestures referring to various objects and manipulations of them, instead pointed (for less than one second) to the object to be manipulated and then issued only the familiar hand-gesture indicating what to do with it. On this new problem, the dolphins responded correctly 80 percent of the time when the object pointed at was on either side of the dolphin (accuracy was only 40 percent when the object was behind the dolphin). They then inserted pointing gestures instead of object-indicating hand-gestures into previously learned, three-part commands that told the dolphins to take one object to the location of another. Here performance dropped to 68–77 percent correct, but on this more complex problem, chance accuracy would have been only 17 percent, given the number of objects in the dolphins' pool. When Tshudin and colleagues (2001) similarly introduced both pointing gestures and gazing at the objects to be manipulated, half of their six dolphins performed above chance on their first 12 trials with pointing, and two performed above chance in response to gazing alone. With further training, four of the six dolphins improved to above-chance performance in response to pointing gestures, and all six improved to above-chance performance in response to gazing.

§8.4(F). TOOL USE AND CONSCIOUS PLANNING FOR THE FUTURE

The manufacture of tools suggests conscious planning for the future insofar as there is displacement between the contexts of manufacture and of use. Tool *use* has now been documented in a wide range of animals in the wild, including chimpanzees, dolphins, elephants, various corvids, and others, although New Caledonian crows and chimpanzees "are the only nonhuman vertebrates where all populations show routine tool use. . . . In all other animal tool users, either there are insufficient data to assess tool use frequency or tool use is known to be absent from many populations" (Kacelnik et al. 2006, p. 519). Tool *manufacture*, however, is very, very rare. Although some enculturated great apes (including Kanzi) have learned to manufacture crude stone cutting tools (Savage-Rumbaugh & Lewin 1994, chapter 8), in the wild only chimpanzees, orangutans, and New Caledonian crows are known to manufacture tools by any process more complex than "detaching or subtracting objects from each other" and only the crows engage in "multistep" tool manufacture (Kacelnik et al., 2006, p. 520).

Apart from tool manufacture, transporting tools or saving them for future use involves the same displacement of tool selection and use. Boesch and Boesch (1984) report that chimpanzees transport hammers for cracking nuts in the wild, and three species of primates have been tested to see if they would save tools for future use. Mulcahy and Call (2006) addressed this question in bonobos and orangutans. They first trained the apes to use tools to retrieve rewards from a test apparatus. They then gave them access to a collection of suitable and unsuitable tools, but not the test apparatus, to see if they would carry an appropriate tool with them and use it later. They began by giving the apes access to the tools while the test apparatus was visible but unaccessible, and then ushered the apes into an adjacent waiting room and allowed the subject to watch as all remaining tools were removed. One hour later, the apes were let back into the testing room and given access to the test apparatus. In this phase of the study, the apes took tools with them most of the time, and brought the appropriate tools into the test room most of the time. The experimenters then changed the procedure, ushering the apes out into a different room that was used for sleeping, and increasing the time delay from one to 14 hours. Only two apes were tested this way (one bonobo and one orangutan), but the results were impressive. "Neither ape took any tools in the first trial. The orangutan took suitable tools in all 11 remaining trials, which she brought back and used in 7 trials to get the reward. The bonobo took suitable tools in 8 of the remaining 11 trials, which he always brought back to the testing room to get the reward" (p. 1039). I know of only one study of tool transport in a species of monkey. Jalles-Filho et al. (2001) trained capuchins (a New World monkey species that is known to use tools in the wild) to retrieve food rewards by breaking open stationary boxes with stones. When they moved the stones 15 meters away, only one out of their 13 subjects solved the problem by transporting a stone to the box. The authors speculate that fear of losing the food to conspecifics might have prevented the behavior emerging in more subjects, however, since the one capuchin only succeeded after the others had lost interest and wandered off (p. 373).

In yet another innovative experiment with scrub jays, Nicola Clayton and colleagues recently demonstrated that they can plan similarly far ahead. In what they call their "'planning for breakfast' experiment," they used cages consisting of a row of three separable compartments. The end compartments each had caching trays, and the middle compartment had a food bowl. Leading up to the experiment, the birds were subjected to the following routine. In the morning, they were confined to one of the two end compartments for an hour, during which time they were either fed (the "breakfast" compartment) or not fed (the "no-breakfast" compartment). In the evenings, they were given access to all three compartments for 90 minutes, during which time powdered pine nuts, which they could eat but not cache, were available in the food bowl in the middle compartment. After six days of this routine (during which they experienced three mornings in the "breakfast" compartment and three in the "no-breakfast" compartment), they were tested by presenting them with whole pine nuts rather than powdered pine nuts during the evening session. This gave them, for the first time, an opportunity to cache food in the end compartments where they would be confined the next morning. "The birds anticipated their hunger the next morning by storing significantly more pine nuts in the caching tray in the "no-breakfast" compartment (16.3 ± 1.8; mean ± s.e.m.) than in the "breakfast" compartment (5.4 ± 1.8; mean ± s.e.m.) (paired t-test with 7 degrees of freedom, $t_7 = 3.01$, $P = 0.02$)." To rule out "the possibility that the differential caching was due to a propensity to cache in places associated with hunger" they performed a variation on this experiment that they called "the 'breakfast choice' experiment" (Raby et al. 2007, p. 919). In this variation, the birds were fed in each morning compartment, but they were fed different foods in each compartment (kibbles in one, peanuts in the other), and the experimental condition presented them with one of these two kinds of food to cache the night before. In this situation, the birds cached significantly more peanuts in the "'kibbles-for-breakfast' compartment" and significantly more kibbles in the "'peanuts-for-breakfast' compartment" (p. 920).

§8.4(G). SUMMARY

There is good evidence that both monkeys and apes use at least some aspects of ToM (understanding, at a minimum, what others can and can't see), and there is some anecdotal evidence of apes and dolphins employing higher order representations involving both their own and others' mental states. Anecdotal field reports also suggest ToM use in both elephants and dolphins, and in the one controlled experiment performed on them to date, elephants performed similarly to great apes. The one innovative study performed on them suggests that scrub jays use ToM. And, relatedly, enculturated dolphins and apes both are able to understand the pointing gesture.

Regarding the other indicator of forward-looking autonoetic consciousness considered in this section, planning for the non-immediate future, both bonobos and orangutans can successfully select and "save" a tool for use the next morning, and scrub jays have been shown to selectively cache food in places where they have learned that they won't be fed the next day.

§8.5. Candidates for Near-Personhood

In the preceding chapter, I described two arguments in favor of what I call the autonoetic consciousness paradigm. This is the view that individuals with a robust, conscious sense of their own past and future deserve some kind of special respect vis-à-vis the merely sentient. The first of those arguments turns on individuals having the abilities to consciously recall the past and consciously anticipate the future, the second on their having desires for the non-immediate future. From a sentientist perspective, having either of these capacities gives an individual's life special moral significance vis-à-vis the life of an individual that "lives entirely in the present."

Although it is obvious that persons (as I have defined them) have these two abilities, the empirical evidence reviewed in this chapter provides some good evidence that at least some non-persons have the first ability, and there is at least some evidence that a few non-persons may also have the second ability.

§8.5(A). The Usual Suspects

The great apes, cetaceans, and elephants figure prominently in what I sometimes call the "Discovery Channel mentality" of the contemporary United States, which holds them up as examples of animals deserving special concern and respect. Although popular culture is an unerring guide to neither values nor facts, the evidence presented in this chapter suggests that common sense may very well have it right in this case.

Although research on episodic memory in non-humans is in its infancy, innovative experimental designs like Clayton and colleagues' caching study and variations on the "element of surprise" approach can pose problems that animals cannot solve without having time-stamped, "episodic-like" memory; and since great apes, cetaceans, and elephants are all mammals, PET scan imaging could, at least in principle, be used to strengthen the argument by analogy for episodic memory in these "usual suspects" by showing that there is similar lateralization of function in the prefrontal cortex while solving such problems.

A number of chimpanzees, one orangutan, one gorilla, one elephant, and some dolphins have all passed the Gallup spot test of mirror self-recognition, and bonobos and other elephants have exhibited self-directed behaviors after becoming accustomed to mirrors. So there are good reasons to believe that the usual suspects all have autonoetic consciousness in the present.

Finally, on the forward-looking element of autonoetic consciousness, the most striking examples of using ToM (anecdotal evidence that they use third- and fourth-order representations) occur in the great apes and dolphins; there is some evidence of ToM use in elephants; and both chimpanzees and orangutans have showed some ability to consciously plan for the non-immediate future, as evinced by saving or transporting tools for later use.

So it is not surprising that a variety of authors have argued for treating great apes as persons (Cavalieri & Singer 1993), "borderline persons" (DeGrazia 1996, 2006), or "quasi-persons" (Tooley 1983; Miller 1994); and others have made similar claims about dolphins and elephants (Herzing & White 1998; Poole 1998).

§8.5(B). SOME OTHER CONTENDERS

As research relevant to the three elements of autonoetic consciousness continues, however, some other—sometimes surprising—contenders are emerging.

Just as relative development of the neocortex is roughly correlated with overall intelligence in mammals, relative development of the hyperstriatum is roughly correlated with overall intelligence in birds (Stettner & Matyniak 1980). Parrots and corvids, which are generally recognized as the most intelligent types of birds, both have a highly developed hyperstriatum. The corvid family consists of the ravens, crows, magpies, jays, and nutcrackers. Scrub jays, in particular, have figured prominently in the research reviewed in this chapter, starring in Clayton and colleagues' innovative research on episodic-like memory, ToM, and planning for the future. Although Irene Pepperberg's African grey parrots have performed as well as the great apes in language studies, no research relevant to assessing them for autonoetic consciousness of the past and future is available, and the one published study of mirror self-recognition in them did not clearly establish it in parrots.

Studies of ToM use have produced positive results in both Old and New World monkeys, although none has exhibited self-directed behaviors in front of mirrors, and there are no data on episodic memory in these species.

Finally, Eichenbaum and colleagues' innovative analysis of error rates on odor list recognition problems suggests that rats have episodic memories, although there are no data on mirror self-recognition or ToM in rodents.

§8.5(C). PRELIMINARY CONCLUSIONS

Working with the currently available evidence, then, the case for near-personhood is strongest in the case of the usual suspects: great apes, cetaceans, and elephants. The evidence presently available also makes a good case for scrub jays. In the case of Old and New World monkeys, rats, and parrots, there is in each case at least some evidence of one of the three elements of autonoetic consciousness. One day we may have strong evidence for episodic memory and some kind of planning for the future in a much wider range of mammals and birds. As with the question of which animals can feel pain (chapter 4), any conclusions that we draw at this time are preliminary and are subject to disproof by future findings. As Tom Regan likes to say, "Wherever you draw the line, draw it in pencil." However, what I called "the Rumsfeld response" (§5.4[A]) applies here too: *you have to make ethical decisions with the evidence you have, not the evidence you might want or wish to have.*

In light of the evidence reviewed in this chapter, the two arguments discussed in the preceding chapter support some kind of special moral significance for the lives of at least the usual suspects: great apes, cetaceans, and elephants; plus one surprising contender: the corvids. For in each of these four cases, we have good evidence that they have capacities that are at least analogous to persons' profoundly long-term and complicated aspirations for the future and to persons' detailed, conscious memories of the past. Their more limited sense of their past allows them to consciously remember pleasant and unpleasant events, even if they do not think of these events as episodes in a story line in the way that persons do. And their more limited sense of the future allows them to form more complicated, longer-term desires than merely sentient animals are capable of, even if they do not think of their futures in terms of a story line in the way that persons do. Thus these near-persons' lives have special moral significance vis-à-vis those of the merely sentient.

Episodic memory and theory of mind are both clearly adaptive cognitive capacities, so as noted in §8.1, convergent evolution may have produced autonoetic consciousness of the past and future in widely separated species, or it could be that versions of them are present in all mammals or even in all mammals and all birds. If that turns out to be the case, it could upset the commonsense picture that places special significance on the lives of "the usual suspects." On the other hand, it may turn out, as noted in §1.7, that no sentient individual literally "lives entirely in the present," and that the difference between "the usual suspects" and the individuals that I am assuming are "merely sentient" is a matter of degree. In that case, we should then be guided by the principle that a continuum of autonoetic consciousness grounds a continuum of moral significance (cf. Poole 1998, pp. 107–08).

§8.6. "Near-Persons" versus "Borderline" and "Quasi-Persons"

Finally, a few words about my choice of the expression "near-person." David DeGrazia (2006) uses the term "borderline persons" to refer to "post-infancy great apes and dolphins" that have not achieved the linguistic feats of Kanzi, Koko, and some language-trained dolphins. He argues that:

> On the whole, great apes and dolphins are fairly well endowed with personhood-relevant properties. Yet, with a few exceptions [meaning Kanzi et al.], they are not so well endowed with these traits to qualify clearly as persons. Normal human children, by comparison, are robustly competent in language, clearly capable of introspective awareness—having knowledge of their own feelings, desires, and beliefs—and more likely to show signs of autonomy. My suggestion, then, is that normal, post-infancy great apes and dolphins are *borderline persons*. Given the vagueness of the concept of personhood, that is, there is no definite yes-or-no answer to the question of whether they are persons. (1997, p. 46)

DeGrazia thus arrives at something like the tripartite distinction that I draw among persons, near-persons, and the merely sentient.

I prefer the term "near-persons," however, when talking about animals for which there is good evidence that they have autonoetic consciousness of the past and future. For these animals are not on the periphery of the paradigm space of "person" as I define it—they are entirely outside it. As we have seen in this chapter, what is "borderline" is the evidence for autonoetic consciousness in some non-human animals. But even if the evidence were overwhelming, so that we were certain that they had autonoetic consciousness, these animals would still not be "borderline" persons under my working hypothesis that having a biographical sense of self requires complex language abilities (§6.4). For we have no good evidence that any non-human animal (including Kanzi et al.) uses language in such sophisticated ways (§6.5).

For similar reasons, I prefer "near-person" to the term "quasi-person," which Michael Tooley uses: "In discussing moderate positions on abortion . . . I mentioned the possibility of a person-making property that admits of degrees, and that, moreover, is morally significant even when present to a lesser extent than is required if something is to be a person. Let us say that an entity is a *quasi-person* if and only if there is such a property, which that entity possesses, but not to the extent required to make it a person" (Tooley 1983, pp. 407–408). On my stipulative definition of "person," there are no known examples of non-human animals that are quasi-persons in Tooley's sense. The evidence reviewed in this chapter might ground "quasi-*near-personhood*" for some of "the contenders" discussed in the preceding section, because there is some evidence of autonoetic consciousness in these animals, while we have none in the case of any others (besides the usual suspects). As argued in §6.5, however, we have no good evidence that any non-human animal meets the necessary condition for having a biographical sense of self assumed in §6.4.

Also, note that while "quasi" can mean "nearly or almost," it can also have the connotation of "in appearance only" or "seemingly but not really." I prefer not to call the usual suspects "quasi-persons," because autonoetic consciousness really does give their lives special moral significance vis-à-vis the lives of merely sentient animals. The arguments in chapter 7 show that they really can benefit (and suffer harm) in ways that the merely sentient cannot, not that it just "seems" so. And since autonoetic consciousness is a necessary component of a person's autobiographical consciousness, it brings any animals that have it genuinely closer to being persons.

Under my working hypothesis about language and personhood, there is only one case in which I think it appropriate to refer to individuals as "borderline" or "quasi-persons," and that is young children who have not yet developed a biographical sense of self. These individuals will (if they progress normally) become persons, but nothing in the animal language studies discussed in chapter 5 suggests this about any non-human animal.

{ PART III }

Formulating ILS Rules for Persons, Near-Persons, and the Merely Sentient

{9}

Replaceability, Population, and Marginal Cases

§9.1. Revising ILS Rules

In this final part of the book, I discuss a range of considerations that arise when we think about how to formulate ILS rules for persons, near-persons, and the merely sentient. I might speak of *revising* ILS rules instead of *formulating* them, because no generation formulates a complete system of intuitive-level rules from scratch. From a Harean perspective, all normal individuals, as they are maturing, internalize some complex set of ILS rules that more or less reflect the common morality of their cultures, and much of that process is beyond the individual's conscious control. As adults, individuals may then consciously work on internalizing ILS rules of a personal morality that distinguishes them from their background culture, and they may consciously work on revising the common morality and the laws of their culture. But no one really formulates a system of ILS rules from scratch—as Mill put it: "all rational creatures go out upon the sea of life with their minds made up on the common questions of right and wrong" ([1861] 1957, p. 31).

As was emphasized in chapter 3, a complete system of intuitive-level rules would include:

1. laws,
2. codes of professional ethics,
3. rules describing a shared "common morality" of one's society or culture, and
4. the rules of each individual's "personal morality."

The first two types of rules are typically written down or otherwise explicitly codified, and debate about revising them may be quite formalized: legislatures and professional societies may adhere strictly to parliamentary procedure when revising laws and codes of professional ethics. At the same time, much more subtle processes affect the lead-up to these formal proceedings. Members of a professional society may talk informally for years about the need for a revision before their code of ethics is taken up at a national meeting, and legislators' votes are often determined more by informal discussions with their constituents (and lobbyists) than the arguments presented during the formal debate. Changes in both laws and codes of professional ethics do generally come down to a formal vote on an explicitly stated revision, however.

When it comes to revisions in common and personal moralities, in contrast, the process is almost never as explicit and formal as a parliamentary debate. For one thing, the rules of common morality may be strictly uncodifiable, and those who have properly "internalized" them may still not be able to state them explicitly (§3.3[D]). This will be true if these rules are stored in something like the "prototypical cases" that capture the "statistically defined central tendencies" of the cases used in moral education (§3.4). So much more subtle forces than formal debate, culminating in a vote on a proposed rewording, will be at work as the common morality of a culture changes over time. In the next chapter (§10.5), I will illustrate one way in which this is so.

There I will also emphasize how Harean thinking does not provide a once-and-for-all answer to the question: "What counts as humane, sustainable agriculture?" From a Harean perspective, a set of ILS rules is good to the extent that it approximates the set that an archangel would prescribe for the target population (§3.2). A Harean archangel would choose and amend ILS rules using explicitly utilitarian critical thinking. As a result, as was emphasized in §3.3(A-B):

1. The best ILS rules for a society will vary with background ecological, technological, economic, and other conditions; and
2. There is a conservative bias or inertia, because various transition costs must be taken into account, including both economic costs and the existence of background consumer preferences and cultural traditions.

At the same time, however, as was emphasized in §3.3(C):

3. Critical thinking can call for radical change in the long haul, with the exercise of describing what seem (in the present) like utopian proposals helping to change background consumer preferences and cultural traditions.

And, as emphasized in §§4.5 and 4.7–4.11:

4. When it seems clear that it will maximize aggregate happiness and one can trust one's judgment that this is so, sound critical thinking can lead to violations of existing laws and the rules of common morality and professional ethics.

For these four reasons, a Harean answer to the specific question posed in the next chapter—What counts as humane, sustainable agriculture?—will be complex and highly contextualized. The same will be true of other questions in animal ethics, as will be emphasized in the sequel (Varner forthcoming).

The remainder of this chapter addresses some very general questions that arise when formulating or revising ILS rules governing persons, near-persons, and the merely sentient. In §9.3 through §9.6, I discuss the notion of "replaceability" and some related questions about population policy. Animals are sometimes said to be "replaceable" in a way that persons are not; the concept is used to help describe the special moral value that persons' lives have. The ways that utilitarians like Singer and Hare have framed the question of replaceability in turn raise issues about humans' obligations to reproduce themselves and about what the optimal populations of animals would be.

The final section of this chapter returns to the problem of marginal cases, providing a more detailed Harean response than was initially offered in §7.6. First, however, the next section discusses a distinction that can be drawn between two categories of ILS norms, a distinction that Hare only implicitly recognized.

§9.2. ILS Principles versus ILS Rules

As Marcus Singer (who is no relation to Peter Singer) draws the distinction, moral *principles* "hold in all circumstances and do not admit of exceptions," they "are always relevant whereas rules are not"; while moral *rules* vary across circumstances and can conflict with one another, moral principles "do not vary with changes in circumstances or conditions"; and "it is impossible for moral principles to conflict" (1961, pp. 327–328). So conceived, there are a limited number of norms that might qualify as ILS principles rather than ILS rules. In Harean terms, I suggest: an ILS norm is a *principle* if a Harean archangel could prescribe it across all cultures, and neither critical thinking nor behavior prescribed by good ILS rules would normally violate it.

I say "could" prescribe such principles across all cultures, rather than "would," because as we will see, such principles are formal, reflecting basic commitments of Harean critical thinking, rather than giving substantive guidance as ILS rules do. So the value of including them in a set of ILS rules is mainly a matter of general moral education; they remind us of the fundamental commitments of critical thinking. Relatedly, I say that such principles would not "normally" be violated by behavior prescribed by good ILS rules, because critical thinking does contradict those rules in non-normal situations, which such rules are not designed to cover.

Hare made no distinction between principles and rules, but in chapter 9 of *Moral Thinking*, he discussed "formal justice" and "the right to equal concern and respect" in a way that suggests that they are candidates for ILS norms that would qualify as principles rather than rules. According to Hare, each of these just restates the logical requirement that moral judgments are universalizable:

"We are being formally unjust if we make different moral judgements about identical cases, and act on them" (p. 157); and "the right to equal concern and respect" is "nothing but a restatement of the requirement that moral [norms] be universalizable" (p. 154). (Hare uses the word "principles" here, but in the present context, "norms" is the better word, since it covers both rules and principles.)

On the same pages, Hare states that all substantive conceptions of justice and all substantive rights claims are confined to the intuitive level. As described in the foregoing paragraph, "the right to equal concern and respect" and "formal justice" are purely formal in that by themselves, they do not tell us what to do. As expressions of the universalizability requirement, they place constraints on the judgments that we can consistently make, but like the principle of utility itself, they can lead us to particular conclusions only when combined with certain facts (p. 157). So these two norms would seem to be, like the principle of utility itself, expressions of what it means to do critical-level thinking. Nevertheless, Hare says that the formal principle of justice (and, because they are both just expressions of the universalizability requirement, also the right to equal concern and respect) "applies at all levels," referring to all levels of moral thinking (p. 158).

So Hare appears to be saying that a good set of ILS rules will include a norm of formal justice, or a right to have similar cases treated similarly. But since this is a logical requirement of all moral judgments, such an ILS norm would not be overridable, as intuitive-level *rules* are.[1] So here we seem to have Hare advocating inclusion of a *principle* among our ILS norms.

Something like "Everyone has the right to have similar cases treated similarly" *does* seem to be taught as part of the common morality of most cultures. In fact, the golden rule seems to be used in most or all cultures to do just this. As noted in §4.2, "How would you feel if Johnny did that to you?" is an easy way to get a child to understand and acknowledge the logical requirement of universalizability, and asking questions like "What if you didn't like rough and tumble play, the way Harold here doesn't like rough and tumble play, how would you then feel about this situation?" is an easy way to improve on a child's moral thinking from a utilitarian point of view.

---

[1] Hare initially says that all moral judgments have all three logical properties of critical-level thinking: universalizability, overridingness, and (in his special sense) prescriptivity (*MT*, p. 24). Since ILS rules are overridable, however, he later qualifies that by saying that a judgment is moral if it either: (1) has all three of those logical properties, including overridingness; or (2) it is treated as overridable in certain circumstances, but its selection for inclusion among one's ILS rules is justified "on the basis of critical thinking, in that this was a principle whose general acceptance would lead to people's actions and dispositions approximating to the greatest extent to the deliverances of a perfectly conducted critical thinking" (p. 61). So ILS rules are universalizable and prescriptive, but overrideable. An ILS *principle*, however, would not be overridable.

So we have at least three candidates for principles—in Marcus Singer's sense—to be included among one's ILS norms. The two suggested by Hare are equivalent, since they both express the need to conform to the logical requirement of universalizability. And the third is, according to Hare, a common-sense expression of universalizability plus the other two logical requirements (overridingness and, in Hare's special sense, prescriptivity). Having such principles among one's ILS norms does not, by itself, give any substantive guidance, but it is useful to the extent that it reminds us of the logical requirement of universalizability and helps us teach children about this requirement.

Another candidate for an ILS norm that would qualify as a principle would concern the taking of sentient life. As a first approximation, consider: "Don't kill sentient animals." A perfectly pain- and stress-free death still removes one sentient being from the world and, if that individual's continued life would have been pleasant, removing it diminishes aggregate happiness. Just as obviously, however, killing sentient beings can be justified from a utilitarian perspective in a broad range of cases: if the individual's continued life would be miserable, for instance; if killing that individual will, under the circumstances, allow us to preserve the lives of some aggregate of other sentient individuals; and, given that persons' lives have greater moral significance than those of the merely sentient, presumably persons are morally justified in killing merely sentient animals for food, if this is the only way they can survive and flourish under the circumstances.

Probably the best we can do, as far as formulating the relevant ILS principle goes, is this: *Don't kill sentient animals unnecessarily*. As Susan Finsen (1990) notes, by "necessary" scientists defending research on animals just mean "required to achieve the intended result," whereas the context demands that we understand it to mean "both required to achieve the intended result *and morally justifiable*." Here it is meant in the latter sense. What is necessary in this sense will, from a Harean perspective, change with changing circumstances, so a Harean archangel could prescribe this principle across all cultures, while prescribing very different moral rules governing slaughter and other practices (as illustrated in chapter 10).

This principle, which embodies a general attitude of respect for sentient life, is a good one for every culture to include among its ILS norms. Like a right to formal justice or the golden rule, by itself this principle yields no substantive conclusions. Including it among our ILS norms is nevertheless useful in the same way as including the golden rule. Just as the golden rule reminds us of the logical requirement of universalizability and helps us teach about it, the principle, "Don't kill sentient animals unnecessarily" reminds us that all cultures should be concerned about how they treat non-human but sentient animals, and it can help us teach children about the importance of this.

Unlike the golden rule, I can't say that this principle seems to be widely articulated in the common moralities of many societies. It may be in the background more often than at first appears, however. For as Jean Kazez notes in *Animalkind*, indigenous cultures have commonly embraced myths about hunting, such as that animals voluntarily give their lives to respectful hunters, or that they don't "really" die, which ensures an unending supply of meat. Kazez finds "no more plausible" the modern idea that animals "chose" domestication. Both ideas, she suggests, are salves for consciences uneasy about humans' relationships with animals. Ancient and modern civilizations have all realized, she says, that "Killing an animal is not like pulling a carrot out of the ground" (2010, pp. 16, 18).

§9.3. The Replaceability Argument for Slaughter-based Agriculture

Certainly humans have faced circumstances in which killing animals for food was "a necessity," for instance among the pre-contact Inuit, and perhaps among all hunter-gatherers, where meat is an integral component of the diet. Whether or not humans living in affluent, contemporary societies need meat or animal by-products in their diets in order to thrive is a complex question. Some nutritionists doubt that a vegan diet, which excludes all animal by-products like milk and eggs, can be nutritionally adequate, at least for people with high metabolic needs like growing children and pregnant and lactating women, and for people with various food allergies. Other nutritionists believe that a vegan diet can be perfectly healthy for these individuals, and some have even claimed that heavy reliance on dairy products actually increases the incidence of osteoporosis among the elderly. (For an overview of where these issues stood in the mid-1990s, see the special edition of the *Journal of Agricultural and Environmental Ethics,* which contains relevant articles by several nutritionists as well as several philosophers; Comstock 1994).

Given the relative moral significance attributed to the lives of persons and the merely sentient in the autonoetic consciousness paradigm (§7.1), presumably human persons would be justified in killing merely sentient animals for food if that were necessary for survival—even Tom Regan's "the rights view" permits that (1983, pp. 301–304, 324). The more interesting question is: *Where it is not necessary to preserve human life or health, can slaughter-based agriculture nevertheless be morally justified?*

Hare answered this question in the affirmative in the only paper he ever published that directly addressed issues in animal ethics. In "Why I am only a Demi-Vegetarian" (*EB*, chapter 15, originally published in 1993, 12 years after *Moral Thinking*), he began by distinguishing two kinds of arguments against eating meat: "The first concerns the wrongness of *killing* animals, regardless of whether this involves suffering or not. The second . . . concerns the wrongness of *causing suffering* to animals, whether or not one kills them" (p. 225). Focusing first on the question of killing, Hare then argued as follows:

> For utilitarians like Singer and myself, . . . [the question is] whether the entire process of raising animals and then killing them to eat causes them more harm overall than benefit. My answer is that, assuming, as we must assume if we are to keep the "killing" argument distinct from the "suffering" argument, that they are happy while they live, it does not. For it is better for an animal to have a happy life, even if it is a short one, than no life at all. . . . [W]hat we ought to be doing is to maximize the amount of quality-adjusted life-years or QALYs of sentient beings. And I do not believe that we should be doing this if we refrained from eating animals. The reason is that if we gave up eating animals the market for meat would vanish, and no more animals would be raised for meat-production. Such animals as there were would be either wild, or kept as pets, or kept for other economic purposes such as milk, eggs, and wool production. And it is certain that there would be very many fewer domestic animals than at present live. This thought gives me pause when I walk the fields around my home in England, and see a great many apparently happy animals, all destined to be eventually eaten. As it happens, they are mainly pigs, who would certainly not be kept except for the bacon market. (pp. 226, 227–228)

The nineteenth-century British philosopher Henry Salt sarcastically referred to this as "the logic of the larder," because, he said, "it implies that the real lover of animals is he whose larder is fullest of them" ([1914] 1976, p. 188).

Contemporary philosophers refer to the underlying reasoning as "the replaceability argument," which can be summarized as follows:

1. We are obligated to maximize aggregate happiness.
2. Animals' sufferings and enjoyments contribute to aggregate happiness.
3. But when an animal that is leading a pleasant life is slaughtered humanely and replaced with another that leads an equally pleasant life, there is no loss in aggregate happiness.
4. At the same time, human happiness is increased in various ways by the practice of humane slaughter.

C. So the practice of slaughtering animals for food can be justified as long as the animals live pleasant lives and are slaughtered humanely.

Obviously, a lot is built into that qualification about "pleasant lives" and "humane slaughter." How we are to conceive of and evaluate animal welfare in production agriculture and standards for humane slaughter will be discussed in the following chapter. Here I am focusing, as Hare does in the passage quoted above, on the argument that slaughter-based agriculture is wrong, not because it involves unnecessary suffering, but because it involves unnecessary killing. The replaceability argument turns on the assumption that the value lost to the world by the death of an individual can be fully compensated for by the birth of another, similarly happy individual.[2]

§9.4. Singer on Replaceability

The replaceability argument suggests that utilitarian reasoning can indeed be used to justify the practice of humane, slaughter-based agriculture, even if humans do not need to eat meat to survive or to be healthy. Nevertheless, in the first edition of *Animal Liberation*, Peter Singer rejected the replaceability argument, describing it as "somewhat bizarre," and as having "a ludicrous air about it" (1975, pp. 240–241). There (pp. 259–260, note 16) he claimed that Henry Salt had identified a fatal flaw of the argument in his 1892 book, *Animals' Rights*. Salt claimed that the argument commits "the gross absurdity of talking of non-existence as a state which is good, or bad, or in any way comparable to existence" ([1892] 1980, pp. 62–63), and Singer agreed:

> Its flaw lies in the implication that we confer a favor on a being by bringing it into existence. . . .[3] A little thought will reveal the error in this reasoning. On whom do we confer the favor of existence? On the nonexistent animal, unborn and unconceived? But this is absurd. There are no such entities as nonexistent beings, waiting around in limbo for someone to bring them into existence. Once a being exists, we have an obligation to avoid making that being suffer unnecessarily, but we have no obligation to nonexistent beings. The very term "nonexistent being" is self-contradictory. Therefore we can neither benefit nor harm a nonexistent being. . . . Moreover if to bring a being into existence is to benefit it, then presumably to decide not to bring a being into existence is to harm it. But there is no "it" to be harmed by this decision. In this area it is easy to talk nonsense without realizing it. (1975, p. 241)

Thus, in the first edition of *Animal Liberation*, Singer sided with Henry Salt and said that the replaceability argument is fatally flawed because it makes no sense to compare nonexistence to existence and to say that a nonexistent being

[2] In his reply to Hare's "Why I am only a Demi-Vegetarian" essay, Singer argues that the logic of the larder "is, at best, not proven," because the wild animals that would live on land currently occupied by cattle might "experience more pleasure than the cattle they replace" (1999, pp. 327–328). As emphasized in §10.6, however, pastured cattle coexist with a variety of "field animals," and replacing cattle with row crops might result in more killing (and associated suffering).

[3] The language omitted here is as follows: "and that we thereby obtain a right to treat the being with less than equal consideration." But the proponent of the replaceability argument need not make this additional—clearly unsound—claim.

can benefit from being brought into existence or be harmed by not being brought into existence.

In the second edition of *Animal Liberation*, however, Singer repudiated Salt's criticism of the replaceability argument. There he wrote:

> In the first edition of this book, I rejected [the replaceability argument] on the grounds that it requires us to think that bringing a being into existence confers a benefit on that being—and to hold this, we must believe that it is possible to benefit a nonexistent being. This, I thought, was nonsense. But now I am not so sure. (My unequivocal rejection of this view is, in fact, the only philosophical point made in the earlier edition on which I have changed my mind.) After all, most of us would agree that it would be wrong to bring a child into the world if we knew, before the child was conceived, that it would have a genetic defect that would make its life brief and miserable. To conceive such a child is to cause it harm. So can we really deny that to bring into the world a being who will have a pleasant life is to confer on that being a benefit? To deny this, we would need to explain why the two cases are different, and I cannot find a satisfactory way of doing that. (1990, p. 228)

Then, when the second edition of *Practical Ethics* appeared three years later, Singer allowed that "it is at least arguable that no wrong is done if the animal killed will, as a result of being killed, be replaced by another animal living an equally pleasant life. Taking this view involves holding that a wrong done to an existing being can be made up for by a benefit conferred on an as yet nonexistent being." And he added that, assuming that the animals in question are "not self-conscious," "the replaceability argument appears to justify" the practice of humane, slaughter-based agriculture (1993, pp. 132–133).

In assuming that the animals are not self-conscious, Singer is assuming that they do not meet his stipulative definition of "persons," and beginning with the first edition of *Practical Ethics*, Singer has tried to account for the special moral significance of persons' lives in several different ways. Each of these has involved arguing that, while it might be appropriate to treat merely sentient animals as replaceable,[4] it is not appropriate to treat persons as replaceable.

In the remainder of this section, I describe Singer's attempts to block the replaceability argument's application to persons and how each of them fails, but first I should briefly note how Singer's conception of personhood compares and contrasts to that adopted in chapter 6 of this book.

---

[4] I say "it *might* be appropriate to treat merely sentient animals as replaceable," because Singer immediately qualified his endorsement of the replaceability argument. He said "it is important to realise [*sic*] how limited it is in its application," arguing that "It cannot justify factory farming, where animals do not have pleasant lives" and that "at the level of practical moral principles [ILS rules?], it would be better to reject altogether the killing of animals for food, unless one must do so to survive. Killing animals for food makes us think of them as objects that we can use as we please" (pp. 133–134).

§9.4(a). Singer on Personhood

In *Practical Ethics*, Singer initially defines "persons" as:

1. "[R]ational and self-conscious" beings (1979, p. 76 [= 1993, p. 87]),

but he goes on (especially in the second edition) to characterize persons in a variety of ways, saying that:

2. Persons are "highly future-oriented in their preferences," in contrast to merely conscious beings who have "no more than a preference for the cessation of a state of affairs that is perceived as painful or frightening" (1993, p. 95);
3. A person is "aware of himself as a distinct entity, existing over time" (1979, p. 96 [= 1993, p. 117]);
4. A person can "aspire to longer life" (1979, p. 102 [= 1993, p. 125]);
5. Persons have "the conception of themselves as living beings with a future" (1979, p. 102 [= 1993, p. 125]);
6. Persons "see [themselves] as existing over time" (1993, p. 125);
7. A person has "a life that is biographical, and not merely biological" (1993, p. 126);
8. "In contrast [to persons], beings who are conscious, but not self-conscious, more nearly approximate the picture of receptacles for experiences of pleasure and pain, because their preferences will be of a more immediate sort. They will not have desires that project their images of their own existence into the future. Their conscious states are not internally linked over time" (1993, p. 126); and
9. Persons have "forward-looking desires that extend beyond periods of sleep or temporary unconsciousness, for example a desire to complete my studies, a desire to have children, or simply a desire to go on living, in addition to desires for immediate satisfaction or pleasure, or to get out of painful or distressing situations" (1993, p. 127).

Singer makes no distinction between persons and what I call near-persons, and from the above characterizations it seems clear that, while the paradigm case of a person in his sense would be a normal adult human being with a biographical sense of self, Singer would also categorize as persons a number of animals that I would classify as only near-persons.

§9.4(b). "Non-Replaceability" and "The Prior Existence View"

In the first edition of *Practical Ethics*, Singer rather tentatively suggested that the non-replaceability of persons could be ensured by applying what he called "the prior existence view" to persons, while endorsing "the total view" for merely sentient individuals (1979, p. 103). According to "the prior existence view," one's duty is to maximize the happiness of individuals that already exist, or will come to exist in the future no matter what you decide.[5] On "the total view," by contrast, one ought to maximize happiness, *simpliciter*. Applying the total view to non-persons would, he said, make the replaceability argument for humane, slaughter-based agriculture apply to them (p. 100), but by sticking to the prior-existence view regarding persons, "we shall hold that it is wrong to kill any [person] whose life is likely to contain, or can be brought to contain, more pleasure than pain" (p. 99). That is, if we adopt the prior existence view when doing utilitarian thinking about persons, then we won't be able to justify killing an already existing person by showing that this will cause another person, who would not otherwise have existed, to come into existence.

In the second edition of *Practical Ethics*, Singer abandoned this approach to treating persons as non-replaceable, citing a thought experiment due to Derrick Parfit (1976, pp. 373–374). Compare these two cases:

1. A woman is three months pregnant. Her fetus has a developmental defect that will significantly diminish its quality of life, but clearly without making that life less than worth living. A pill is readily available which, if taken every day for the next three weeks, will cure this defect, without any side effects on either her or the fetus.
2. A woman who has been taking birth control pills for years plans to stop them and become pregnant. If she does not take pills for the next three-week cycle to prevent conception, then she will become pregnant with a fetus having a developmental defect precisely as serious as the fetus in case one, but a defect for which there is no possible treatment. However, if she takes the three week cycle of pills and then stops, she will become pregnant next month and the resulting fetus will be normal.

If neither woman proceeds with the three-week cycle of pills, two equally disabled persons will result, but, Singer argues, only on the total view has the second woman done anything wrong. For on the prior existence view, only the person resulting from the first case can later complain that she or he has been harmed, since the person who would result from the second case could only have come into existence with the defect. In contrast, on the total view we can say that it is equally wrong for either of the women not to take the three-week

[5] Singer explains, in a bibliographic note, that he substitutes "prior existence" for Parfit's term "person-affecting," because "the view has no special reference to persons, as distinct from other sentient creatures" (1979, p. 225; 1993, p. 365). But that is an odd thing to say when stipulating, as Singer was suggesting in the first edition of *Practical Ethics*, that the prior existence view should be applied only to persons.

cycle of pills, because the second woman could have maximized utility by bringing a different possible person into existence (Singer 1993, pp. 123–125).

Thus, Singer argued, applying the prior existence view to our thinking about persons would lead to some counterintuitive conclusions. More fundamentally, I think it would be *incoherent* to endorse the total view with regard to the merely sentient while rejecting it with regard to persons, because this would implicitly devalue personhood. For if the total view applies to the merely sentient while the prior existence view applies to persons, then all possible merely sentient beings can count in the ledger against the continued life of a happy person, while possible persons cannot. This means that while the killing of a happy person could never be justified in light of its bringing into existence an aggregate of persons all of whom lead happy lives, the killing of a happy person *could* be justified in light of its bringing into existence an aggregate of merely sentient individuals all of which lead pleasant lives. Thus, this account of replaceability leads to the conclusion that persons are replaceable with respect to the merely sentient but not with respect to other persons, and this is at odds with the whole use of "person" as a label for individuals whose lives have greater moral significance than those of the merely sentient.

§9.4(c). "Non-Replaceability" and Long-Term Preferences

In his preface to the second edition of *Practical Ethics*, Singer abandoned the foregoing account, announcing that he had

> dropped the suggestion—which I advanced rather tentatively in the fifth chapter of the first edition—that one might try to combine the "total" and "prior existence" versions of utilitarianism, applying the former to sentient beings who are not self-conscious and the latter to those who are [i.e., persons]. I now think that preference utilitarianism draws a sufficiently sharp distinction between these two categories of being to enable us to apply one version of utilitarianism to all sentient beings. (1993, pp. x–xi)

Here Singer seemed to be claiming that persons are not replaceable because they have long-term preferences for the future, including a general preference for continued life, whereas merely sentient animals are "stuck in the present."

It is not clear, however, that persons are not replaceable according to preference utilitarianism. For as H. L. A. Hart noted in a 1980 review of the first edition of *Practical Ethics*:

> although a person's wish to live is a reason for not killing him it is not a conclusive reason. Singer himself states that a person's wish to live might be in conflict with the preferences of other persons and may be outweighed by

> them. So, to take a crude example, if the death of one person who wished to live was necessary to save the lives of ten others who wished to live, Preference Utilitarianism would support the killing of that one person. The fact that in some sense each person's life plans are different from those of others would not affect the calculation of what in such cases would maximize overall preference satisfaction, for presumably in such calculations conflicting preferences would be weighted only by number and intensity, not by their content.
>
> If preferences, even the desire to live, may be outweighed by the preferences of others, why cannot they be replaced by new preferences created to take their place? The fact that preferences may be outweighed shows that for Preference Utilitarianism persons *are* in a sense mere receptacles. (Hart 1980, pp. 29–30)

So what else could a utilitarian mean by the claim that persons are not replaceable?

§9.4(D). "NON-REPLACEABILITY" AND PREFERENCES AS "DEBITS"

In an August 1980 letter to the *New York Review of Books*, Singer floated a response to Hart that he subsequently abandoned. There Singer described unsatisfied preferences as "debits" in "the moral ledger." On this view, a humanely killed person's unsatisfied preferences can never be canceled out by bringing other individuals with similar preferences into existence, since the satisfaction of their preferences only cancels out the debits made by bringing their own preferences into existence.

However, this response to Hart entails a version of negative utilitarianism with very pessimistic implications. Negative utilitarianism is the view that we are obligated to minimize negative utility rather than maximize positive utility. But if the unsatisfied preferences of persons count as negative utilities, then, as Michael Lockwood pointed out to Singer,

> . . . it must invariably be *wrong* to bring new human beings into existence, or to refrain from aborting foetal ones. For every child that emerges into full consciousness will inevitably acquire certain desires or preferences which will turn out to be frustrated—desires, moreover, that would not have existed at all had the child never been conceived, or had its life been extinguished at a sufficiently early stage. (Lockwood 1979, p. 164, italics in original)

This is in line with Schopenhauer's conclusion that "existence is certainly to be regarded as an error or mistake, to return from which is salvation" ([1844] 1958, volume II, p. 605).

§9.4(E). "NON-REPLACEABILITY" AND "LIFE'S UNCERTAIN VOYAGE"

Singer later abandoned the debit view of preferences in light of these problems for negative utilitarianism (Singer 1987a, p. 166). In the same essay, titled "Life's Uncertain Voyage," he used planning for and executing a long, ambitious journey as a metaphor for understanding the special significance of persons' lives, a metaphor that he continued to use in the second edition of *Practical Ethics* and that he claims helps us make sense of the non-replaceability of persons.

To illustrate the metaphor, Singer asks us to consider the life of a man who loves walking in the mountains and learns that a trek through Nepal to the Thyangboche monastery in the Himalayas would be the ultimate mountain trek, albeit a hazardous and, in several ways, a very costly one. He sketches four versions of the story, in which the journey is variously accomplished, undertaken unsuccessfully, or abandoned in light of the sacrifices it would entail. Singer ranks these outcomes as indicated in table 9.1.

In the second edition of *Practical Ethics*, he notes that the uncertain journey metaphor explains his intuitive judgment in Parfit's two pregnancy cases (§9.4[B]). For either woman to knowingly forgo the pills and produce a handicapped child would be wrong, insofar as "they both quite unnecessarily send out voyagers with fewer prospects of making a successful journey than other voyagers whom they might have placed at the starting line" (Singer 1993, p. 131). In the "Uncertain Voyage" essay, he emphasizes how the metaphor also illuminates comparative judgments about the loss of fetal, child, and adult life. Since "the decision to embark on the journey of creating a life" usually requires little effort at first (1987a, p. 160), a lost pregnancy does not seem as serious as

TABLE 9.1. Singer on "Life's Uncertain Voyage"

Versions of the story	Singer's ordinal rankings (1 = best)	Analogously valuable lives
V1 Dreamed of voyage completed in spite of many hardships.	1	Bertrand Russell dies at 97
V2 Voyage ends due to mishap prior to goal and is never later completed.	4	F.P. Ramsey dies at 26
V2' Voyage ends due to mishap prior to goal but is later completed.	[2]	
V3 Voyage is dreamed of but never begun due to disillusionment.	3	[Jimmie R., with amnesia after 19, or a "replaced" dead child]
	3	
V4 Voyage is never dreamed of to begin with.	3	[Aborted fetus]

Source: Based on Singer (1987a), although statements in brackets are inferred from, rather than explicit in Singer's text.

the loss of an older child, and "To die soon after the development of mental continuity" is not as tragic as the death of a young adult (p. 163). Further, the metaphor also illuminates judgments about how tragic death is toward the end of life. He notes that Stanley Benn's death seemed less tragic because he had just completed his capstone work, Bertrand Russell's death at 97 was hardly tragic at all, but the death of promising young philosopher F. P. Ramsey at 26 was like version #2 of the Himalayan trekker story.

Finally, Singer claims that the uncertain voyage metaphor helps us to understand the non-replaceability of persons. He points out that planning for a journey is a large part of the enjoyment:

> Here, as with the pleasure that we get from bringing up children, much of the enjoyment derives from the fact that this is a forward-looking activity, and we are savouring the pleasures of anticipation. Whatever the balance of pleasures and pains of being a parent, much of the positive value we derive comes from the fact that all this activity is goal-directed, and the goal is to rear a child to maturity. Once the children themselves understand that they will grow to be adults, they too will join in this purposive activity, and share in its goal. (p. 158)[6]

It should be obvious why I find this account attractive: it nicely maps onto the arguments I gave, in chapter 7, for thinking that persons' lives have special moral significance. The ability to make complex plans for the very remote future allows our lives to go better or worse for us in the ways that Singer describes using his journey metaphor. And—he should have added—our ability to tell stories about past journeys can also add value (and tragedy) to our lives. (In light of his reliance on this metaphor, I find it puzzling that Singer has not recognized an intermediate category between persons and the merely sentient, as explained in §11.2.)

However, Singer once again draws a much stronger conclusion, and one that is decidedly non-utilitarian. He writes:

[6] Singer adds that, although from "the *inward* perspective" a life journey only begins when a child acquires "the concept of a continuing self [i.e., becomes a person]" (1987a, p. 162), "from the *outward* perspective . . . The life of a being may represent a journey for others—for instance the parents—even if it is not a journey from the internal perspective of that being itself" (pp. 160, 162, 165). In a chapter of the sequel (Varner forthcoming), I will argue that pet ownership is undervalued, and I think this observation by Singer may help to explain (and justify) the special value that pet owners attribute to their animals. I assume that my cat Nanci did not have a biographical sense of self, but *I* can include her life in the narrative that constitutes my biographical sense of self. In this respect, pets differ from similarly cognitively impoverished farm animals that usually live only a short time, and this difference makes it natural (and appropriate) for pet owners to value their pets and grieve their deaths in ways that a farmer typically does not (although farmers sometimes do with those production animals that live fuller lives).

> It seems plausible to hold that . . . the value that is lost by a person's death after years of effort and striving towards a goal, is not going to be replaced by bringing a new person into existence, for then . . . there will be two lots of strivings for only one lot of fulfilment. Even if we do hold that there is some positive value in bringing into the world a being whose preferences will be satisfied, we can plausibly deny that the positive value of bringing a new being into the world is ever going to make up for such a loss. (p. 169)

But surely we cannot, not without abandoning the aggregative, maximizing quality essential to utilitarianism. As we saw earlier, from the fact that creating a single person with a similar, satisfied preference does not compensate for the unsatisfied preference of a humanely killed person, it does not follow that there is no possible number of created persons that could, in the aggregate, justify the killing on utilitarian grounds.

§9.5. Hare on Replaceability

For his part, Hare consistently rejected the Salt-style objection to the replaceability argument. Salt argued that it is impossible to judge that existence is preferable to nonexistence, but in "Why I am only a Demi-Vegetarian," Hare said flatly: "To this we can answer that happy existing people are certainly glad that they exist, and so are presumably comparing their existence with a possible non-existence; so what the objection says is impossible actually happens" (1993, p. 226). Moreover, Hare implicitly agreed with Salt that people who accept the replaceability argument should admit that "Also, the argument must apply to mankind" (Salt [1914] 1976, p. 185). For Hare argued that his account of the logic of moral discourse committed him to the total view and, as Singer pointed out, this commits us to thinking of individuals as replaceable. In this section, I first describe why Hare's analysis commits him to the total view, and I then spell out the parallel conclusions that Hare drew about animal and human population policies.

§9.5(a). How the Total View Is Entailed by the Logic of Moral Discourse

Singer writes as if deep features of utilitarianism—such as whether we ought to seek to maximize aggregate happiness *simpliciter* ("the total view") or to maximize the aggregate happiness of individuals who already exist or will come to exist in the future no matter what we do ("the prior existence view")—are to be determined by appeal to various moral intuitions. Singer initially suggested adopting the prior existence view with regard to humans as a way of capturing the intuition that persons are not replaceable in the way that merely sentient animals are; and he subsequently abandoned it in light of its counterintuitive implications for the two pregnancy cases, due to Parfit.

According to Hare, however, this gets things backward. As emphasized in §§1.4 and 1.5, Hare eschews appeals to moral intuitions and appeals instead to the logic of moral discourse to show that utilitarianism is the ultimate basis for all moral thinking. He then explains the moral intuitions that we have in terms of what ILS rules we have internalized, arguing that these intuitions are good ones to the extent that those ILS rules approximate the ones that an archangel would teach us.

As described in chapter 2, Hare's theory holds that all moral judgments have three logical properties: they are universalizable, overriding, and (in Hare's special sense) prescriptive. Hare's argument for the total view (over the prior existence view, which Singer initially preferred) focuses on the first property, universalizability: "There is no way of framing a properly universal principle which prescribes for actual cases but does not similarly prescribe for non-actual cases which resemble the actual cases in all their universal properties and differ from them only in the roles played in them by particular individuals" (*MT*, p. 114). That is, if "a properly universal" principle must prescribe for all relevantly similar cases, then any properly universal principle applies to all purely hypothetical (possible but not actual) cases that are relevantly similar.[7]

This *does* seem to be a logical feature of moral judgments. For in moral arguments, we *do* freely appeal to hypothetical cases. "I understand the principle that you're endorsing," we commonly say, "but what would you say about a case where . . .," at which point we feel free to sketch a hypothetical case,

---

[7] Hare did emphasize that certain implications of the prior existence view are counterintuitive. For instance, in "Abortion and the Golden Rule" (which was originally published in 1975, six years before *Moral Thinking*), Hare reckoned that

> it would be strange if there were an act whose very performance made it impossible for it to be wrong. But if the objection were correct [that we can't have duties to possible people], the act of aborting a possible person would be such an act; by preventing the existence of the object of the wrongdoing, it would remove its wrongness. This seems too easy a way of avoiding a crime. (*EB*, p. 164)

And again:

> We can do harm to and wrong succeeding generations by using up all the world's resources or by releasing too much radioactive material. But suppose that this not merely made them miserable, but actually stopped them being born (e.g. that the radioactive material made everybody sterile all at once). . . . It seems strange to say that if we behaved only a little worse, so that the next generation was half the size it would have been, we had done badly for that generation, but that if we behaved much worse, so that the succeeding generation was reduced to nil, we had not done badly for it at all. (*EB*, pp. 165–166)

These are appeals to intuition, and they are offered in support of the total view over the prior existence view, but Hare's core reason for endorsing the total view is the appeal to the logic of moral discourse described in the text.

often one that is hopelessly unrealistic and, therefore, strictly hypothetical (§2.3[A] and §4.8). That we commonly employ such wildly hypothetical cases in moral argumentation suggests that we assume, as Hare argues, that the universalizability requirement ensures that moral principles must apply to non-actual cases.

Hare argues that this logical feature of moral judgments entails the total view, so that, as he puts it in an essay on population policy ("Possible People," which was originally published in 1988), "where we have a choice between bringing someone into existence and not doing so, the interests of that possible person have to be considered" (*EB*, p. 67). This led him to reach conclusions about human population policy that are broadly parallel to his conclusions about animals.

§9.5(B). HARE ON HUMAN POPULATION POLICY

Hare argues that his theory implies conclusions about human population policy that many people find strongly counterintuitive, at least in the form in which Hare initially formulates them. The first concerns what Derick Parfit (1986) calls "the repugnant conclusion." As Hare puts it, Parfit's point is that "we can *conceive* of a population vastly increased in size, but with a quality of life for all barely above the level at which life is just worth living," in which the "lower quality of life is compensated for by the increased numbers enjoying it" (*EB*, p. 68, italics in original). Thus utilitarian thinking about human population policy appears to lead to the conclusion that average happiness should be sacrificed for the sake of total happiness, as illustrated in table 9.2. Under the assumptions embodied in that table, greatest average utilitarianism would call for halting population growth at 200, whereas greatest total utilitarianism would call for increasing it all the way to 1,600—this is something like Parfit's "repugnant conclusion." Comparing the results of the 200 and 400 population levels illustrates the counterintuitive implications of greatest average utilitarianism, and comparing the results of the 200 and 1,600 population levels illustrates the counterintuitive implications of greatest total utilitarianism.

TABLE 9.2. Total versus Average Utilitarianism

Population:	100	200	400	800	1,600	3,200
Average happiness of individuals in population:	10	10	9.99	7	4	1
Total happiness (population x average):	1,000	2,000	3,996	5,600	6,400	3,200

Hare italicized the word "conceive" in his restatement of Parfit's repugnant conclusion in order to set up his response:

> I do not wish to dispute this. I shall be arguing, rather, that *in practice* enough disutilities would be created by the transition to this end-state, and by the circumstances that would then obtain, to cancel the balance of advantage. The conclusion which most people find repugnant is one about what we ought now to do in our actual situation . . . that we ought now to take steps to increase the population until the end-state, which I have just agreed is conceivable, is actually realized. And when the question is what we ought to do, the distinction between what is conceivable and what is practicable is of the highest importance. (*EB*, p. 68)

Hare then emphasizes the following two points (these are paraphrases of two sections of his essay):

1. Due to diminishing marginal utility and the human tendency toward envy, the ideal distribution of happiness will be moderately equal. In order to provide for both incentives and philanthropy, however, it will not be absolutely equal. But if distribution is unequal and average utility is near the point at which life becomes not worth living, then there will be some people below the break-even point, and for them it would be rational to either commit suicide or rebel, either of which would reduce total utility (*EB*, pp. 78–81).
2. So before increasing population we ought to approximate the ideal of a moderately egalitarian distribution of happiness and then increase population while holding the distribution constant. This suggests that we may already have reached or at least approached the limit in societies like Britain, the United States, and Australia, which are moderately inegalitarian and have an appreciable number of people whose lives are just barely worth living. If so, then total utilitarianism does not call for such a counterintuitive population policy after all (*EB*, pp. 81–83).

In terms of the numbers in table 9.2, Hare is effectively arguing that the total happiness numbers become inaccurate prior to the population's increasing from 1,600 to 3,200, so that a total utilitarian would have to endorse a conclusion close to that reached by greatest average utilitarianism.

Another initially counterintuitive conclusion that Hare reaches is that we have a prima facie duty to procreate and that, therefore, elective abortion is prima facie wrong.[8] In "Abortion and the Golden Rule" (which was originally

[8] In the same essay, Hare argues that "the potentiality principle," which is often invoked by opponents of abortion and infanticide, follows from the golden rule (*EB*, p. 156), which just expresses Hare's universal prescriptivism (*EB*, pp. 153, 156). "[T]he potentiality principle says that if it would be wrong to kill an adult human because he has a certain property, it is wrong to kill an organism (e.g. a foetus) which will come to have that property . . . if we do not kill it" (*EB*, pp. 154–155).

published in 1975, six years before *Moral Thinking* appeared), Hare argues that abortion of a profoundly disabled fetus can be justified when it is sure to be replaced by a normal one. But, generally, replacement is much less likely than birth, and therefore "abortions in general ought to be avoided." Because the odds of a single coitus resulting in birth are so much lower than the odds of normal birth once a fetus exists, however, Hare concludes that a prohibition on contraception is not justified. Still, there is "a general duty to produce children," and while contraception is not as bad as abortion, on Hare's view it is not as much *less* bad as "common opinion" would have it (*EB*, pp. 158–160).

To contemporary liberal ears, this is certainly a counterintuitive set of conclusions. Like all sets of ILS rules, however, Hare recognizes that population policy must be tailored "for particular societies and families" (*EB*, p. 163). And this means that here, as in the case of Parfit's repugnant conclusion, the final picture is not as counterintuitive. For Hare says that we might be justified in adopting as ILS rules (presumably *laws*) both a defeatable presumption against abortion and a totally pro-choice policy regarding contraception (*EB*, p. 161). There can, he says, be good reasons for having a child later rather than sooner or vice versa: for example if one believes that one will be married later (and that this will give the child a better start), or if a woman is about to turn 40, when the risk of Down's syndrome is said to rise (*EB*, p. 182). So Hare says that the exceptions to his preferred abortion policy would include those "that many people would now allow" (*EB*, p. 159). He even characterizes as having "a great attraction" an ILS rule (presumably a *law*) allowing the practice of infanticide on newborns with severe disabilities for some short time after birth (*EB*, p. 162). Compare "The Abnormal Child" (originally published in 1973), where he says that "A hard-headed utilitarian might . . . say that in cases of suspected abnormality we should let the child be born, operate if appropriate, and then kill the child if the operation resulted in a very severe handicap, and have another child instead" (*EB*, p. 188). So Hare's final conclusions regarding ILS rules on contraception, abortion, and killing severely disabled newborns are very close to those currently embodied in the laws of the United States.

I think that Hare is correct to conclude that, at least in modern societies, the best ILS rules of common morality and law will leave decisions about procreation up to individuals. In an ecologically "marginal" society, a Harean could justify an ILS rule requiring maximum procreation. Given his position on the repugnant conclusion described above, however, Hare should endorse at most an ILS "replacement" rule for contemporary Western nations, something like: *Everyone has a duty to procreate at a replacement level (about 2.2 children per couple).* But even then, if some people love having large families and others prefer to be childless, the better justified ILS rule would leave still more latitude. Perhaps the best rule would refer to overall government policies rather than obligations of individuals, for example: *Government policies (including, but not limited to taxation, public schooling, immigration, etc.) ought to be arranged so as to approximate replacement-rate reproduction within the country, without dictating family size.* But now childless people like me are off the hook! As long as there are plenty of folks who want to have children, and some who want to have lots of them, the rest of us can go through life without raising any. Of course, we are taxed every year to support in various ways the child-rearing of others: much of our property taxes go to support free public education, and our income taxes must be somewhat higher than they would need to be if there were no exemption for dependent children. But this can be construed as a way of fulfilling our part of a societal duty to procreate.

§9.5(C). HARE ON ANIMAL POPULATION POLICY

For analogous reasons, if there is a general duty to bring into existence happy animals, then—at least in affluent modern societies—this duty is probably best expressed in an ILS rule directed at the government, which does not require each and every individual to raise animals, but allows humane animal husbandry and slaughter by those inclined toward animal husbandry, and pet ownership by those who are so inclined.

To underscore the epistemological basis of this conclusion, first consider the final paragraph of "Why I am only a Demi-Vegetarian":

> Lastly, the question of how the numbers of various species should be apportioned within the available resources and ecological limits is an immensely difficult moral question, whose resolution requires *both* an understanding of some rather deep philosophical issues which I have dealt with . . . *and* a grasp of the enormously complicated ecological issues which affect the answer. Nobody that I know of has mastered both these groups of issues; few philosophers have even begun to master the ecology, and few ecologists and other environmentalists have even begun to understand the logic of the arguments, and the pitfalls which beset them. So there is everything to be said for the two disciplines getting together. (*EB*, pp. 234–235, italics in original)

Note also this passage from the final paragraph of "Possible People":

> I have . . . considered only the population of humans. A defensible population policy would have to take account of the interests of sentient nonhuman animals. But to formulate this, we should have to make difficult judgements about the relative quantity of life of humans and other animals in the circumstances that alternative polities would produce. This problem I thought too intractable for brief discussion in this paper. It could be that leaving room for the right numbers of animals in each species, whatever those are, would reduce the amount of land available for humans. I have no wish to make light of these difficulties . . . (*EB*, p. 82)

Hare's theory implies that there is some optimal mix of human and animal populations that we ought, in an ideal world, to maintain. But only an archangel could know what that mix is. So there is some ideal, but unspecifiable number of sentient animals that human beings should support. Similarly, the human population should be increased to some ideal, but unspecifiable size. In the face of this uncertainty, and given the side effects of infringing on various personal liberties, real-world inhabitants of modern, liberal democracies will do best to leave decisions about animal procreation up to the farmers and pet owners involved (as long as they adhere to legal standards for animal welfare), just as we do best to leave decisions about human procreation up to the couples involved (as long as they adhere to legal standards of child welfare).

§9.6. Non-Replaceability as a Feature of ILS Rules

From a Harean perspective, it is fairly obvious both (1) why Singer was bound to fail in his various attempts to account for the intuition that persons are not replaceable in the same basic way that merely sentient animals are, and (2) how to account for that intuition. Regarding the first point, Singer sought to ground the non-replaceability of persons in explicitly utilitarian thinking. As Hare acknowledges, however, explicitly utilitarian thinking makes everyone—including human persons—replaceable. Regarding the second point, the solution is to admit this and to argue instead that good ILS rules governing our relations with persons will treat them as non-replaceable. Given that Singer announced, in the preface to the second edition of *Practical Ethics*, that he had made "two significant changes" from the first edition, one of which was that he had begun using Hare's distinction between the intuitive and critical levels of moral thinking (1993, p. x), it is somewhat surprising that Singer has never noticed this.

In this section, I first spell out how and why persons should be regarded as non-replaceable in a good set of ILS rules. I then ask whether or not near-persons should also be regarded as non-replaceable. The answer to that question turns out to be much more complex, as illustrated briefly in the following chapter and more extensively in the sequel (Varner forthcoming).

§9.6(a). Justifying Non-Replaceability for Persons

In an essay on "The Abnormal Child" (which was first published in 1988, seven years after *Moral Thinking* appeared), Hare seems to say flatly that the replaceability argument could not apply to existing adults and children:

> Certainly it would be wrong to kill an adult in order to replace him or her with some other person who might be born. This is because the existing

> adult has desires (above all the desire to live) which will be frustrated if he is killed. That is the reason why we have a general ban on killing adults. And this applies even to young children. Whether it applies to neonates, who do not have the desire to live, is a controversial question which there is no room to discuss here. It certainly applies to children from a very early age. But it clearly does not apply to foetuses; so at any rate *that* reason for saying that foetuses ought not to be killed lacks force. (*EB*, p. 182)

But we must understand him to be talking here about intuitive-level rules. To see why, suppose that the following were all true:

1. There are only two sorts of adult people: huge, unhappy ones who require much to eat; and tiny, very happy ones who need very little to eat.
2. The human population is currently at the maximum that can be supported at an 8-to-2 ratio of huge people to tiny people; and
3. The huge people could be killed without causing them any fear or pain and replaced with a larger number of tiny people, without the tiny people's sensibilities being affronted.

Treating this as what I call "a strictly hypothetical case" (§2.3[A] and §4.8), the replacement policy would clearly maximize aggregate happiness and Harean critical thinking would have to endorse it. This completely unrealistic case shows that adult humans really *are* replaceable from the perspective of critical thinking.

Such fantastic cases are not taken into consideration when selecting ILS rules, however, and a number of general facts about human nature and the human condition imply that good ILS rules would tend to treat human beings as non-replaceable. The fact (as Hare mentions in the above quotation) that normal adult humans have desires for the future that will go unsatisfied if they are killed (I won't say, as Hare does, that their desires "will be frustrated" if they are killed, for reasons given in the footnote on Aristotle in §6.2) is a reason for thinking that good ILS rules will not treat persons as replaceable, a reason that does not apply to the merely sentient. But there are other, more significant reasons.

For instance, unlike the huge and tiny people depicted in assumption #3 above, real-world human persons are capable of understanding what is afoot and suffering in ways that the merely sentient are not. They also tend to have sensibilities that would in fact be affronted by human replacement policies. Notice, for instance, that human cultures universally condemn cannibalism (at least when practiced on members of one's own culture), except under the most extreme conditions. From a Harean perspective, this suggests that a ban on cannibalism is an important feature of any society's ILS rules. At the same time, the hesitance of many people to condemn starving people for eating their fellow human beings who have already died of starvation or from accidents (as in the actual cases of the Donner party in California during the winter of 1846–1847 and the Uruguayan soccer team in the Andes Mountains in 1972) illustrates that even the most important ILS rules are subject to being overridden by critical thinking under unusual conditions. From a Harean perspective, people who insist on condemning people for eating human flesh under such extreme circumstances have (as Hare used to put it) gotten "stuck" thinking at the intuitive level.

More generally, part II of this book extended Hare's theory by showing why it should recognize a three-way distinction among "persons" (defined as individuals with a biographical sense of self), "near-persons" (defined as individuals with a fairly robust, conscious sense of their own past, present, and future, but no biographical sense of self), and "the merely sentient" (defined as individuals that are sentient but lack any robust, conscious sense of their own past and future). For the reasons given in chapter 7, the lives of near-persons are *more morally significant* than those of the merely sentient, and the lives of persons are *more morally significant* than those of either near-persons or the merely sentient. This makes it appropriate to take special care in our dealings with them (§1.8), and this means that good ILS rules will specify different standards for respectful treatment of near-persons vis-à-vis the merely sentient, and for respectful treatment of persons vis-à-vis both near-persons and the merely sentient.

In Harean terms, this is what justifies a kind of anthropocentrism in ethics—not what I call (1998, p. 121) "valuational anthropocentrism," which denies all moral standing to non-humans, but "axiological anthropocentrism," which gives pride of place to certain interests that only human beings have. Outside science fiction, the only persons we encounter are human beings, and all normal adult humans are persons, so as noted in §1.9, the fact that one meaning of "person" is "human being" just adds to the usefulness of the term in formulating ILS rules governing treatment of individuals who have a biographical sense of self.

One way of embodying special respect for persons in ILS rules is to treat them as having certain non-utilitarian rights—"trump cards against utilitarian arguments." In *Moral Thinking*, Hare said that rights are "an immensely important element" in intuitive-level thinking (p. 155), and in an essay that first appeared in 1983, he argued that among the rights "which will be entrenched in the principles of a good psychiatrist" will be patients' rights to confidentiality, liberty, and informed consent (*EB*, pp. 24–26). He also argued that under almost all realistic circumstances, a ban on human slavery would be incorporated in any good set of intuitive-level rules (*MT*, p. 167). Like the right to vote, none of these rights could meaningfully be attributed to any non-human animals, and embodying them in the ILS rules of common morality, law, and professional ethics is another appropriate way of expressing the special respect that persons are due under the autonoetic consciousness paradigm.

§9.6(B). SHOULD NEAR-PERSONS BE REGARDED AS NON-REPLACEABLE?

For the foregoing reasons, I conclude that good ILS rules will treat persons as not replaceable, while treating merely sentient animals as replaceable. But what about near-persons? For the reasons given in chapter 7, near-persons are also due some form of special respect vis-à-vis the merely sentient. Should we, then, conclude that good ILS rules will also treat near-persons as non-replaceable in something like the ways in which good ILS rules treat persons as non-replaceable? I believe there is no simple answer to this question.

To this question, Harean utilitarianism does not yield a cut-and-dried, once-and-for-all answer. As I will argue in the next chapter, in some circumstances humans would be justified in killing near-persons for food, even if contemporary societies are not. Similarly, in the sequel (Varner forthcoming) I will argue that some kinds of biomedical research on near-persons would be justified, and this, too, involves treating them as replaceable. There I will also argue that biologists managing wild populations of near-persons should treat their charges as replaceable, because the consequences of attributing them a right to life would be so horrible. On the other hand, I will argue that some pets and working animals should *not* be thought of as replaceable by their owners, even if the animals in question are not good candidates for near-personhood.

§9.7. The Problem of Marginal Cases, Again

This book is focused on questions about human treatment of animals, but a discussion of the question, "What makes humans special?" was needed in chapters 6 and 7 to justify the common intuition that normal human beings *are* special, morally speaking. The so-called "problem of marginal cases" was briefly discussed at the end of chapter 7, but a more detailed discussion of the problem is appropriate here.

Jan Narveson first used the expression "argument from marginal cases" to describe the argument that because we believe that certain forms of respect are due "non-paradigmatic cases of human beings," we should conclude that similar forms of respect are due certain animals (1977, p. 164). I use the expression "*problem* of marginal cases" to describe the related problem that arises for any view that, like the one defended in this book, ascribes special moral significance to personhood and defines personhood in terms of cognitive capacities that not all human beings have.

Evelyn Pluhar (1995, pp. 64–65) distinguishes two versions of the argument from marginal cases. What she calls "the categorical version" goes like this:

1. Beings who are similar in all important morally relevant respects are equally morally significant.
2. Non-humans exist who are similar in all important morally relevant respects to marginal humans.
3. Marginal humans are maximally morally significant.
4. *Therefore*, the non-humans who are similar in all important morally relevant respects to marginal humans are also maximally morally significant.

This version assumes that we are obligated to treat marginal humans as "maximally morally significant." What Pluhar calls "the biconditional version" does not assume this:

1. Beings who are similar in all important morally relevant respects are equally morally significant.
2. Non-humans exist who are similar in all important morally relevant respects to marginal humans.
3. *Therefore*, those non-humans who are similar in all important morally relevant respects to marginal humans are maximally morally significant if and only if marginal humans are maximally morally significant.

But this second version of the argument does express the position that it is impossible to ground different levels of "moral significance" for marginal humans and animals unless some important morally relevant *dissimilarities* can be identified between the two. And that entails that *if* you are going to attribute "maximal moral significance" to both normal and marginal humans, *then* you must attribute it also to various animals.

§9.7(A). SINGER ON MARGINAL CASES

Pluhar points out that in *Animal Liberation*, Singer relies on the biconditional version. This is accurate as a first approximation, but I think that a careful look at what Singer says in a key passage supports a more nuanced reading, so what he says is worth quoting at length:

> To avoid speciesism we must allow that beings who are similar in all relevant respects have a similar right to life—and mere membership in our own biological species cannot be a morally relevant criterion for this right. Within these limits we could still hold, for instance, that it is worse to kill a normal adult human, with a capacity for self-awareness and the ability to plan for the future and have meaningful relations with others, than it is to kill a mouse, which presumably does not share all of these characteristics; or we might appeal to the close family and other personal ties that humans have but mice do not have to the same degree; or we might think that it is the consequences for other humans, who will be put in fear for their own lives,

> that makes the crucial difference; or we might think it is some combination of these factors, or other factors altogether.
>
> Whatever criteria we choose, however, we will have to admit that they do not follow precisely the boundary of our own species. We may legitimately hold that there are some features of certain beings that make their lives more valuable than those of other beings; but there will surely be some nonhuman animals whose lives, by any standards, are more valuable than the lives of some humans. A chimpanzee, dog, or pig, for instance, will have a higher degree of self-awareness and a greater capacity for meaningful relations with others than a severely retarded infant or someone in a state of advanced senility. So if we base the right to life on these characteristics we must grant these animals a right to life as good as, or better than, such retarded or senile humans.
>
> This argument cuts both ways. It could be taken as showing that chimpanzees, dogs, and pigs, along with some other species, have a right to life and we commit a grave moral offense whenever we kill them, even when they are old and suffering and our intention is to put them out of their misery. Alternatively one could take the argument as showing that the severely retarded and hopelessly senile have no right to life and may be killed for quite trivial reasons, as we now kill animals.
>
> Since the main concern of this book is with ethical questions having to do with animals and not with the morality of euthanasia I shall not attempt to settle this issue finally. I think it is reasonably clear, though, that while both of the positions just described avoid speciesism, neither is satisfactory. What we need is some middle position that would avoid speciesism but would not make the lives of the retarded and senile as cheap as the lives of pigs and dogs now are, or make the lives of pigs and dogs so sacrosanct that we think it wrong to put them out of hopeless misery. What we must do is bring nonhuman animals within our sphere of moral concern and cease to treat their lives as expendable for whatever trivial purposes we may have. At the same time, once we realize that the fact that a being is a member of our own species is not in itself enough to make it always wrong to kill that being, we may come to reconsider our policy of preserving human lives at all costs, even when there is no prospect of a meaningful life or of existence without terrible pain. (1990, pp. 19–20)

As Pluhar notes, when Singer says "this argument cuts both ways," he is invoking the biconditional version of the argument from marginal cases, but in the final two sentences of the above passage he still seems to imply that *some* different status for marginal humans and non-human animals would be appropriate. Pretty clearly, Singer thinks that it would be appropriate to both increase the moral significance that we assume attaches to various animals' lives and to decrease the moral significance that is commonly attributed to humans' lives. But I also sense that he wants to say that *some* marginal humans' lives would be attributed both less moral significance than those of normal humans and more than those of animals.

§9.7(B). A HAREAN RESPONSE TO THE PROBLEM

A Harean perspective helps us understand the complicated sense in which differential treatment of marginal humans and animals can be justified in utilitarian terms. Note that, in the first paragraph of the long passage quoted above, Singer says that without being guilty of speciesism "we could still hold" any of the following views about what justifies recognizing individuals as having a right to life:

1. that having "a capacity for self-awareness and the ability to plan for the future and have meaningful relations with others" is what grounds it,
2. "or we might appeal to the close family and other personal ties that humans have but mice do not have to the same degree,"
3. "or we might think that it is the consequences for other humans, who will be put in fear for their own lives, that makes the crucial difference."

A "souped-up" version of (1) is the view of personhood (having a biographical sense of self) that I defended in chapters 6 and 7, and I have argued (in §6.5) that we have no good evidence that any non-human animal is a person.

Marginal humans are a diverse lot, ranging from anencephalic babies through various forms of cognitive impairment, dementia, and comas, both reversible and irreversible.

1. Some of these (e.g., high-functioning Down syndrome people and certainly high-functioning autistics) will have a biographical sense of self and therefore qualify as persons.
2. Some (e.g., low-functioning Down syndrome people and some demented or senile people) may lack a biographical sense of self, but will still qualify as near-persons because they have a fairly robust conscious sense of their own past and future.
3. Others (e.g., the profoundly demented or senile) may be at best merely sentient.

Can we justify adopting different standards of treatment of marginal humans in category 2 and category 3 vis-à-vis animals? I think that we can.

Note, first, that Pluhar's statement of the argument from marginal cases uses the term "maximally morally significant." She does this in order to avoid pitching the argument in terms of any particular view's chosen terms; she says that it "implies either basic moral rights, including the right to life, or a strong presumption against killing" (1995, pp. 63–64). But these different ways of unpacking "maximal moral significance" make a difference. For our legal system and common morality already treat marginal humans as less than "maximally morally significant" in terms of what "basic rights" we ascribe to them. We treat some marginal humans more paternalistically than normal humans, for instance: we don't allow them to conduct their own defenses in court, we don't allow them to make their own medical decisions, we don't allow them to live alone, and so on. In such ways, we deny both marginal humans and animals various legal rights that we would respect in normal humans.[9]

At the same time, however, with very few exceptions, our laws and common morality do not allow us to kill unwanted marginal humans. Aside from passive termination of severely disabled newborns, it is increasingly common for doctors to give terminal patients doses of painkillers that become "unintentional" fatal overdoses. Ballot initiatives permitting physician-assisted suicide passed in the states of Oregon (1994) and Washington (2008), and the Montana Supreme Court has ruled (*Baxter v. Montana*, decided December 31, 2009) that while the state's constitution does not include a right to assisted suicide, "state law protects doctors in Montana from prosecution for helping terminally ill patients die" (Johnson 2009). The Netherlands, Belgium, Luxembourg, and Switzerland all allow physician-assisted suicide. All of this reflects, I think, the fact that common morality and the law seem to be in transition regarding killing humans when, as Singer puts it in the passage quoted above, "there is no prospect of a meaningful life or of existence without terrible pain."

On the other hand, without exceptions, law and common morality never allow us to raise marginal humans for food, or to experiment on them. That is, our laws and common morality embody a "strong presumption against killing" all human beings, but not animals that are at best merely sentient, just as some marginal humans are. Can this be defended? I think that it can be, for the following reasons.

In addition to having "a capacity for self-awareness and the ability to plan for the future and have meaningful relations with others" ground a right to life (in our ILS rules), in the passage quoted above, Singer allows that two other non-speciesist reasons are available:

2. "we might appeal to the close family and other personal ties that humans have but [some animals] do not have to the same degree,"
3. "or we might think that it is the consequences for other humans, who will be put in fear for their own lives, that makes the crucial difference."

[9] It might be said that what I am doing in this section is arguing for the adoption, at the intuitive level of moral thinking, of a "species norm" view, according to which the kind of treatment that an individual deserves is determined by what would be appropriate treatment of normal members of its species. This is how Carl Cohen appears to handle the problem of marginal cases (1986, p. 866). I would resist this characterization, however, in light of the differential legal rights of normal and marginal humans described in this paragraph.

Regarding (2) and (3), these are "indirect" reasons that a Harean utilitarian can give for treating marginal humans differently from similarly cognitively endowed non-human animals. To call them "indirect" reasons is to emphasize that these reasons for treating the marginal humans a certain way don't turn on the effects on the marginals themselves, but rather effects on other, normal humans.[10] Regarding (2), marginal humans are usually members of families who care deeply about them and fervently want to see them treated differently from the way that animals are commonly treated. And regarding (3), since all of us are at risk of becoming marginals ourselves, we would be justifiably fearful of policies that cheapened the lives of marginal humans.

Notice, however, that these considerations are contingent on what the facts are, and if human nature were very different, then these considerations might not favor marginal humans. Suppose, for instance, that humans had evolved some kind of symbiotic relationship with another species, so that humans almost never had contact with each other but they had lifelong relationships with members of that other species. If, as a result, we were "hard-wired" to care deeply about cats but we had no feelings at all for our fellow humans, then consideration (2) would not apply. Obviously, real-world humans are not like this, but this underlines how contingent is this utilitarian line of defense against the problem of marginal cases. Many people find it counterintuitive that the moral significance of various individuals should depend on such contingent features of their social environment or evolutionary history. Hare's theory can explain why people find this counterintuitive, however: if it's true that the best ILS rules for modern societies would treat both persons and marginal humans as non-replaceable or as having non-utilitarian rights, then a properly socialized individual will have such intuitions.

[10] Williams emphasizes how "peculiar" he finds utilitarians' invocation of such indirect effects: "The attitude or tone involved in invoking these effects may sometimes seem peculiar; but that sort of peculiarity soon becomes familiar in utilitarian discussions, and indeed it can be something of an achievement to retain a sense of it" (1973, p. 101). But, as I emphasized in §4.3, prominent utilitarians have commonly held that common sense thinking about ethics is "unconsciously" or "instinctively" utilitarian. Doubtless Williams would say that this too is a "peculiar" thought, but current law and common morality turn out, I think, to roughly reflect what a Harean archangel would prescribe for nations in the developed West; compare Hare on population policy as described above in §9.5(B).

{ 10 }

A Harean Perspective on Humane Sustainable Agriculture

§10.1. "Humane Sustainability"

In this chapter, I illustrate the application of Harean utilitarianism to the debate over what constitutes humane sustainable agriculture. My goal is to illustrate how nuanced and contextualized is a Harean perspective on humans' use of animals for food, and how subtle forces influence changes in law and common morality over time. In the sequel (Varner forthcoming), I will provide a more extensive treatment of this same topic, while extending the discussion to other areas of humane sustainable communities, including pet ownership and working animals, scientific research on animals, and wildlife/ecosystem management.

"Sustainability" is applied to a burgeoning list of things, including but by no means limited to "sustainable communities" and "sustainable agriculture." The Brundtland Commission (United Nations, 1987) defined "sustainable development," but now we hear of "sustainable forestry," "sustainable resource management," "sustainable economies," "sustainable building," "sustainable homes," "sustainable tourism," and on and on. Sustainability is now almost universally regarded as a morally necessary feature of social systems. However, "sustainability" means different things in different contexts.

Everyone agrees that a social system is sustainable only if it is structured in such a way that it can be used into the indefinite future. This is the *descriptive* element of sustainability. As Paul Thompson has emphasized (1995, chapter 7), even here there are variations by context: what counts as "the system," and how long is "the indefinite future"? But, as Thompson also notes, sustainability always includes—at least implicitly—another, *normative* component. This consists of a value commitment that allows the user of the term "sustainability" to choose among or prioritize various systems that are each sustainable in the descriptive sense. Various authors have spelled out what sustainable systems would look like if various versions of the normative component were endorsed.

For instance, Wes Jackson's (1985) vision of sustainable agriculture based on restoring a continuously harvestable prairie to the Great Plains includes normative assumptions about the appropriate scale of human communities, and some "sustainable development" plans for third world nations have been criticized for being driven by first world assumptions about what is most valuable in those nations' environments.

By "humane sustainability," I mean thinking about sustainability with the well-being of the affected animals as a primary focus. The focus cannot be exclusively on animal welfare, because "sustainability" is applied to human social systems, which exist to serve human interests. In a vision of *humane* sustainability, however, there would be a strong emphasis on animal welfare. From a Harean perspective, it makes sense to distinguish among what I call "prelapsarian," "contemporary," and "utopian" visions of humane sustainability.

One place to look for examples of descriptively sustainable systems is among ancient societies that survived relatively unchanged for millennia, such as those of Native Americans. I call these *prelapsarian visions* of sustainability because environmentalists often venerate pre-modern societies and assume that these descriptively sustainable communities also met whatever normative criteria are implicit in their conception of sustainability. On the other hand, such communities certainly appear morally suspect from the perspective of humane sustainability when their animal husbandry and slaughter practices are compared either to what contemporary scientific euthenics and contemporary engineering make possible or to what might someday be possible.

By *contemporary visions* of humane sustainability, I mean attempts to improve animal welfare using contemporary science and engineering, but constrained, at least generally, by contemporary consumer preferences and economic realities. From a Harean perspective, such visions concede—as I emphasized in §3.3(B)—that utilitarian thinking has a conservative bent or inertia, but at the same time they work to achieve meaningful welfare reforms.

Finally, what I call *utopian visions* involve bracketing such concerns about "social engineering" and economic and technological feasibility. Such proposals are either not currently economically and technologically feasible, or they imagine that consumer preferences could be altered to suit the envisioned system. At the same time—as I emphasized in §3.3(C)—like Mill's writings on slavery and feminism, such utopian proposals give concrete expression to the ideals of utilitarianism, and over time they can "shift the goal posts" of common morality, making achievable what once were utopian proposals.

In later sections of this chapter, I will describe in some detail examples of each of these three kinds of visions. First, however, the next section briefly describes four contrasting ways in which animal scientists conceive of "animal welfare," the degree to which animal scientists currently agree on various targets for reform, and how these map onto the three visions of humane sustainability discussed in the remainder of the chapter.

§10.2. "Animal Welfare" in the Animal Science Literature

In this context, "animal scientists" refers to scientists who study animals in production agriculture, specifically. This includes but is not limited to what is sometimes called "the science of animal welfare." Animal scientists are themselves deeply divided over how to conceptualize and/or operationalize "animal welfare," and over the extent to which questions about animal welfare are amenable to scientific study.

Three prominent approaches to assessing animal welfare emphasize measures of productivity, physiology, and behavior. The first holds that so long as agricultural animals are producing (meat, milk, eggs, etc.) and reproducing at a normal rate, their welfare is not compromised. In physiological terms, welfare is considered adequate unless and until certain physiological indices—including both nutritional measures and hormones associated with stress—stray from their normal ranges. In behavioral terms, welfare is considered adequate if the animals are able to engage in the full range of their "natural" behaviors and they exhibit no behavioral signs of poor welfare, for example pointless "stereotyped" behaviors, aggression, gnawing on bars or stall walls, and so on.

Obviously, these three approaches will lead to different conclusions in various cases. For instance, productivity and profitability can sometimes be maintained under what appear to be poor welfare conditions. Bernard Rollin states flatly that veterinarians in large animal practice "frequently face dilemmas arising out of obligations to the client's economic interests conflicting with obligations to the animals" (1999, p. 86), as examples of which he gives a farmer refusing to treat a broken leg in a sow that is about to farrow (p. 115) or to pay for surgery for painful eye cancer in a cow that is about to calve (p. 81). Rollin also states that in the egg industry,

> Although . . . it has long been known that the stocking of fewer birds per cage leads to greater production per bird, it is nonetheless more economically efficient to put a greater number of birds into each cage, accepting lower productivity per bird but greater productivity per cage. In other words, though each hen is less productive when crowded, the operation as a whole makes more money with a high stocking density: chickens are cheap, cages are expensive. (1995a, p. 119)

The physiological and behavioral approaches can also pull in opposite directions. For instance, when sows are kept for extended periods in gestation stalls where they cannot turn around, their welfare may not be compromised according to the purely physiological criterion; but on behavioral grounds, their welfare is compromised, because they cannot engage in various natural behaviors and they do things like endlessly chewing on the bars of their crates. On the other hand, certain behaviors that seem quite rewarding and presumably make positive contributions to animal welfare, such as sexual activity, exercise, and certain forms of play, are accompanied by increased levels of stress hormones. According to the purely physiological criterion, these activities decrease animal welfare, while on the "natural behavior" criterion, we are diminishing animals' welfare if we prevent these activities. (This has led some researchers to distinguish between "*dis*tress" and "*eu*stress.")

A fourth school of thought puts conscious mental states at the forefront of thinking about animal welfare. While acknowledging how difficult it is to study animal consciousness scientifically, Ian Duncan has been a prominent advocate of this approach among animal scientists. In this vein, Duncan stresses that "health, which is surely the most obvious of the physical needs, is not necessarily a prerequisite for welfare. Of course, usually there will be a close relationship between physical health and welfare," but unless and until unhealthy conditions such as tooth decay and cancer impinge on the individual's conscious states, he argues, welfare is not adversely affected (Duncan & Petherick 1991, p. 5018). Relatedly, Duncan speculatively describes conscious "wants" and "desires" as "evolved cognitive representations of [animals'] needs," that "motivate behavior in a more flexible way than do reflexes" (Duncan 1996, p. 31).

Much of this debate among animal scientists can be traced back to the Brambell Report, which was commissioned by the British government in the wake of Ruth Harrison's popular book *Animal Machines* (1964). As Joy Mensch notes:

> The members of the Brambell Committee not only identified particular production practices that raised concerns about welfare and made recommendations about those practices, but also provided a conceptual foundation for future research. Two concepts emerged that have proven pivotal to discussion about farm animal welfare. The first of these is that suffering is the primary consideration in evaluating the welfare of animals in intensive confinement systems. The second is that animals have behavioral "urges"; that is, behaviors that they need to perform to ensure their welfare and that are frustrated under conditions of confinement. (Mensch 1998, pp. 91–92)

Mensch argues that, in fact, animal scientists' attention has been narrowly focused on suffering and pain, and on a limited range of desires, to the neglect of positive mental states and broader questions about cognition in farm animals (pp. 95–96). She urges animal scientists "to make an ethical leap and formulate a broader operational definition of animal welfare" (p. 94) that would incorporate "positive experiences" (p. 95).

Despite these deep disagreements about the very concept of "animal welfare," David Fraser suggests that, at this point in time, animal scientists can agree on three propositions, arranged in descending order of general acceptance as follows:

> First, there appears to be widespread consensus that a high level of welfare implies freedom from suffering in the sense of intense and prolonged pain, fear, distress, discomfort, hunger and thirst. More specific applications of this principle enjoy less agreement. For example, farmers and animal protectionists may disagree about whether the pain inflicted by castration and branding is sufficiently intense and prolonged to seriously detract from animal welfare.
>
> Second, . . . it seems widely agreed that a high level of welfare requires a high level of biological functioning, including freedom from debilitating diseases, injury, malnutrition and reductions in normal growth. Again, more specific issues are less well resolved. For example, does the exclusion of disease in a confined, minimum-disease herd enhance welfare enough to compensate for the curtailment of freedom that the system requires?
>
> A third principle, perhaps less universally accepted, is that a high level of welfare implies that the animals should have positive experiences such as comfort and contentment, and the pleasure of normal activities such as play, exploration and non-aggressive social behavior. The specific application of this principle involves very disparate viewpoints, including disagreement about how spacious and natural an environment must be in order to satisfactorily accommodate natural behavior. (Fraser 1995, p. 112, citations omitted)

Fraser's hierarchy provides a general framework for assessing proposed reforms from a Harean perspective.

Obviously, for a utilitarian it is conscious mental states that matter, so an adequate conception of "animal welfare" must focus on animal consciousness, and it must be admitted that positive conscious states improve animal welfare, just as suffering detracts from animal welfare. Insofar as animal scientists rely on utilitarian arguments in general, and the replaceability argument in particular (§9.3), they should agree with both of these points. That said, however, the study of animal consciousness is in its infancy, as Duncan acknowledges, and this is especially true with farm animals, as Mensch's observation about cognition studies of farm animals emphasizes.

So from a Harean perspective, when it comes to farm animal welfare, we are far from archangels. Under present circumstances, then, the easiest reforms to justify would be those that eliminate "intense and prolonged" suffering, since as Fraser notes in his first proposition above, there is a "widespread consensus that a high level of welfare implies freedom from suffering in the sense of intense and prolonged pain, fear, distress, discomfort, hunger and thirst." Presumably, many reforms that eliminate "debilitating diseases, injury, malnutrition and reductions in normal growth," as described in Fraser's second proposition, would benefit farmers and would thus be more readily justified from a Harean perspective, despite its conservative bent or inertia (§3.3[B]).

The more difficult reforms to justify would be those that require farmers to go further and provide animals with "positive experiences" and a range of natural behaviors in a "spacious and natural" environment.

Generally, only what I call "prelapsarian visions" of humane sustainable agriculture will allow "intense and prolonged" suffering. "Contemporary visions" appropriate to affluent, industrialized nations will tolerate no such suffering in normal production practices, and they will, to the extent consistent with current economic and technological realities, move in the direction of providing production farm animals with a range of natural behaviors. "Utopian visions" propose radical alternatives that simultaneously meet all three of the goals on Fraser's list. Although the conservative bent or inertia of utilitarianism militates against requiring their implementation in the near future, I will emphasize in §10.5 how individuals, by approximating such utopian visions in their personal moralities, can help move the goal posts of common morality and make their implementation in future generations more readily justifiable from a Harean perspective.

§10.3. Prelapsarian Visions

What I call "prelapsarian visions" of sustainability look to ancient societies that survived relatively unchanged for millennia, such as those of Native Americans. As I noted earlier, I call these "prelapsarian" visions because environmentalists often venerate such societies and assume that their descriptively sustainable practices also met whatever norms are implicit in their conception of sustainability. At the same time, however, their animal husbandry and slaughter practices can look morally suspect, compared either to what contemporary scientific euthenics and contemporary engineering make possible or to what might someday be possible.

The truth often lies in the middle, however, because Hare's theory allows for what I characterized in §3.3(A) as "a benign form of relativism." For just as technological advances may have made it inappropriate for Inuits to continue practicing infanticide, technological advances may make it inappropriate for people in affluent, developed nations to use animals in ways that would have been appropriate in earlier times.

For a dramatic illustration, consider the ways in which Native Americans were reported to hunt bison at the time of their first contact with Europeans. This included killing individual animals, which were easy to approach from downwind, but also driving whole herds over cliffs, into natural or artificial enclosures, or onto thin ice. In these drives, anywhere from a few dozen to more than a thousand bison would be killed at one time, and in the higher numbers it was impossible to utilize the carcasses efficiently. Eyewitnesses across decades of the nineteenth century reported Native Americans leaving hundreds of carcasses to rot, sometimes butchering only the cows, or taking only favorite parts such as tongues and humps, or fetuses. These radically inefficient drive-based techniques may well have been descriptively sustainable in pre-contact times, since the decline of the bison is thought to have begun only after Native Americans began trading bison meat and hides with European immigrants, who shipped them out to large non-native markets by railroad (Krech 1999, pp. 138ff).[1]

Such inefficient hunting strategies may have been morally justifiable as well, given the extreme conditions under which pre-contact Native Americans lived. For the bison provided pre-contact Native Americans with more than meat; they "provided over one hundred specific items of material culture" from various parts of the animals, ranging from winter clothing and tipi covers to ropes and glue (Krech 1999, p. 128). And to reliably harvest enough animals to provision an entire tribe, Native Americans lacking firearms and horses may have needed to rely on inefficient drive hunts rather than trying to kill individual animals. From a Harean perspective, then, inefficient hunting of merely sentient animals would presumably be justified for persons living under the Native Americans' ecological and technological circumstances. That is, good ILS rules for persons living under such extreme environmental conditions and technological limitations would permit such inefficient methods of hunting. So the drive hunts of pre-contact Native Americans may well have been an example of humane sustainability *for persons in their circumstances.*

But that should not blind us to the fact that the Native Americans' drive hunts were not just inefficient but caused an enormous amount of suffering that obviously could be avoided using more modern hunting or slaughter techniques. Bison driven onto thin ice died by drowning; those driven off cliffs would often be stunned rather than killed and left with broken legs or backs; and Europeans who witnessed the slaughter of those driven into enclosures were shocked to see Native Americans butchering still-conscious bison (Krech 1999, pp. 128–135). So even if pre-contact Native Americans were morally

[1] Although there are no reliable estimates of bison and Native American populations prior to the nineteenth century, in 1800 there are estimated to have been 120,000 Native Americans and 30 million bison on the Great Plains. Krech estimates that per capita, the Native Americans needed about six bison per year and that 120,000 Native Americans could have sustainably harvested that number (720,000) annually from a herd of 30 million (1999, p. 137). But given how inefficient the drive hunts were, it seems likely that the Native Americans would have killed far more than the number of animals they needed each year.

Krech also reports that many Plains tribes believed that "when buffaloes disappeared for the season, they went to lake-bottom grasslands, and that when they reappeared they came from those habitats." So even when no bison were in evidence, these Native Americans believed that the underworld was still teeming with them. For this reason, a Cheyenne once told an Army colonel—in the 1870s when commercial slaughter of bison was in high gear—that "however recklessly the white men slaughter, they could never exterminate them" (Krech 1999, p. 148).

justified in using them, surely modern humans would not be justified in using such inefficient and relatively inhumane hunting techniques.

For another example, consider that the practice of eating great apes, cetaceans, and elephants—"the usual suspects" for near-personhood—is now widely believed to be unjustified, and it is widely believed that these animals deserve fairly strong legal rights. I believe that both of these claims are supported by a Harean perspective, at least in modern societies. This does not mean, however, that from a Harean perspective, it was always wrong to kill them for food. Even Tom Regan's worse-off principle implies that humans are justified in killing non-human animals for food if that is the only way to survive (Regan 1983, pp. 301–304, 324). So just as pre-contact Inuits were, perhaps, justified in using infanticide for sex selection (§3.3[A]), some communities of persons living under pre-modern environmental and technological conditions may have been justified in killing near-persons for food, even if today's Inuits are justified in doing neither. Similarly, in pre-modern times, good ILS rules might have allowed people to kill great apes for food, even though that would be ruled out by good ILS rules for modern societies.

This illustrates how, from a Harean perspective, "showing appropriate respect for animals" may mean something very different in pre-industrialized societies from what it means in a wealthy, industrialized state. As I said in §3.3(A), this is a "benign" form of relativism, because full-blown cultural ethical relativism is the view that what justifies the members of a society in believing an ethical rule and acting accordingly is the bare fact that it is accepted in their society. In Hare's theory, by contrast, members of a society are justified in believing and acting according to the rules of their common morality to the extent that those rules approximate the ones that clear critical thinking would lead them to choose, and as the society's circumstances change, reformers can use critical thinking to urge changes in the common morality.

§10.4. Contemporary Visions

In what I call "contemporary visions" of humane sustainability, attempts are made to improve animal welfare using contemporary science and engineering, but are constrained, at least generally, by contemporary consumer preferences and economic realities. From a Harean perspective, such visions concede—as I emphasized in §3.3(B)—that utilitarian thinking has a conservative bent or inertia, but at the same time, they work to achieve meaningful welfare reforms.

In this section, I describe three examples of such contemporary visions of humane sustainability. First, I describe some of the ways in which slaughter practices were reformed during the twentieth century. These reforms became feasible as new slaughter technologies were developed, and they reflected a general shift in public perceptions of the slaughter industry that was significantly influenced by popular writings. I then describe some certification programs that are thought to improve farm animal welfare in various ways but without imposing significant, non-voluntary costs on consumers and/or producers. Finally, I describe the cage-free movement and European-style reforms in poultry production. Such reforms are aimed at changing birds' environments on the assumption that they are better off if they are able to perform various "natural behaviors." Although such reforms remain controversial among U.S. producers, the European Union (EU) nations have committed to moving in this direction.

§10.4(A). SLAUGHTER REFORM IN THE TWENTIETH CENTURY

Although Upton Sinclair's 1906 novel *The Jungle* is mainly credited with inspiring the Pure Food and Drug and Beef Inspection Acts, which were meant to ensure a safe food supply to humans and were passed into law just six months after its initial publication,[2] everyone remembers his descriptions of hog slaughter at the time:

> They had chains which they fastened about the leg of the nearest hog, and the other end of the chain they hooked into one of the rings upon the wheel. So, as the wheel turned, a hog was suddenly jerked off his feet and borne aloft. . . . Neither squeals of hogs nor tears of visitors made any difference to them; one by one they hooked up the hogs, and one by one with a swift stroke they slit their throats. There was a long line of hogs, with squeals and lifeblood ebbing away together; until at last each started again, and vanished with a splash into a huge vat of boiling water. ([1906] 1981, pp. 34–45)

And cattle slaughter at the time:

> Along one side of the room ran a narrow gallery, a few feet from the floor; in to which gallery the cattle were driven by men with goads which gave them electric shocks. Once crowded in here, the creatures were prisoned, each in a separate pen, by gates that shut, leaving them no room to turn around; and while they stood bellowing and plunging, over the top of the pen there leaned one of the "knockers," armed with a sledge hammer, and watching for a chance to deal a blow. The room echoed with the thuds in quick succession, and the stamping and kicking of the steers. The instant the animal had fallen, the "knocker" passed on to another; while a second

[2] Sinclair, however, "intended it less as an exposé of the meat industry than as an argument for socialism, to which he had recently been converted," writing "I aimed at the public's heart, and by accident I hit it in the stomach" (Dickstein [1981] 2002, p. 50). The novel describes the dehumanizing effects of early-twentieth-century American industrialism on a newly arrived Lithuanian immigrant. He gladly takes a job in the Chicago stockyards, only to be ground down by the exploitation, suffering debilitating injuries, and alternately playing a part in machine politics and campaigning for socialism.

man raised a lever, and the side of the pen was raised, and the animal, still kicking and struggling, slid out to the "killing bed." (p. 38)

The above depictions are essentially accurate.[3] In large, early-twentieth-century slaughterhouses, "knocking boxes" were used to loosely restrain cattle so that they could be hit in the head to stun them before slitting their throats to bleed them out. Under those conditions, inaccurate blows were bound to happen a significant percentage of the time. And the hoisting of unstunned pigs that Sinclair described was not outlawed until the Humane Methods of Slaughter Act of 1958 (Title 7, Chapter 48 of the U.S. Code).

The Act defined slaughter of livestock as "humane" when "animals are rendered insensible to pain by a single blow or gunshot or an electrical, chemical or other means that is rapid and effective, before being shackled, hoisted, thrown, cast, or cut" (section #1902)[4]; and the 1978 amendments to the Act gave USDA inspectors the authority to shut down slaughter plants because of observed cruelty violations (although animal protection groups emphasize that this provision has been used irregularly over the years).

Work by animal scientist and popular author Temple Grandin indicates that, by contrast, in a well-run slaughter plant using current equipment, a very small number of animals suffer from inadequate stunning. Here I will discuss cattle, specifically, but similar improvements have been achieved in hog slaughter. Large slaughter operations in the United States today generally use curved, walled paths approaching a kill chute that employs a double-track conveyor system to restrain the animals while a penetrating captive bolt pistol is applied to the head. The pistol uses compressed air to fire a pinky-finger sized bolt through the animal's skull into the brain. The overall design of the system serves several functions simultaneously. First, it keeps the animals calm. The curved, walled paths approaching the kill chute are similar to those the animals have previously experienced, and prevent them from seeing human handlers or other potentially stressing sights. At the kill chute, the two-track conveyor system holds the animals on either side of their brisket, a

---

[3] Before writing *The Jungle*, Sinclair "spent seven weeks in Chicago living with workingmen in the packing plants" (Thompson 1970, p. 7), funded by the socialist journal *Appeal to Reason*. Conservatives complain to this day that the novel inaccurately depicted food safety-related conditions and practices, but Doubleday, Page and Company agreed to publish it only after confirming to their own satisfaction Sinclair's general portrait of the packing industry, and the Roosevelt administration's Neill-Reynolds Commission "confirmed the essential truth of the way *The Jungle* depicted slaughterhouse conditions" (Thompson 1970, p. 12).

[4] The Act exempted kosher slaughter from this provision by stipulating that "slaughtering in accordance with the ritual requirements of the Jewish faith or any other religious faith that prescribes a method of slaughter whereby the animal suffers loss of consciousness by anemia of the brain caused by the simultaneous and instantaneous severance of the carotid arteries with a sharp instrument" also counts as humane (section #1902).

grip that Grandin says has a calming effect comparable to that of a strong hug on an autistic child. The chute is shielded so that the person using the pistol cannot be seen except by the animal being stunned, and that at only the last moment. When the pistol is applied to an animal, it is already supported by the conveyor, so that the animal behind it only sees its legs go limp. Holding them on the conveyor also prevents bruising, and increases efficiency. For as soon as an animal is stunned, a worker attaches a shackle to one of its rear legs from below. Then, as the conveyor continues moving forward, the stunned animal is automatically hoisted by that rear leg for immediate bleeding (which is also important for meat quality).[5]

Working as a consultant for McDonald's, Grandin found very low percentages of inadequate stunning in cattle presented to the kill chute operator in such systems. Among steers and heiffers, which are the mainstay of cattle slaughter, she found that only 0.16 percent showed any sign of regaining consciousness after stunning. Among cows and bulls slaughtered in similar facilities, however, just over 1 percent did. (These are older, unneutered animals that have been used for dairying or breeding, respectively; and thicker skulls in bulls and some breeds like Brahmins can make stunning more difficult.) She also found that about 5 percent of cattle were only rendered insensible on the second shot (2002). In a parallel study of electrical stunning of hogs, Grandin reported similarly low rates of inadequate stunning (2001).

These low rates of inadequate stunning are not ideal. In Harean terms, the question is which option would you choose if you had to successively "stand in the shoes" of the animals being slaughtered (§2.2). In terms of the statistics reported above, the question would be specifically: *Would you rather undergo the experience of thousands of animals being slaughtered where in every case you "never knew what hit you," or in 5 percent of which you experienced the first of two shots from the stun gun, and in 0.16 percent of which you "came to" while being hoisted and bled?* I think the answer is obvious: *Never* knowing what hit you would be better than consciously experiencing it any percentage of the time. Just as clearly, however, the observed rates of inadequate stunning are a dramatic improvement over what Upton Sinclair described in 1906.

I noted above that similar improvements in slaughter practices have been achieved with swine. Swine are usually stunned electrically, because restraining them is more difficult than restraining cattle, but systems have been designed

---

[5] The information in this paragraph is based on conversations I had with Grandin while visiting a Colorado slaughterhouse in 1991. For what it's worth, as I watched cattle being slaughtered there for upward of an hour, the operator fired a second shot from the pistol about 5 percent of the time, as reported in the following paragraph. But in only one case do I recall seeing an animal move in a way that suggested that it was conscious before it was shot the second time: I saw one animal rear slightly after the first shot, as if it was trying to get off the conveyor. I inferred that all of the other second shots by the operator were "insurance."

that stun swine electrically without needing physical restraint. However, the Humane Methods of Slaughter Act covers a limited range of animals. For instance, it does not apply to birds, which are slaughtered in far, far greater numbers than mammals, nor does it cover fish. And it explicitly exempts Jewish and Islamic ritual slaughter, in which animals lose consciousness as a result of exsanguination, from the requirement to stun animals before they are hoisted or cut (section #1902).

Ritual slaughter and the slaughter of birds and fish will be discussed in the sequel (Varner forthcoming). Slaughter also occupies only the final moments of these animals' lives, and there are numerous other welfare issues in production animal agriculture. (For a comprehensive overview, see Rollin 1995a.) All I have intended to illustrate here is how legal enforcement of a welfare standard, coupled with technological innovation, have significantly improved slaughter techniques in the United States over the last century, while the affected industries have continued to thrive.

§10.4(B). CERTIFICATION PROGRAMS

In "Why I am only a Demi-Vegetarian," Hare says that he is "pretty certain that one of the most important reforms to agitate for, if one is the agitating sort, is a law requiring full disclosure of the sources of all foodstuffs" (*EB*, p. 230). Mandating "full disclosure" in all cases could be more costly than Hare seems to realize, for instance in restaurants (which might have to revise their menus constantly) and under proposals to require labeling of all foods containing certain kinds of ingredients (which some producers have claimed would stigmatize their products). Depending on the details of the proposal, those kinds of costs could be significant. But where voluntary certification programs are concerned, those kinds of costs are, if not irrelevant, at least less likely to be significant.

One example is the "Certified Humane Raised & Handled" program (http://www.certifiedhumane.org/). Participating producers are audited annually (although the program reserves the right to make unannounced audits), and changes in production practices that might affect animal welfare must be cleared by the certification program. Although specific standards are developed for each species by a scientific advisory board,

> [t]he most basic criteria require that animals have clean and sufficient food and water; a humane environment under conditions and care that limit stress; are able to express normal behaviors and live in an appropriate and comfortable environment that includes sufficient space, proper facilities, shelter, a resting area, and company of their own kind; and have a healthy life, benefitting from disease and injury prevention and rapid diagnoses and treatment. In addition, the standards require that managers and staff be

thoroughly trained, skilled, and competent in animal husbandry and welfare, and have good working knowledge of their system and the animals in their care.

In early 2011, the program's web site listed 26 cage-free egg operations and 6 dairy operations, but only one operation each under beef, bison, chicken, pork, and veal.

A particularly ambitious certification program came from the Pork Industry Institute at Texas Tech in 2001. (Unless noted otherwise, all references to the Sustainable Pork Project are based on: http://www.depts.ttu.edu/porkindustry-institute/Sustain_Web/SP_guidelines%20Dec%2001.htm, accessed December 20, 2007.) To be certified as producing "sustainable pork" by the Sustainable Port Project, hog farmers would have to be "animal friendly," "environmentally friendly," "worker friendly," and "community friendly." For instance, certified producers:

1. Must use no sub-therapeutic antiobiotics, gestation crates or tethering on the animals (relatedly, the standards also state that "If no scientific literature exists, or if the scientific literature is ambivalent on a given subject, but consumer perception is that a component of the system is preferred, then an attempt will be made to satisfy the consumer perception of the best standard");
2. They must provide health insurance for all full-time employees; and
3. Participating farms are encouraged to hold "Field days, open houses, producer seminars and talks to local school students and community betterment groups."

Although certification does not require outdoor housing, a series of studies by the Pork Industry Institute also documented the animal welfare and environmental benefits of such systems.

The system studied involved the use of a square area of land divided into 12 angular paddocks that are connected via a circular, central exchange area. The paddocks can be inexpensively created using single strand electric wire fencing 60 centimeters high. Each paddock is about 0.4 hectares in area, and by transferring pigs from paddock to paddock, the system can be used to maintain vegetative ground cover continuously, when used in favorable climates and topographies. This can prevent the pigs' wastes from posing any environmental hazard to groundwater or neighboring fields (Rachuonyo 2001), thus avoiding one of the most significant environmental hazards of intensive swine production systems: the need to maintain large waste collection tanks.

Housing animals in such extensive, outdoor systems allows them to engage in a range of natural behaviors, and reduces some of the problems that arise in the densely stocked, indoor confinement operations that are currently the norm in the United States. For instance, stereotypic behaviors and tail biting are less common. Raising pigs outdoors does have its disadvantages from an animal welfare perspective, however. For instance, piglet mortality is higher than when gestation crates are used, and predation can be a problem. The animals also have to deal with extremes of heat and cold. For both of those reasons, outdoor systems tend to be less economically efficient as measured by feed conversion ratios, but outdoor systems can be cheaper to build on a per animal basis, and "if well managed, an outdoor system can be as competitive as an indoor system" (Johnson 2001, p. 93).

Unfortunately, the ambitious Sustainable Pork certification project didn't catch on. One farm applied for certification, but before it could make the improvements required for certification, it went out of business. In mid-2010, the animal scientist in charge of the project told me: "I think it is too early for such a program—but I believe its time will come" (John McGlone, personal e-mail, July 15, 2010).

§10.4(c). Banning Battery Cages for Laying Hens

Another example of a contemporary vision of humane sustainable agriculture involves the move toward cage-free housing in the egg industry. Although still something of a niche market, numerous cage-free egg producers are competing successfully in local markets, and several brands are proving competitive on regional and even national scales. Controversy remains among animal scientists over whether these systems are, from an animal welfare perspective, superior to conventional battery cage systems, but a consensus has emerged in the EU on phasing out the battery cage systems.

Current egg production systems are commonly divided into several categories. "Battery cage" systems house four or more hens in a cage, usually made of wire mesh, with as little as 53 square inches of space per bird (that's equivalent to an area of 7.3 x 7.3 inches). These are also referred to as "conventional" systems, because, since the U.S. poultry industry intensified after World War II, the large majority of layers (today over 90 percent) have been housed in such systems (Scanes et al. 2004, table 13.11, and p. 213).

"Furnished" cage systems include various "enhancements" in the cages, such as "nesting boxes" (although these are usually just plain enclosures—hens prefer to lay eggs in such enclosed areas), perches, and areas to dust bathe.

"Cage-free" systems include both aviaries and "free-range" systems. In aviaries, a large enclosure houses the birds, usually with perches and nesting boxes provided. Some aviaries are constructed in barns, with the boxes and perches added to the walls. Others are constructed in special buildings furnished with rows of stacked perches and nesting boxes with narrow walkways between them. In "free-range" systems, rather than being confined indoors, the hens have substantial free access to outdoor areas.

Although surveys show that consumers generally believe that hens suffer from overcrowding in them, battery cage systems have some real, measurable advantages from an animal welfare perspective. Generally, mortality and morbidity are lowest in battery cages and higher in furnished-cage and non-cage systems (LayWel Project 2006, p. 27). This is because the hens have less exposure to disease vectors than in cage-free systems, and fewer opportunities to injure themselves than in aviaries and furnished cage systems. The manager of an aviary that I visited, for instance, estimated that annual flock mortality in a well-run cage system would be just 3–4 percent, whereas he estimated that in his aviary it would probably be up to 10 percent, although that was in part because, as a certified organic facility, it could not use various medicines and supplements. (I cannot reveal the company or the location of the facility that I visited.)

Of course, some increased risk of death and disease might be "worth it" from a moral perspective. For, when doing explicitly utilitarian critical thinking, we are trying to maximize "quality adjusted life years" or QALYs (*EB*, p. 227). So, in addition to increases in mortality and morbidity, any improvement in experiential welfare that is achieved in furnished cages and non-cage systems must be taken into account.

More welfare-related research has been done on laying hens than on any other farm animal, and more of this research has been done in the European Union than in the United States (Rollin 1995a, p. 117). Still, when a consortium of the EU and several of its member states pooled their data in the LayWel report (http://www.LayWel.eu), differences in the design of various studies and the types of data collected in them limited the statistically significant results (LayWel Project, pp. 6ff, 15). Nevertheless, the LayWel Report concluded that "the welfare of laying hens is severely compromised" in battery cages by the "discomfort and abnormal behaviour which are inherent to the system" (p. 27), and recommended that all hens be provided with nesting boxes and perches, and that more research be done on varieties of substrates for foraging and dust bathing (p. 35).

So although battery cages do have some welfare-related advantages, and much relevant research remains to be done, the EU nations have all committed to phasing out battery cages by 2012 (Blokhuis et al. 2007, p. 103). By announcing it a decade before the ban took effect, the E.U. reduced the adverse economic impacts on producers. These impacts could be further reduced by including "tax incentives, such as accelerated depreciation for new and remodeled structures" as recommended by a 2008 Pew Commission report that called for eliminating not only battery cages, but also restrictive gestation and farrowing crates for swine (Pew Commission on Industrial Farm Animal Production 2008, p. 85). In such ways, what might otherwise be radical, utopian reforms can be made practicable, contemporary visions of humane sustainable agriculture.

§10.5. Utopian Visions

In chapter 3, I emphasized that utilitarian thinking has a degree of conservative bias or inertia, since economic interests, cultural traditions, and existing consumer preferences must all be taken into account when assessing policy changes. I also emphasized, however, that a utilitarian analysis can call for radical changes, at least in the long haul, and that utopian writings can help prepare societies for such changes. For example, Mill argued against slavery and in favor of extending voting rights to women at a time when race-based slavery seemed natural to many people and women's suffrage seemed unthinkable. But across time, educational efforts, including Mill's own writings, changed background cultural attitudes in ways that reduced the costs of changes in common morality and laws, effectively "shifting the goal posts" of common morality.

Similarly, some visions of humane sustainability can be read as challenges to deeply entrenched cultural traditions and economic interests, and people who strive to live up to these ideals in the present can be understood as attempting to reeducate the public by their example. Thus, from a Harean perspective, self-styled "animal activists" can be understood as living lifestyles that it would be unethical to force upon society at large at present, but which might become feasible as social norms in future generations. In Harean terms, such animal activists have internalized, as their personal morality, a set of ILS rules that require more of them than the common morality of their present society. On a Harean analysis, this makes their personal commitments supererogatory as described in §3.2(B), so it is no wonder that they sometimes consider their moral commitments heroic. In the long run, however, such activists can influence entrenched cultural assumptions and, over time, reduce the transition costs of reform. Through demonstrations, popular and philosophical literature, films, advertisements, political campaigns and so on, they sometimes succeed in shifting the goal posts of their culture's common morality.

Singer's *Animal Liberation* can be interpreted as an example. In his preface to the second edition, Singer said that "To reread the original preface of this book is to return to a world half forgotten." A generation has now passed since the book was first published in 1975, and I think it fair to say, as Singer added in that second edition preface, that "A new consciousness exists about the need to extend sympathies for dogs and cats to pigs, chickens, and even laboratory rats" (1990, p. viii). I also think it fair to say that Singer's famous book has been a very significant catalyst in that change. The book has been referred to as "the Bible" of the modern animal rights movement, PETA sometimes gives out copies of the book as membership premiums, and the number of copies in print has been treated as a reflection of how public concern about animal issues has grown since 1975 (Jasper & Nelkin 1992, pp. 37–38). I cannot count the number of people who, upon learning that I am a philosopher who studies animal issues, have told me that reading *Animal Liberation* led them to significantly change their diets, or at least caused them to think about—and sometimes agonize over—traditional, and still widely accepted, uses of animals.

In this section, I discuss three examples of "utopian visions" of humane sustainable agriculture, although my first example isn't really a proposal, per se, and it could just as well be treated as a contemporary vision.

§10.5(A). DEMI-VEGETARIANISM

I discuss demi-vegetarianism here, however, because it concerns the symbolic significance of dietary choices. Traditional foods and etiquette surrounding meals are central to all cultures, and for this reason dietary choices can set off discussion—sometimes heated discussion, as many a college student has learned upon returning home a newly converted vegetarian. For this very reason, however, individuals' dietary choices can, over time, exercise a subtle but significant influence on background cultural traditions and consumer preferences by critical reflection on accepted practices. In this way, the diffused dietary choices of many individuals can ultimately have the same kind of impact as celebrated publications like Singer's *Animal Liberation*. In "Why I am only a Demi-Vegetarian," Hare recognized this "propaganda or publicity value" of dietary choices (*EB*, p. 230), but he emphasized that those he called "demi-vegetarians" have an important advantage over complete vegetarians.

The term "demi-vegetarian" is apparently more widely used in England than in the United States. Hare defined it thus: "It means someone who, while not being a full vegetarian, let alone vegan, eats little meat, and is careful what kinds of meat he (or she) eats. Usually the selection is on dietetic grounds (lean meat rather than fat, fish rather than meat, etc.); but no doubt moral considerations come in too" (*EB*, pp. 224–225). In the mid-1990s, when I did a critical review of the scientific literature on vegan diets (Varner 1994b, 1994c, and 1994d), nutrition scientists were standardly using the following terminology:

> "*lacto-ovo* vegetarians" eat "meat by-products," like dairy products and eggs, but no flesh;
> "*vegans*" eat neither flesh nor any of these meat by-products;
> "*pesco*-vegetarians" eat fish (and maybe other seafood) but neither mammals nor birds; and
> "*semi*-vegetarians" eat relatively small amounts of various kinds of meat.[6]

[6] The term "fruitarian" was also in common usage among nutritionists in the 1990s, referring to someone who eats only foods that can be obtained without killing plants. (Fruitarians assume that this includes nuts and seeds, which suggests that they deny the potentiality principle—see footnote 8 to §9.5[B].) More recently, I have seen the terms "*pollo*-vegetarian" and "*crusto*-vegetarian" used to describe (respectively) those who eat chicken and those who eat crustaceans (which would include shrimp, crawfish, crabs, and lobsters) but no other flesh.

Since nutritionists already use the term "semi-vegetarian" to describe individuals who use "dietetic grounds" to select small quantities of meat to eat, I suggest that we reserve the word "*demi*-vegetarian" for:

> *People who eat limited quantities of various animal by-products, as well as limited quantities of meat, poultry, and/or seafood, using "moral considerations"—specifically, animal welfare considerations—to determine which products to include in their diets.*

Others have used the term "critical omnivory" for similarly motivated dietary commitments.

In his essay on demi-vegetarianism, Hare stressed the economic influence of the practice (*EB*, pp. 230–231):

> Demi-vegetarians have one very powerful advantage over full vegetarians when it comes to influencing the market. Since they eat little meat, but do eat some, they would be in a quite strong position to influence animal husbandry practices if they could get enough information to enable them to make use of their power. By selecting carefully the meat they did buy, they might help to cause those practices to be abandoned which occasion most suffering to animals.

He did allow that "It might be urged in favour of full vegetarianism that to make a stand in this simple way against meat eating will bring home to people how awful some animal husbandry practices are, and so lead to public revulsion against them and thus possibly to legislation forbidding them." But he noted that widespread vegetarianism is "certainly likely to reduce" the numbers of farm animals, which flies in the face of the replaceability argument (§9.5). The question of which has greater "propaganda value"—demi-vegetarianism or veganism—is an empirical one. While there is no decisive evidence one way or the other, certain statistics suggest that Hare may be right.

The percentage of vegetarians found in a population varies widely, depending on whether the respondents are asked simply "Are you a vegetarian?" or some variation on "How many times a week/month/etc. do you eat beef/chicken/fish/etc.?" This is because people who eat semi-vegetarian diets often self-identify as "vegetarian," *simpliciter*. Although the number of "true" vegetarians in both the United Kingdom and the United States remains small, it is growing linearly, while the number of semi-vegetarians and those who say that they "avoid red meat" is growing at an accelerating rate. (All information in this paragraph is based on Sabaté et al. 2001, pp. 6–10.)

How much of this growth is attributable specifically to concerns about animal welfare and/or animal rights is hard to say, as illustrated by a 1992 survey of practicing vegetarians: only 15 percent cited "animal rights" as their reason, in contrast to 46 percent who cited health reasons. Five percent cited "ethics"

(Sabaté et al. 2001, p.8), but these could as well have been concerned with *environmental* ethics or human justice–oriented issues.

Various other surveys suggest, however, that demi-vegetarianism as defined here (a commitment to semi-vegetarianism based on animal welfare considerations) could take hold of the consumer psyche and have a significant economic impact on production practices in the future. For instance, NOP polls conducted for the Royal Society for the Prevention of Cruelty to Animals and Animal Aid in 1997 and 1998 found that while only "36 percent of non-vegetarians would consider becoming a vegetarian if they felt animal farming was cruel," "80 percent claimed to prefer to buy food that is cruelty free and good for the environment," and "69 percent were willing to pay more for a product that comes from a humanely reared animal" (http://www.vegsoc.org/info/statveg90.html, accessed April 14, 2009).

The prefix "demi-" is used to describe half-sized items, as in "demitasse," and things of a mixed nature, as in "demi-vegetarian," but also things of an *imperfect* nature, as in "demigod." Accordingly, some vegetarians use the term "demi-vegetarian" disparagingly. Also, some vegetarian groups define "demi-vegetarian" so as to exclude poultry and red meat, essentially equating it with "pesco-vegetarian." I need to emphasize, however, that Hare intended neither to disparage demi-vegetarianism nor to define it so as to exclude eating mammals and birds, and that is how I intend it.

The commitment to demi-vegetarianism can be manifest in diverse ways: eating eggs and turkey, but only from "free-range" birds; eating red meat, but only from extensively reared and humanely slaughtered cattle or bison; eating fish, but only from an aquaculture operation that uses a sophisticated system to kill them without causing them stress or suffering;[7] eating dairy products, but only from an operation that pastures the cows; eating only fish, birds, or deer that one kills oneself; and so on.

For his part, in reply to Hare's "Why I am only a Demi-Vegetarian" essay, Singer acknowledges that some audiences will be more receptive to demi-vegetarianism than they will be to "full" vegetarianism (let alone veganism). For that reason, he allows that "It is probably a good thing if different people and organizations promote each of these goals" (1999, p. 325), but he says that "No more finely grained dietary stance could have been as effective a symbol of my rejection of our conventional attitudes and practices towards animals" (p. 327).

---

[7] In the controlled environment of a fish farm, individual fish can be slaughtered humanely and economically using electrical and percussive stunning systems analogous to those used with cattle and swine. For instance, one system involves fish swimming up a narrowing channel to a point at which "an automated blow can be delivered to the head, immediately rendering the fish unconscious" (Braithwaite 2010, p. 182). Another involves electrically stunning fish in small groups that can be quickly bled out while still unconscious before another group is allowed into the stunning tank.

§10.5(B). THE BUFFALO COMMONS PROPOSAL COUPLED WITH IN-SITU SLAUGHTER

An example of a truly utopian vision that I personally find very compelling is inspired by the foregoing demi-vegetarianism vision. In *Bring Back the Buffalo! A Sustainable Future for America's Great Plains* (1996), Ernest Callenbach proposed that dwindling populations in rural areas of the Great Plains be supported by a combination of electric wind farms and bison ranching on a truly grand scale. Callenbach claims that native prairie grasses are well adapted to the natural grazing habits of bison, which "tend to bite off the top parts of grass rather than tear away the near-ground growing shoots as do cattle and sheep," and that bison move almost constantly, grazing an area heavily, but then leaving it alone for a long time before returning (p. 24). This causes bison on "suitably large unfenced ranges" (p. 29) to naturally achieve the effects of Allen Savory's highly management-intensive (and fence-intensive) rotational grazing system for cattle (Savory & Butterfield 1999, parts VIII and IX). "With bison and their companion grazers," Callenbach claims, "it seems almost certain that the best management is no management" (p. 29).

Apart from finding Callenbach's proposal highly appealing aesthetically, I believe that bison rearing and slaughter could be made tremendously humane, given their natural behavior and resilient health, and modern carcass-processing technologies. For although bison are generally more difficult to herd and handle than cattle, and they have a keen sense of smell, Native Americans were able to kill individual buffalo by approaching from downwind, and European explorers commonly had to shoo the animals out of their way (Krech 1999, pp. 124, 128). I remember seeing footage of bison that continued to graze casually while animals around them were shot from a short distance away, and so I suppose that experts shooting (perhaps with silencers on their guns) could selectively harvest animals with minimal disruption to the herd. Mobile processing facilities developed for use with African game, with reindeer herds in Scandinavia (Hoffman & Wiklund 2006, p. 201), and with deer in Scotland (Sharman 1983, p. 502) could be adapted to hygienically process the bison carcasses immediately after they are shot. Obviously, some of the meat would be bruised as the animals fell, but that seems like a small cost to bear in eliminating all of the welfare concerns raised by handling and transport.

The Wild Idea Buffalo Ranchers are a consortium of bison ranchers in South Dakota who claim to do just this. Their web site (http://wildideabuffalo.com/wild-idea-buffalo-practices, accessed February 1, 2011) states that they "Respectfully, and humanely harvest their buffalo in the field, causing zero stress to the animal." This does add to the price of the meat. In the documentary film *Facing the Storm: Story of the American Bison* (High Plains Films 2011), founder Dan O'Brien says in an interview: "We harvest our buffalo in the field, which adds to the price. It's a lot more difficult. It's a lot easier just to put them in trucks, haul them to a slaughter plant, put them in a long chute where they hit them over the head going through. Sure that's easy, that's what the beef guys do."

The "buffalo commons" proposal is not utopian because it would replace meat from cattle with bison meat. Bison meat is still a "niche market" product, and it costs more than beef. It has about half the calories and fat of beef, and one-third less cholesterol (according to http://exoticmeats.com/, accessed December 28, 2007), however, so it is conceivable that health-conscious consumers might make a large shift toward bison in the near future.

Callenbach's proposal *is* utopian, however, in its grand scope. For he imagines turning vast expanses of the Great Plains into unfenced bison range, with bison fenced out of remaining small towns, which could involve enormously expensive land transfers. Callenbach borrowed the term "buffalo commons" from a 1987 essay by Deborah Epstein Popper and Frank J. Popper, who quoted approvingly the following approach to such a buyout:

> Bret Wallach, a University of Oklahoma geographer and MacArthur fellow, has suggested that the Forest Service enter into voluntary contracts with Plains farmers and ranchers, paying them the full value of what they would cultivate during each of the next 15 years but requiring them not to cultivate it. During this time, they would instead follow a Forest Service-approved program of planting to reestablish the native shortgrasses. Afterwards, the service would, as part of the original contract, buy out their holdings except for a 40-acre homestead. (1987, p. 17)

Doing this on any great scale would obviously be tremendously expensive, but it would also face great resistance from landowners. The previously mentioned film *Facing the Storm* includes footage of Plains residents characterizing the Poppers as "nuts," "a couple of interlopers from Rutgers," and as needing armed guards when they first traveled to talk in the area.

Nevertheless, the buffalo commons land-use plan, when coupled with the in situ slaughter option I have described, provides a vision of humane sustainability that can be seen as inspiring to contemporary consumers and producers. Just describing it goes some small way toward moving the goal posts of our common morality.

§10.5(C). HUMANE FRANKENFOODS

Sometimes welfare improvements and consumer preferences go hand in hand; for instance, if it is true both that free-range, organically raised animals are better off than factory farmed ones and that they taste better (although I doubt that blind taste tests confirm this). But sometimes consumer preferences and welfare improvements don't go hand in hand.

Suppose, for instance, that "we could genetically engineer essentially decerebrate food animals, animals that have a merely a vegetative life but no experiences" as imagined by Bernard Rollin in *The Frankenstein Syndrome* (1995b, p. 193). In a similar vein, in *Vexing Nature,* Gary Comstock asks us to imagine genetically engineered "living egg machines, transgenic animals genetically engineered to convert feed and water into eggs more efficiently than any of their evolutionary ancestors, layer hens," animals with "no ability to perceive or respond to any information it might receive if it had eyes, ears, or a nose," which each produce one cholesterol-free egg every 12 hours (2000 p. 152). These thought experiments were in turn presaged in the 1952 science fiction novel *The Space Merchants*. There a machine called "Chicken Little" was depicted as part of dystopian future where advertisers controlled popular thought and everyone ate the same chicken-like substance produced by the machine from pond-grown algae (Pohl & Kornbluth 1952, pp. 76, 87).

The foregoing examples were put forth as thought experiments or science fiction, but commercially viable sources of in vitro meat may be just around the corner. According to a *New York Times* article, it may soon be possible to grow meat in vitro on a commercial scale:

> Scientists at NASA and at several Dutch universities have been developing the technology since 2001, and in a few years' time there may be a lab-grown meat ready to market as sausages or patties. In 20 years, the scientists predict, they may be able to grow a whole beef or pork loin. A tissue engineer at the Medical University of South Carolina has even proposed a countertop device similar to a bread maker that would produce meat overnight in your kitchen. (Robin 2005)

There is also a non-profit organization, called New Harvest, which is dedicated to developing "cultured meat" (http://www.new-harvest.org/default.php, accessed July 19, 2010), and PETA offered a $1 million "reward" for the first in vitro chicken meat successfully marketed in at least 10 states by June 30, 2012 (http://www.peta.org/feat_in_vitro_contest.asp, accessed July 19, 2010). And Adam Shriver (2009) draws on recent research on the dissociation of the affective and informational components of the pain-processing system in mammals that was discussed in §5.4(B) to seriously propose developing genetically modified animals that are capable of functioning "normally" within intensive production systems, while having muted capacities for pain and other forms of suffering.

Any of these new technologies would be attractive from an animal welfare perspective, insofar as they would either entirely eliminate concerns about animal welfare from the meat-production process or (in the case of Shriver's proposal) significantly attenuate them. The proposed products would probably not be acceptable to many contemporary consumers, however. For even if the meats were judged equivalent in blind taste tests, the so-called "yuck factor" would kick in for consumers who both knew about the new production process and found it revolting, or at least "aesthetically abhorrent" (Rollin 1995b, p. 193). If one were contemplating outlawing traditional production techniques in favor of such new technologies, these negative reactions of consumers would be a genuine cost that must be taken into account in a utilitarian analysis of the proposal, along with costs to families invested in current production technologies.

Reliance on in vitro meat technologies would also impoverish the world from a utilitarian perspective insofar as it would entirely eliminate the sentient animal from the meat-production process. In that respect, another recent development is more attractive, although "the yuck factor" may still keep it in the category of "utopian" visions of humane sustainability. Using traditional breeding strategies rather than genetic engineering, researchers have developed a strain of blind hens as a solution to the problems of feather pecking among intensively reared chickens. Confined chickens tend to peck at others' feathers and combs, which can cause pain, and the resulting wounds tend to attract more pecking, which can ultimately lead to cannibalism. Capitalizing on rare but natural genetic mutations, traditional breeding has been used to produce populations of blind chickens that do not suffer from these problems.

> The blindness in one experimental strain of layers is caused by an autosomal recessive mutation. A study comparing blind and sighted chickens (Ali & Cheng 1985) showed that the blind chickens do not have problems with feather pecking, comb damage or cannibalism. Nor do they seem to have any additional problems measurable by normal standards applied in animal welfare research. Cortisol levels do not differ between the two groups. The blind chickens are able to find food and water. Their food intake is about 25 per cent less than the sighted chickens but their body weight remains the same. This is thought to be due to the reduced activity of the blind chickens in combination with the reduced heat loss of their intact plumage. In addition, the blind chickens produce about 13 per cent more eggs.
>
> Of course, being blind may deprive the hens of some pleasures. However, there is good reason to think that these hens are well adapted to their situation and may find pleasure by many other means. Assuming that the blind hens do not suffer in any other way, it seems clear that, by utilitarian standards, they live a better life than their sighted counterparts. (Sandøe et al. 1999, pp. 321–322)

Although these specialized strains of chickens have been available since 1985, they have not yet been used in commercial operations. Perhaps it would cost early adopters more to stock a facility with such specialized breeds, but a law requiring poultry operations to use such breeds would level the competitive playing field and would probably also quickly lower the cost of such breeds.

So the cost to producers and consumers of requiring this welfare innovation would probably be minimal, while the welfare advantages could be significant. The only thing that keeps this welfare innovation in the utopian category is consumer preferences. As Sandøe and colleagues put it, "many people will think that this is a completely wrong approach to improving animal welfare," because "we should change the housing systems to fit the animals, not the animals to fit the housing systems" (1999, p. 222). From the perspective of Harean critical thinking, this reasoning makes the mistake of adopting the prior-existence view rather than the total view. For in explicitly utilitarian thinking one should, for the reasons given in §9.5, take into account the interests of all possible sentient beings, and weigh these against the interests of those that already exist or will come into existence under current policies. The "we should change the housing systems to fit the animals, not the animals to fit the housing systems" stance assumes, in effect, that only chickens like those already in existence should be brought into existence. But insofar as we can make significant improvements from an animal welfare perspective by replacing current strains of chickens with blind strains, we have grounds for changing the consumer preferences. This simple reform remains a utopian vision of humane sustainable agriculture, however, because consumer preferences cannot be changed by the waving of a philosophical wand.

§10.6. Davis's Argument for Omnivory and the Special Value of Ruminants

In this final section, I describe a current debate that illustrates how an animal ethics issue can be complicated by how detailed is the information needed to apply the principle of utility to it. My focus here is one of the most popular and influential arguments in favor of vegetarian diets, one that is generally understood to have nothing to do with animal welfare, specifically. The argument turns on the claim that people can be fed more efficiently on crops, rather than having them eat meat from animals that were raised, or at least "finished," on grains that were grown as row crops. This argument certainly applies to much of the contemporary Western meat industry. As Hare notes in "Why I am only a Demi-Vegetarian," however, this argument's application is limited, because "there are substantial, though not enormous, areas of the world which are suitable for pasture but unsuitable for crops" (*EB*, p. 223). Hare gives, as specific examples, farms he knows of in Jamaica and England, but there are also large arid and cold areas where ruminants can turn sparse grasses into meat, but growing row crops would not be feasible. Aldo Leopold ([1923] 1979) cautioned that it only took a century for cattle grazing in the Southwestern United States to significantly degrade the carrying capacity of that arid region, but to the extent that it is possible raise meat this way without degrading the carrying capacity of the land, raising ruminants for meat can *decrease* the cultivated acreage needed to support a human population. Where the region's people are poor and nutritionally stressed, as in many developing nations, this can be an especially important source of food.

Even in developed nations with strong environmental movements, however (which generally oppose grazing on arid and other landscapes unsuited for crop production), meat from ruminants may have advantages from an animal welfare perspective, as animal scientist Steven Davis has emphasized in a controversial paper. In just four or five pages of a 2003 essay in the *Journal of Agricultural and Environmental Ethics*, Davis argues that if the goal is to kill the fewest animals, then when we take into account the effects of agricultural practices on "field animals"—rodents, rabbits, various birds, and other animals that live or feed in farm fields—an omnivorous diet that includes some products of "pasture-forage-ruminant systems" would be preferable to a vegetarian diet. Discussing Davis's argument and the debate that it has engendered provides a fine illustration of how extremely detailed information is sometimes needed to apply the principle of utility correctly.

Davis argues that a lot of field animals are killed when tractors and combines are run through the fields and by the harvesting itself, which removes cover that protects animals from predators. By contrast, pastures either do not need to be harvested at all or they require "fewer passages through the fields with tractors and other farm equipment" (Davis, p. 390), and ruminants kill almost no field animals while grazing. Davis grew up on a farm himself, and he remembers "seeing mice, gophers, and pheasants in the field that were injured or killed every time we worked the fields" (p. 388). In contrast, "The predominant feeling among wildlife ecologists is that no-till agriculture will have broadly positive effects on mammalian wildlife" (citing 390, quoting Wooley et al., 1984). Davis estimates that half as many "animals of the field" would die, and that would make up for the increased number of animals slaughtered for human consumption. With almost no empirical research available on the effects of harvest practices on field animals, Davis relies on just two published studies. (The dearth of relevant studies is noted by Davis, p. 389; Tew & Macdonald 1993, pp. 279, 281; and Fraser 1995, p. 181.)

The first, Tew and Macdonald 1993, studied the effects of harvesting wheat and barley, and subsequent stubble burning, on field mice in England. Using a capture-mark-recapture calculation, they estimated that the population of field mice grew rapidly as a result of both breeding in the field and immigration from adjoining wood lots, and then dropped abruptly from 25/hectare to 5/hectare at harvest (p. 280, figure 1, and p. 281). To estimate how much of the population decline was due to emigration and how much to harvest-induced mortality, they radio-collared 33 mice and tracked them during harvest operations. They found that only one mouse was killed by the combine harvester, and two by subsequent stubble burning, meaning that only 9 percent of their radio-collared mice were directly killed by harvest operations. However, they found that 17 (52 percent) of the mice "disappeared within one week of harvest." Of these, "conclusive" evidence indicated that 9 (27 percent) had been killed by predators, and the authors speculate that most of the others were, too. Relatedly, they noted that:

> tawny owls changed their hunting behaviour at harvest. Prior to harvest, they rarely hunt farmland. When they do, they concentrate particularly on the grassy banks bordering hedgerows since the crop prevents them from stooping on prey. Immediately following harvest, however, the owls fly low over the fields and stoop onto prey both in the stubble and directly into the straw lines left by the combine. (p. 282)

In sum, Tew and Macdonald found that while only 9 percent of field mice were killed when English wheat and barley fields were harvested, at least 27 percent and as much as 52 percent were displaced from the fields and killed by predators.

The other study cited by Davis (Nass et al. 1971) examined the fate of Polynesian rats in Hawaiian sugarcane fields. These are harvested only once every 22 to 38 months, so they provide especially good habitat for wild animals. Of a total of 35 rats fitted with radio collars prior to the harvesting of two fields, two collars failed and the fate of 13 could not be determined, either because their collars were never recovered or the collars were found empty with no evidence of how the rats died. Of the 22 whose fate could be determined, 14 (64 percent) were killed by the operation of field equipment (either directly or after they were buried in their burrows and apparently suffocated), three (14 percent) were swept up by the equipment and hauled off to the mill where they died, and three (14 percent) were displaced to neighboring fields where they were killed by mongooses. Thus only 2 out of 22 of the tracked rats (9 percent) appeared to have survived the harvest.

Noting that these two studies cover only two species of field animals and only a single field operation (harvesting), Davis concludes that it is "reasonable" to assume that 60 percent of all field animals are killed annually as a result (direct or indirect) of crop farming practices (2003, p. 389). He then cites a USDA estimate that there are currently about 120 million hectares in crop production in the United States. Assuming the same density for all field animals per hectare as Tew and Macdonald's field mice (25/hectare), Davis estimates that "1.8 billion animals would be killed annually to produce a vegan diet for the USA" on land currently in crop production (p. 390). If half of that 120 million hectares were converted to pasture production, and if switching to pasture production reduced by half the number of field animals killed annually per hectare, then only 1.35 billion field animals would be killed annually. Obviously, more ruminants would be slaughtered under this proposal, but, Davis notes, only about 37 million ruminants are slaughtered annually in the

TABLE 10.1. Estimates of the Total Number of Animals Killed Annually (Both During Field Operations and Slaughtered for Food) in U.S. Agriculture

Assumptions:		
	A1. Hectares currently in crop production in U.S.	120 million[a]
	A2. Wild animals killed annually during crop production	15 per hectare[b]
	A3. Wild animals killed annually during pasture production	7.5 per hectare[c]
	A4. Ruminants currently slaughtered annually in U.S. agriculture	37 million[a]
Resulting estimates:		
	Wild animals killed annually during current crop production = (A1 x A2)	1.8 billion
	Wild animals killed if half of acreage currently in crops were converted to pasture = [(A1)/2 x A2] + [(A1)/2 x A3]	1.3 billion
	Total killed if slaughter rate doubles = 1.3 + (A4 + [A4 x .5])	1.424 billion
Sources of assumptions:		
[a] USDA figures.		
[b] Assuming 25 animals/hectare (the density of field mice in Tew and Macdonald 1993) and an annual mortality rate of 60 percent.		
[c] Assuming that pasture production kills half as many field animals as crop production.		

Source: Davis (2003)

United States, a small fraction of the number of field animals that he estimates are killed annually during crop production. So, even if slaughter of ruminants *doubled* under his favored scenario, that would add only a little over two-thirds of a billion deaths, leaving the total still several billion short of the number of animals that he estimates are killed annually under the present system. Davis concludes that a system in which humans consume beef and dairy products produced by pasturing cattle would cause fewer total animal deaths than a system in which humans all went vegan. (Table 10.1 summarizes Davis's analysis.)

Davis couches his argument in terms of minimizing death, because he pitches it as a response to Tom Regan's rights view. For similar reasons, however, philosopher George Schedler has argued that "utilitarian vegetarians are obligated to substitute for some of the commercially harvested vegetables they consume the meat of grazing animals and any home-grown vegetables they can produce" (2005, p. 500). He argues that such "ethical meat eating" maximizes both total and average utility vis-à-vis both the status quo and universal vegetarianism. It increases aggregate happiness vis-à-vis the status quo by eliminating factory farming, and it increases happiness vis-à-vis universal vegetarianism insofar as both (1) it reduces the amount of commercial crop production, which reduces the amount of suffering experienced by field animals for the reasons given by Davis, and (2) it increases happiness by bringing into existence grazing animals that lead good lives while allowing human beings to enjoy eating meat.

Obviously, Davis's argument piles assumption upon assumption, as Andy Lamey emphasizes. Lamey concludes that "we do not know enough to make even the rough calculations that Davis offers" (2007, p. 339). He also notes that several aspects of beef production "make it potentially more hazardous to human beings than crop harvesting," including that the manure used to fertilize no-till pastures carries various diseases, cattle kill about 24 people per year in the United States (in attacks by bulls, by crushing, and when cars hit them), slaughterhouses are particularly dangerous places to work, and cattle account for 19 percent of humans' methane emissions, which contribute to global warming (p. 340).

More philosophically speaking, Gaverick Matheny's reply notes that Davis focuses exclusively on how many animals are killed, ignoring questions about how well wild and domestic animals' lives go overall. For his part, Matheny speculates that a wild animal that perishes in a harvester lives a better life up to that point than a grass-fed cow or steer (2003, p. 508).

Matheny's main criticisms are again on the empirical side, however. He argues, first, that Davis should have considered the number of animals killed *per consumer*. Since a given amount of acreage in crops can feed more people, Matheny argues that converting current crop land to pasture would raise the number of animals killed per consumer. Assuming that 1,000 kilograms of protein can be produced on "as few as 1.0 hectares planted with soy and corn, 2.6 hectares used as pasture for grass-fed dairy cows, or 10 hectares used as pasture for grass-fed beef cattle," he estimates that on Davis's assumptions about how many field animals are killed yearly per hectare, "a vegan-vegetarian would kill 0.3 wild animals annually, a lacto-vegetarian would kill 0.39 wild animals, while a Davis-style omnivore would kill 1.5 wild animals" (2003, p. 507). Relatedly, Matheny emphasizes that Davis fails to consider the number of wild animals that would be prevented from existing under his proposal: "[R]uminant production uses ten times as much land as crop production to yield the same amount of food. Thus, as long as the combined number of wild animals on nine wild acres plus one cultivated acre is greater than the number of animals on ten grazed acres, a vegan-vegetarian will allow the greatest number of wild animals to exist" (p. 510). Lamey, too, notes that Davis ignores facts related to the relative inefficiency of meat-based diets in comparison to vegan diets (p. 339).

This is, I conclude, a case in which the utilitarian analysis of an issue hinges on a large number of empirical questions that simply cannot be answered with any certainty at this point in time. All that can be said, I think, is that a reasonable case can be made for eating small quantities of meat from extensively reared ruminants as a form of demi-vegetarianism (§10.5[A]).

§10.7. Conclusion

In one chapter it is impossible to survey all of the welfare issues in contemporary animal agriculture (for a comprehensive overview, see Rollin 1995a). The range of welfare reforms discussed in §10.4 and §10.5 has, however, served to illustrate how nuanced and contextualized a Harean perspective will be. And the discussion of Davis's argument for omnivory in §10.6 illustrates how detailed can be the information needed to apply the principle of utility correctly.

As illustrated in this chapter, Harean utilitarianism does not produce simple, clear-cut, once-and-for-all answers on such complicated issues. But if the best minds of our era find such issues challenging to sort through, any moral theory that did would be suspect for that very reason. In my estimation, Harean two-level utilitarianism mirrors the complexities, uncertainties, and hazards of the moral life as we actually encounter them better than any alternative theory. It reveals rather than obscures the complexities of such issues. Because such issues are so complex, diverse moral agents sincerely concerned about animal welfare do not automatically come to consensus on a vision for humane sustainability, and a number of contemporary and utopian visions will always be under discussion.

{11}

Harey Animals

§11.1. Singer on Harean, Two-Level Utilitarianism

The previous two chapters have illustrated various ways in which a Harean perspective on animal ethics issues is nuanced and contextualized. In this final chapter, I compare and contrast such a "Harey" perspective with the views of the best-known philosopher writing on animal issues, Peter Singer. As noted in §1.1, Singer wrote his dissertation with Hare at Oxford. Also, the defense of utilitarianism that Singer gives in *Practical Ethics* is similar in important ways to Hare's defense of it, and since the second edition of *Practical Ethics* (1993), Singer has invoked Hare's distinction between the intuitive and critical levels of moral thinking in various contexts. In this section, I survey these similarities between Hare's theory and Singer's writings on utilitarianism, before turning to a comparison of what Singer says about animal ethics with the "Harey" perspective provided in the preceding chapters.

Singer's defense of utilitarianism in the first and second editions of *Practical Ethics* is similar in important ways to Hare's. Noting that "philosophers from the Stoics to Hare and Rawls" have tried to derive moral standards from the notion that "ethical conduct is acceptable from a point of view that is somehow universal," Singer argues that "The universal aspect of ethics . . . does provide a persuasive, although not conclusive, reason for taking a broadly utilitarian position" (1993, pp. 12–13). The argument goes as follows. In a "pre-ethical stage," one decides what to do based solely on how one's own interests are served. When one "begins to think ethically," the universal aspect of ethics requires "recognizing that my own interests cannot count for more, simply because they are my own, than the interests of others. . . . This requires me to weigh up all those interests and adopt the course of action most likely to maximise [*sic*] the interests of those affected" (p. 13). Singer concludes that:

> we very swiftly arrive at an initially utilitarian position once we apply the universal aspect of ethics to simple, pre-ethical decision making. This, I believe, places the onus of proof on those who seek to go beyond utilitarianism. The

> utilitarian position is a minimal one, a first base that we reach by universalizing self-interested decision making. We cannot, if we are to think ethically, refuse to take this step. If we are to be persuaded that we should go beyond utilitarianism and accept non-utilitarian moral rules or ideals, we need to be provided with good reasons for taking this further step. (p. 14)

Note how this argument, similarly to Hare's, relies on an appeal to the logic of moral discourse and makes morality into a kind of "universalized prudence" (§1.5). Singer's argument is simpler, because it relies only on the logical requirement of universalizability and it doesn't depend on Hare's very controversial conception of prescriptivity (§2.5).

The same argument for utilitarianism is repeated in the second edition of *Practical Ethics*, but there he says that he has made "two significant changes" to the form of utilitarianism he embraces:

> The first is that I make use of the distinction drawn by R. M. Hare, in his book *Moral Thinking*, between two distinct levels of moral reasoning—the everyday intuitive level and the more reflective, critical level. The second is that I have dropped the suggestion—which I advanced rather tentatively in the fifth chapter of the first edition—that one might try to combine both the "total" and "prior existence" versions of utilitarianism, applying the former to sentient beings who are not self-conscious and the latter to those who are. (1993, pp. x–xi)

That second change was discussed in §9.4, in the survey of the different ways in which Singer has tried to account for the intuition that human persons are "not replaceable." Following is a survey of the ways that Singer makes use of Hare's distinction between ILS rules and explicitly utilitarian critical thinking in the second edition of *Practical Ethics* and elsewhere.

In response to the objection that a utilitarian should approve of secret killings that would clearly maximize aggregate happiness, Singer notes that such "unusual hypothetical cases" don't go into the formulation of intuitive-level rules (1993, pp. 92–94). This parallels Hare's "template" response to the more unrealistic "test cases" for utilitarianism described in §4.8. Similarly, Singer invokes the critical/intuitive level distinction in reaction to a fanciful case that involves killing "a person who has—though she does not yet realize it—fallen into the hands of homicidal sadists who will torture her to death" (p. 201).

In the chapter of *Practical Ethics* devoted to "Taking Life: The Embryo and the Fetus," Singer notes that the explicitly utilitarian reasoning that leads him to endorse abortion under various circumstances also leads to the conclusion that infanticide would be permissible in various circumstances. One response he considers is this: "recalling Hare's distinction between the critical and intuitive levels of moral reasoning, one could hold that the ethical judgment we have reached applies only at the level of critical morality [and that] for everyday decision-making, we should act as if an infant has a right to life from the moment of birth" (p. 172). From a Harean perspective, this would amount to holding that the best intuitive-level rules would treat newborns as persons, with a non-utilitarian "right" to life, even though critical thinking would endorse euthanasia in some cases. Later in the book (p. 190), however, Singer endorses a window of opportunity for euthanasia of severely disabled newborns similar to what he proposed (with coauthor Helga Kuhse) in *Should the Baby Live?* (Kuhse & Singer 1985, pp. 189–197). Singer describes this as the view that "there should be at least some circumstances in which a full legal right to life comes into force not at birth, but only a short time after birth—perhaps a month" (p. 172). As a thorough-going utilitarian, Singer has expressed skepticism about appeals to "rights" (e.g., 1987b), yet in *Animal Liberation* I count 15 references to "right(s)" in the first two pages. Singer says that this is only "a convenient political shorthand" (1990, p. 8), but from a Harean perspective, it makes perfect sense to articulate various ILS rules, including Singer's stance on euthanasia of severely disabled newborns, in terms of legal rights. For as I said in §3.2(D), laws are a species of ILS rules, and as I noted in §3.1, Hare thought that rights claims are a useful way of building "deontological flavor" into ILS rules. At the same time, these "trump cards" against utilitarian thinking are justified by explicitly utilitarian critical thinking, so a Harean utilitarian need have no qualms about invoking rights.

Finally, on pp. 222–229 of *Practical Ethics*, Singer discusses "five differences that normally exist between killing and allowing to die." He says that these differences "explain" why common morality treats the duty not to kill people as being more weighty than the duty to save lives (e.g., through contributing to famine relief), although he emphasizes that these differences *don't* exist in the case of physician-assisted, voluntary euthanasia (p. 225). He doesn't put it this way, but Singer's point could be expressed in Harean terms as follows: good ILS rules of common morality will prohibit killing people without also requiring people to do everything they can to save lives (e.g., through contributing very heavily to foreign relief efforts), but when it comes to codes of medical ethics, a code allowing doctors to practice active euthanasia under certain circumstances would be superior to one that allowed only passive euthanasia in those same circumstances.

Several years later, in the "Response" he wrote for *Singer and His Critics*, Singer notes that critical-level thinking may lead one to conclude that "asking too much" of people is counterproductive when it comes to famine relief efforts (1999, p. 302), and he discusses the ways in which good ILS rules would allow partiality toward friends, lovers, and family members (pp. 297–301).

§11.2. Comparisons and Contrasts with "Harey Animals"

So while Singer has not flat-out endorsed Hare's theory, he is clearly inclined toward a similar, two-level version of utilitarianism. For that reason, I find it puzzling that Singer has never tried to account for the intuition that persons are not "replaceable" in Harean terms, as described in §9.6. For Hare, the "non-replaceability" of persons is, like rights, a feature of good intuitive-level rules. In terms of explicitly utilitarian critical thinking, however, even persons are replaceable in the sense that we can be justified harming them for the sake of aggregate benefits to others (§9.5). This is another way in which critical thinking reaches counterintuitive conclusions in "strictly hypothetical" cases (§2.3[A] and §4.8). Given certain general facts about human beings and the human condition, however, good ILS rules will embody various non-utilitarian rights for persons, and critical-level thinking will only be used to override these rights in extraordinary circumstances.

Relatedly, in light of the use that Singer makes of the "life's uncertain voyage" metaphor described in §9.4(E), I find it surprising that he has not recognized an intermediate category between persons and the merely sentient. In *Practical Ethics*, having defined "persons" as "self-conscious," he cites evidence from language studies with great apes that he thinks clearly shows that they are self-conscious (pp. 111–112), but he says that, in addition to the great apes, "a case can be made, though with varying degrees of confidence, on behalf of whales, dolphins, monkeys, dogs, cats, pigs, seals, bears, cattle, sheep and so on, perhaps even to the point at which [the class of persons] may include all mammals" (1993, p. 132). But the person setting out on an "uncertain journey" is not just self-conscious; she or he has a particularly rich and complex understanding of a substantial phase of his or her life.

Indeed, there are many parallels between the ways that we describe our lives as "journeys" and the stories that we tell about our lives. Most important, just as a biography unfolds according to the logic of narrative—and for that reason understanding one's life as a story requires something very much like a natural language (§6.4)—understanding one's life as a journey requires a very sophisticated kind of planning for the remote future—and for that reason it seems likely that only humans are capable of understanding their lives as "journeys." As described in §8.4(F), studies suggest that some non-human animals are capable of consciously planning for the future, at least as much as a day out. Planning a day ahead might fairly be described as planning a "trip," but describing life as a "voyage" or "journey" involves planning for a very long-term future. And just as understanding one's life as a biography involves understanding how the "chapters" are meaningfully related to each other, a journey has substantial phases that are understood to build on and prepare the way for one another.

For the reasons given in chapters 6 and 7, we have no good evidence that any non-human animals have the biographical sense of self that would qualify them for personhood on my view. Understanding one's life as a "voyage" requires similarly complex conceptual machinery. To the extent that Singer has settled upon the "life's uncertain voyage" metaphor for understanding personhood, then, he should admit that no animals are persons. For the reasons given in chapter 8, however, we have good reason to believe that some animals have a fairly robust, conscious sense of their own past and future. And for the reasons given in chapter 7, this gives their lives special moral significance vis-à-vis those of merely sentient animals. So a Harean utilitarian like Singer should recognize an intermediate category of "*near*-persons" deserving of special protections in the ILS rules of common morality, law, and professional ethics.

At the same time, I have argued (in chapter 10) that a Harean analysis of proposed agricultural reforms must be sensitive to background ecological, technological, economic, and cultural conditions in the target societies. This means that the ILS rules appropriate to various societies in various eras will vary with respect to slaughter practices, and with respect to how (or even if) they embody special respect for near-persons. In the sequel (Varner forthcoming), I will argue that a Harean analysis of human relationships with pets and working animals should be similarly nuanced, calling for humans to cultivate what I call "domesticated partnerships" (Varner 2002b) with some candidates for near-personhood, but also, in some cases, with animals that are not good candidates for near-personhood.

With regard to merely sentient animals, I have argued that good intuitive-level rules will generally treat them as replaceable (§9.3), and this is what led Hare to endorse the replaceability argument for humane, slaughter-based agriculture in "Why I am only a Demi-Vegetarian." This would seem to be a dramatic contrast with Singer's views, for he is popularly understood to be a staunch advocate of full vegetarianism. That's not quite the whole story, however.

For one thing, in *Practical Ethics*, Singer allows that fish and chickens might be merely sentient (pp. 95, 133—although on the same page he characterizes as a "shaky assumption" the claim that "ducks are not self-conscious"), and in the second edition of the book Singer endorsed the replaceability argument with respect to merely sentient animals (as described in §9.4). This means that he should endorse a sufficiently humane form of slaughter-based agriculture for animals like fish and chickens. He acknowledges precisely this, saying: "As a piece of critical moral reasoning, this argument may be sound." Note, however, that he characterizes it as "a piece of *critical* moral reasoning." He hastens to add that the replaceability argument "cannot justify factory farming, where animals do not have pleasant lives." Further:

> at the level of practical moral principles, it would be better to reject altogether the killing of animals for food, unless one must do so to survive. Killing animals for food makes us think of them as objects that we can use as we please. Their lives then count for little when weighed against our mere wants. As long as we continue to use animals in this way, to change our attitudes to animals in the way that they should be changed will be an impossible task.

> How can we encourage people to respect animals, and have equal concern for their interests, if they continue to eat them for their mere enjoyment? To foster the right attitudes of consideration for animals, including non-self-conscious ones, it may be best to make it a simple principle to avoid killing them for food. (1993, pp. 133–134)

That is, Singer's opposition to slaughter-based agriculture is qualified. With regard to merely sentient animals, he believes that clear critical thinking may show that sufficiently humane forms of it are justified. However, he doubts that modern, intensive agricultural systems meet premise #3 of the replaceability argument as represented in §9.3. And given current tendencies of people to "think of them as objects that we can use as we please," he thinks that a commitment to vegetarianism would be a good ILS rule—not because eating animals like fish and chickens is inherently wrong, but because, in Singer's judgment, killing them for "mere enjoyment" reinforces people's tendency to think of them as counting for nothing, morally speaking.

Some readers will disagree because they think that there are good nutritional reasons for eating meat. As noted in §9.2, some nutritionists doubt that a vegan diet can be nutritionally adequate, at least for people with high metabolic needs like growing children and pregnant and lactating women, but neither dairy products nor eggs can be produced on a very large scale without slaughtering large numbers of animals. This is because dairy cattle must be "freshened" periodically, and contemporary breeds of laying hens are not desirable as broilers. So maintaining dairy herds produces large numbers of offspring that cannot be used as replacement milkers. Indeed, around 1990, the U.S. dairy industry accounted for about one-seventh of all cattle slaughtered in the United States (Varner 1994b, p. 10). And although the technology exists to sex eggs during incubation (Scanes et al. 2004, p. 259), it is still too expensive and, consequently, males of laying strains are destroyed after hatching, usually by grinding them up (Jeffrey Armstrong, personal communication). And anyway, as described in Scanes et al. (2004, p. 259), the sexing procedure is performed at 15 days into the 21-day incubation of chicken eggs, when, according to Scanes et al. (2004, p. 53), the embryos are well-developed and might be able to feel pain.

Others will disagree because they deny that eating animals for "mere enjoyment" has led them to think that anything goes in their treatment of those same animals while alive. Relatedly, what Singer says on this score is at least in tension with something he says about infanticide. In response to the objection that allowing active euthanasia will push us onto a "slippery slope" to genocide, he argues:

> There is, anyway, little historical evidence to suggest that a permissive attitude towards the killing of one category of human beings leads to a breakdown of restrictions against killing other humans. Ancient Greeks

> regularly killed or exposed infants, but appear to have been at least as scrupulous about taking the lives of their fellow-citizens as medieval Christians or modern Americans. In traditional Eskimo societies it was the custom for a man to kill his elderly parents, but the murder of a normal healthy adult was almost unheard of. I mention these practices not to suggest that they should be imitated, but only to indicate that lines can be drawn at places different from where we now draw them. If these societies could separate human beings into different categories without transferring their attitudes from one group to another, we with our more sophisticated legal systems and greater medical knowledge should be able to do the same. (1993, p. 217)

But if he is so confident about contemporary humans' ability to "draw lines" among humans, why is Singer so skeptical of our ability to "draw lines" among animals? One disanalogy is that prohibiting euthanasia would harm the terminally ill who would prefer a quick death, whereas a switch to universal vegetarianism would not harm the animals that would not be born as a result. Nevertheless, Singer's argument that good ILS rules will prohibit even humane, slaughter-based agriculture using merely sentient animals is undermined by his reply to the slippery slope argument against active euthanasia.

Note also that—as described at the end of §10.5(A)—in reply to Hare's demi-vegetarianism essay, Singer adopts a "let a thousand flowers bloom" sort of stance, acknowledging that because some audiences will be more receptive to demi-vegetarianism than they will be to "full" vegetarianism (let alone veganism), "It is probably a good thing if different people and organizations promote each of these goals" (1999, p. 325). In Harean terms this can be described as acknowledging that appropriate ILS rules will be different for different cultures, or that individuals' personal moralities can, legitimately, differ.

§11.3. On to the Sequel

In the first paragraph of his preface to *Moral Thinking*, Hare wrote:

> I offer this book to the public now rather than later, not because I think it needs no improvement, but because of a sense of urgency—a feeling that if these ideas were understood, philosophers might do more to help resolve important practical issues. These are issues over which people are prepared to fight and kill one another; and it may be that unless some way is found of talking about them rationally and with hope of agreement, violence will finally engulf the world. (*MT*, p. iii)

I am painfully aware of how much improvement this book needs. I have stressed throughout how Harean utilitarianism provides a framework for understanding how and why we disagree, however, and I am anxious to get on to the sequel (Varner forthcoming), where I will illustrate this on a broader range of issues in animal ethics, including pet ownership and working animals, scientific research on animals, and wildlife/ecosystem management.

I believe that if the various concerned interest groups could all agree on that framework, then they could discuss these issues more constructively, without so much suspicion, and without vilifying one another. And if Hare was correct, along with Sidgwick, Mill, and Bentham before him, that human societies do tend to track—if only in a more-or-less delayed fashion—the thinking of a Harean archangel (§4.3), then the process will surely move forward more quickly if we are all consciously aware that the trajectory is always "ultimately" utilitarian.

I admitted in §1.1 that you probably wouldn't be converted to Harean utilitarianism by the time you reached this final section. I have tried to illustrate in this book how a Harean framework sheds light on the complexities inherent in the principle of utility, however, and on knotted questions about personhood and the moral status of animals. That effort will continue in the sequel, which will be written for consumption by those who are on the front lines of the animal ethics debates. I do hope that the two volumes, taken together, will convince some of those who approach these issues from diverse constituencies that they "ultimately" have something in common: *a shared commitment to humane sustainable communities*. What puts the "humane" in that phrase is the well-being of the animals affected by our laws, codes of professional ethics, and personal decisions—which means just about all of the animals on earth.

{ BIBLIOGRAPHY }

Ali, A., and K. M. Cheng. 1985. "Early Egg Production in Genetically Blind (rc/rc) Chickens in Comparison with Sighted (Rc+/rc) Controls." *Poultry Science* 64: 789–794.

Allen, Colin. 2004. "Animal Pain." *Nous* 38: 617–643.

Allen, Colin, and Marc Bekoff. 1997. "Consciousness: Essential or Dispensable?" In Colin Allen and Marc Bekoff, eds., *Species of Mind*, pp. 139–159. Cambridge, MA: MIT Press.

Anderson, James R., and Gordon G. Gallup. 1997. "Self-recognition in *Saguinus*? A Critical Essay." *Animal Behavior* 54: 1563–1567.

Aristotle. 1985 [n.d.]. *Nicomachean Ethics, translated by Terence Irwin*. Indianapolis: Hackett Publishing.

Bateson, Patrick. 1991. "Assessment of Pain in Animals." *Animal Behavior* 42: 827–839.

Bentham, Jeremy. [1780] 1948. *Introduction to the Principles of Morals and Legislation*. Wilfrid Harrison, ed. Oxford: Basil Blackwell.

Blokhuis, H. J., T. Fiks van Niekerk, W. Bessei, A. Elson, D. Guémené, J. B. Kjaer, G. A. Maria Levrinoo, C. J. Nicol, R. Tauson, C. A. Weeks, and H. A. Van de Weerd. 2007. "The LayWel Project: Welfare Implications of Changes in Production Systems for Laying Hens." *World's Poultry Science Journal* 63: 101–114.

Boesch, C., and H. Boesch. 1984. "Mental Map in Wild Chimpanzees: An Analysis of Hammer Transports for Nut Cracking." *Primates* 25: 160–170.

Braithwaite, Victoria. 2010. *Do Fish Feel Pain?* New York: Oxford University Press.

Brandt, Richard. 1979. *A Theory of the Good and the Right*. Oxford: The Clarendon Press.

Brown, Donald E. 1991. *Human Universals*. Boston: McGraw Hill.

Butler, Bishop. [1729] 2006. "Fifteen Sermons Preached at the Rolls Chapel." In David E. White, ed., *The Works of Bishop Butler*, pp. 33–146. Rochester, NY: University of Rochester Press.

Byrne, Richard, and Andrew Whiten, eds. 1988. *Machiavellian Intelligence: Social Expertise and the Evolution of Intellect in Monkeys, Apes, and Humans*. New York: Oxford University Press.

Callenbach, Ernest. 1996. *Bring Back the Buffalo! A Sustainable Future for America's Great Plains*. Berkeley: University of California Press.

Campbell, Stephen Michael. 2006. "Phenomenal Well Being." M.A. thesis, Texas A&M University.

Carritt, Edgar Frederick. 1947. *Ethical and Political Thinking*. Oxford: Clarendon Press.

Carruthers, Peter. 2000. *Phenomenal Consciousness: A Naturalistic Theory*. New York: Cambridge University Press.

Carruthers, Peter, and Peter K. Smith, eds. 1996. *Theories of Theories of Mind*. New York: Cambridge University Press.

Cavaliera, Paola, and Peter Singer. 1993. *The Great Ape Project*. New York: Saint Martin's Press.

Claridge, G. Cyril. [1922] 1969. *Wild Bush Tribes of Tropical Africa*. New York: Negro Universities Press.

Clark, Andy. 1998. "Connectionism, Moral Cognition, and Collaborative Problem Solving." In Larry May, Marilyn Friedman, and Andy Clark, eds., *Mind and Morals: Essays on Ethics and Cognitive Science*, pp. 109–127. Cambridge, MA: Bradford Books/MIT Press.

Clark, R. E., and L. R. Squire. 1998. "Classical Conditioning and Brain Systems: The Role of Awareness." *Science* 280: 77–81.

Clayton, N. S., and A. Dickinson. 1998. "Episodic-like Memory During Cache Recovery by Scrub Jays." *Nature* 395: 272–274.

Clayton, N. S., D. P. Griffiths, N. J. Emery, and A. Dickinson. 2001. "Elements of Episodic-like Memory in Animals." *Philosophical Transactions of the Royal Society of London B* 356: 1483–1491.

Coggleshall, R. E., R. B. Leonard, M. L. Applebaum, and W. D. Willis. 1978. "Organization of Peripheral Nerves and Spinal Roots of the Atlantic Stingray, *Dasyatis sabina*." *Journal of Neurophysiology* 41: 97–107.

Cohen, Carl. 1986. "The Case for the Use of Animals in Biomedical Research." *New England Journal of Medicine* 315: 865–869.

Comstock, Gary, ed. 1994. "Might Morality Require Veganism?" Special issue of *Journal of Agricultural and Environmental Ethics* 7(1): 1–124.

Comstock, Gary, ed. 2000. *Vexing Nature? On the Ethical Case Against Agricultural Biotechnology.* Boston: Kluwer Academic Publishers.

Conway, Martin A. 1990. *Autobiographical Memory: An Introduction*. Bristol, PA: Open University Press.

Daniels, Norman. 1979. "Wide Reflective Equilibrium and Theory Acceptance in Ethics." *Journal of Philosophy* 76(5): 256–282.

Davis, Steven. 2003. "The Least Harm Principle May Require That Humans Consume a Diet Containing Large Herbivores, Not a Vegan Diet." *Journal of Agricultural and Environmental Ethics* 16: 387–394.

DeGrazia, David. 1996. *Taking Animals Seriously: Mental Life and Moral Status*. New York: Cambridge University Press.

DeGrazia, David. 1997. "Great Apes, Dolphins, and the Concept of Personhood." *Southern Journal of Philosophy* 35: 301–320.

DeGrazia, David. 2006. "On the Question of Personhood Beyond *Homo sapiens*." In Peter Singer, ed., *In Defense of Animals: The Second Wave*, pp. 40–53. Malden, MA: Blackwell Publishing.

DeGrazia, David, and Andrew Rowan. 1991. "Pain, Suffering, and Anxiety in Animals and Humans." *Theoretical Medicine* 12: 193–211.

Delius, Juan D., and Julia A.M. Delius. 2006. "Intelligences and Brains: An Evolutionary Bird's Eye View." In Edward A. Wasserman and Thomas R. Zentall, eds., *Comparative Cognition: Experimental Explorations of Animal Intelligence*, pp. 555–579. New York: Oxford University Press.

Dennett, Daniel C. 1998. *Brainchildren: Essays on Designing Minds*. Cambridge, MA: Bradford Books.

Dickstein, Morris. [1981] 2002. "Introduction to *The Jungle*." In Harold Bloom, ed. *Upton Sinclair's The Jungle*, pp. 49–59. New York: Chelsea House Publishers.

Dong, W. K., T. Hayashi, V. J. Roberts, B. M. Fusco, and E. H. Chudler. 1996. "Behavioral Outcome of Posterior Parietal Cortex Injury in the Monkey." *Pain* 64: 579–587.

Doty, B. A., C. N. Jones, and L. A. Doty. 1967. "Learning-set Formation by Mink, Ferrets, Skunks, and Cats." *Science* 155: 1579–1580.

Downes, H., D. R. Koop, B. Klopfenstein, and N. Lessov. 1999. "Retention of Nociceptor Responses During Deep Barbiturate Anesthesia in Frogs." *Comparative Biochemistry and Physiology Part C: Pharmacology, Toxicology and Endocrinology* 124: 203–210.

Dreyfus, H., and S. Dreyfus. 1990. "What Is Morality? A Phenomenological Account of the Development of Ethical Expertise." In D. Rasmussen, ed., *Universalism vs. Communitarianism: Contemporary Debates in Ethics*. Cambridge, MA: MIT Press.

Duncan, Ian. 1996. "Animal Welfare Defined in Terms of Feelings." *Acta Agriculturae Scandinavica*, Section A, Animal Science Supplement 27: 29–35.

Duncan, Ian, and J. Carol Petherick. 1991. "The Implications of Cognitive Processes for Animal Welfare." *Journal of Animal Science* 69: 5017–5022.

Dworkin, Ronald. 1977. *Taking Rights Seriously*. Cambridge, MA: Harvard University Press.

Eichenbaum, H., A. Fagan, and N. J. Cohen. 1986. "Normal Olfactory Discrimination Learning Set and Facilitation of Reversal Learning after Medial-temporal Damage in Rats: Implications for an Account of Preserved Abilities in Amnesia." *Journal of Neuroscience* 6: 1876–1884.

Eichenbaum, Howard, Norbert J. Fortin, Ceren Ergorul, Sean P. Wright, and Kara L. Agster. 2005. "Episodic Recollection in Animals: 'If It Walks like a Duck and Quacks like a Duck. . . .'" *Learning and Motivation* 36: 190–207.

Emery, Nathan J., and Nicola S. Clayton. 2001. "Effects of Experience and Social Context on Prospective Caching Strategies by Scrub Jays." *Nature* 414: 443–446.

Fagot, Joël, and Isabelle Barbet. 2006. "Grouping and Segmentation of Visual Objects by Baboons and Humans." In Edward A. Wasserman and Thomas R. Zentall, eds., *Comparative Cognition: Experimental Explorations of Animal Intelligence*, pp. 15–28. New York: Oxford University Press.

Feinberg, Joel. 1974. "The Rights of Animals and Unborn Generations." In William T. Blackstone, ed., *Philosophy and Environmental Crisis*, pp. 43–68. Athens: University of Georgia Press.

Finsen, Susan. 1990. "On Moderation." In Marc Bekoff and Dale Jamieson, eds., *Interpretation and Explanation in the Study of Animal Behavior*, vol. 2, pp. 394–419. Boulder, Colorado: Westview Press.

Fiorito, Graziano, and Pietro Scotto. 1992. "Observational Learning in *Octopus vulgaris*." *Science* 256: 545–547.

Fivush, Robyn, and Catherine A. Haden. 2003. *Autobiographical Memory and the Construction of a Narrative Self: Developmental and Cultural Perspectives*. Mahwah, NJ: Lawrence Erlbaum Associates.

Flanagan, O. 1991. *Varieties of Moral Philosophy: Ethics and Psychological Realism*. Cambridge, MA: Harvard university Press.

Flombaum, Jonathan I., and Laurie R. Santos. 2005. "Rhesus Monkeys Attribute Perceptions to Others." *Current Biology* 15: 447–452.

Frankfurt, Henry. 1971. "Freedom of the Will and the Concept of a Person." *Journal of Philosophy* 68: 5–20.

Fraser, David. 1995. "Science, Values, and Animal Welfare: Exploring the 'Inextricable Connection.'" *Animal Welfare* 4: 103–117.

Fuster, Joaquin. 1997. *The Prefrontal Cortex: Anatomy, Physiology, and Neuropsychology of the Frontal Lobe*, 3rd ed. Philadelphia: Lippincott-Raven.

Gallup, Gordon G. 1970. "Chimpanzees: Self-recognition." *Science* 212: 695–696.

Gallup, Gordon Jr., James R. Anderson, and Daniel J. Shillito. 2002. "The Mirror Test." In Marc Bekoff, Colin Allen, and Gordon M. Burghardt, eds., *The Cognitive Animal: Empirical and Theoretical Perspectives on Animal Cognition*, pp. 325–333. Cambridge, MA: Bradford Books.

Gheerbrant, Alain. 1961. "Introduction" to Garcilasco de la Vega's *The Incas: The Royal Commentaries of the Inca*. New York: Orion Press.

Goodpaster, Kenneth. 1978. "On Being Morally Considerable." *Journal of Philosophy* 75: 308–325.

Grandin, Temple. 2001. "Solving Return-to-sensibility Problems after Electrical Stunning in Commercial Pork Slaughter Plants." *Journal of the American Veterinary Medicine Association* 219: 608–611.

Grandin, Temple. 2002. "Return-to-sensibility Problems after Penetrating Captive Bolt Stunning of Cattle in Commercial Beef Slaughter Plants." *Journal of the American Veterinary Medical Association* 221: 1258–1261.

Grau, James. 2001. "Instrumental Conditioning." In W. E. Craighead and C. B. Nemeroff, eds., *The Corsini Encyclopedia of Psychology and Behavioral Science*, 3rd ed., pp. 767–769. New York: Wiley.

Grau, James. 2002. "Learning and Memory Without a Brain." In Marc Bekoff, Colin Allen, and Gordon M. Burghardt, eds., *The Cognitive Animal: Empirical and Theoretical Perspectives on Animal Cognition*, pp. 77–87. Cambridge, MA: The MIT Press.

Greene, Joshua, Sylvia A. Morelli, Kelly Lowenberg, Leigh E. Nystrom, and Jonathan Cohen. 2008. "Cognitive Load Selectively Interferes with Utilitarian Moral Judgment." *Cognition* 107: 1144–1154.

Greene, Joshua D., Leigh E. Nystrom, Andrew D. Engell, John M. Darley, and Jonathan D. Cohen. 2004. "The Neural Bases of Cognitive Conflict and Control in Moral Judgment." *Neuron* 44: 389–400.

Groot, Adriaan de. 1965. *Thought and Choice in Chess*. New York: Basic Books.

Haidt, J. 2001. "The Emotional Dog and its Rational Tail: A Social Intuitionist Approach to Moral Judgment." *Psychological Review* 108: 814–834.

Hammond, Kenneth R., Robert M. Hamm, Janet Grassia, and Tamra Pearson. 1987. "Direct Comparison of the Efficacy of Intuitive and Analytical Cognition in Expert Judgment," *IEEE Transactions on Systems, Man, and Cybernetics* 17: 753–770.

Hampton, Robert R., and Bennett L. Schwartz. 2004. "Episodic Memory in Nonhumans: What, and Where, Is When?" *Current Opinion in Neurobiology* 14: 1–6.

Hare, Brian, J. Call, B. Agnetta, and M. Tomasello. 2000. "Chimpanzees Know What Conspecifics Do and Do Not See." *Animal Behaviour* 59: 771–786.

Hare, R. M. 1949. "Imperative Sentences." *Mind* 58: 21–39.

Hare, R. M. 1952. *The Language of Morals*. London: Oxford University Press.

Hare, R. M. 1963. *Freedom and Reason*. London: Oxford University Press.

Hare, R. M. 1970. "Meaning and Speech Acts." *Philosophical Review* 79: 3–24.

Hare, R. M. 1973. "Rawls' Theory of Justice—I" and "Rawls' Theory of Justice—II." *Philosophical Quarterly* 23: 144–155 and 23: 241–252.

Hare, R. M. 1981. *Moral Thinking: Its Levels, Method and Point*. New York: Oxford University Press.

Hare, R. M. 1991. "Universal Prescriptivism." In Peter Singer, ed., *A Companion to Ethics*, pp. 451–463. Malden, MA: Blackwell Publishers.

Hare, R. M. 1993. *Essays on Bioethics.* New York: Oxford University Press.

Hargrove, Eugene C. 1985. "The Role of Rules in Ethical Decision Making." *Inquiry* 28: 3–42.

Harlow, Harry F. 1949. "The Formation of Learning Sets." *Psychological Review* 56: 51–55.

Harris, Charles E., Michael S. Pritchard, and Michael J. Rabins. 2000. *Engineering Ethics: Concepts and Cases*, 2nd ed. Belmont, CA: Wadsworth.

Harrison, Ruth. 1964. *Animal Machines: The New Factory Farming Industry*. London: Vincent Stuart.

Hart, H. L. A. 1980. "Death and Utility." *New York Review of Books*, May 15, pp. 27–32.

Hauser, Marc. 2000. *Animal Minds: What Animals Really Think*. New York: Henry Holt.

Hayes, Keith J., and Catherine Hayes. 1951. "The Intellectual Development of a Home-raised Chimpanzee." *Proceedings of the American Philosophical Society* 95: 105–109.

Herman, Louis M. 1986. "Cognition and Language Competencies of Bottlenosed Dolphins." In Ronald J. Schusterman, Jeanette A. Thomas, and Forrest G. Woods, eds., *Dolphin Cognition and Behavior: A Behavioral Approach*, pp. 221–252. Hillsdale, NJ: Lawrence Erlbaum Associates.

Herzing, Denise L., and Thomas I. White. 1998. "Dolphins and the Question of Personhood." *Ethica & Animali* 9: 64–84.

Hobbes, Thomas. [1651] 1958. *Leviathan*. Indianapolis: Bobbs-Merrill Company.

Hoebel, E. Adamson. 1954. *The Law of Primitive Man: A Study in Comparative Legal Dynamics*. Cambridge, MA: Harvard University Press.

Hoffman, L. C., and E. Wiklund. 2006. "Game and Venison—Meat for the Modern Consumer." *Meat Science* 74: 197–208.

Holland, P. C. 1990. "Event Representation in Pavlovian Conditioning: Image and Action." *Cognition* 37: 105–131.

Howe, Mark L., and Mary L. Courage. 1993. "On Resolving the Enigma of Infantile Amnesia." *Psychological Bulletin* 113: 305–326.

Hume, David. [1739–1740] 1978. *A Treatise of Human Nature*, 2nd ed. New York: Oxford University Press.

Hume, David. [1779] 1947. *Dialogues Concerning Natural Religion*. Norman Kemp Smith, ed. Indianapolis: Bobbs-Merrill.

Hyatt, Charles W., Troy Metzler, Brian French, and Deborah Fahrenbruck. 2003. "Mirrors as Enrichment for Asian Elephants (*Elephas maximus*)." *Journal of the Elephant Managers Association* 14(3): 12–16.

Jackson, Wes. 1985. *New Roots for Agriculture*. Lincoln: University of Nebraska Press.

Jalles-Filho, E., R. G. T. da Cunha, and R. A. Salm. 2001. "Transport of Tools and Mental Representation: Is Capuchin Monkey Tool Behavior a Useful Model for Plio-Pleistocene Hominid Technology?" *Journal of Human Evolution* 40: 365–377.

Jamieson, Dale. 1983. "Killing Persons and Other Beings." In Harlan B. Miller and William H. Williams, eds., *Ethics and Animals*, pp. 135–146. Clifton, NJ: Humana Press.

Jamieson, Dale, ed. 1999. *Singer and His Critics*. Oxford: Blackwell.

Jasper, James M., and Dorothy Nelkin. 1992. *The Animal Rights Crusade: The Growth of a Moral Protest*. New York: Macmillan/The Free Press.

Johnson, Anna K. 2001. "Behavior, Physiology and Performance of Sows and Their Litters Produced on a Sustainable Pork Farm." Ph.D. diss., Texas Tech University.

Johnson, Kirk. 2009. "Montana Ruling Bolsters Doctor-Assisted Suicide." *New York Times*, December 31, 2009 (http://www.nytimes.com/2010/01/01/us/01suicide.html).

Johnson, Mark. 1993. *Moral Imagination: Implications of Cognitive Science for Ethics*. Chicago: University of Chicago Press.

Kacelnik, Alex, Jackie Chappell, ben Kenward, and Alex A. S. Wier. 2006. "Cognitive Adaptations for Tool-Related Behavior in New Caledonian Crows." In Edward A. Wasserman and Thomas R. Zentall, eds., *Comparative Cognition: Experimental Explorations of Animal Intelligence*, pp. 515–528. New York: Oxford University Press.

Kahneman, Daniel. 2003. "A Perspective on Judgment and Choice: Mapping Bounded Rationality." *American Psychologist* 58: 697–720.

Kant, Immanuel. [1785] 1948. *The Moral Law: Kant's Groundwork of the Metaphysic of Morals*. Translated by H. J. Paton. New York: Hutchinson's University Library.

Kant, Immanuel. [1788] 1956. *Critique of Practical Reason*. Translated by Lewis White Beck. Indianapolis: Bobbs-Merrill.

Kant, Immanuel. [1797] 1909. "On A Supposed Right to Tell Lies from Benevolent Motives." In T.K. Abbott, trans., *Kant's Critique of Practical Reason and Other Works on Ethical Theory*, pp. 361–365. New York: Longmans, Green.

Kazez, Jean. 2010. *Animalkind: What We Owe to Animals*. Malden, MA: Wiley-Blackwell.

Kellogg, Winthrop N. 1968. "Communication and Language in the Home-Raised Chimpanzee." *Science* 162: 423–427.

Kirsh, D., and Maglio, P. 1992. "Reaction and Reflection in Tetris." In J. Hendler, ed., *Artificial Intelligence Planning Systems: Proceedings of the First Annual International Conference*. San Mateo, CA: Morgan Kaufman.

Kohlberg, Lawrence. 1971. "From Is to Ought: How to Commit the Naturalistic Fallacy and Get Away with It in the Study of Moral Development." In T. Mischel, ed., *Cognitive Development and Epistemology*, pp. 151–235. New York: Academic Press.

Krech, Shepard, III. 1999. *The Ecological Indian: Myth and History*. New York: Norton.

Kuhse, Helga, and Peter Singer. 1985. *Should the Baby Live? The Problem of Handicapped Infants*. New York: Oxford University Press.

LaGraize, S., C. Labuda, R. Rutledge, R. Jackson, and P. Fuchs. 2004. "Differential Effect of Anterior Cingulated Cortex Lesion on Mechanical Hypersensitivity and Escape/avoidance Behavior in an Animal Model of Neuropathic Pain." *Experimental Neurology* 188: 139–148.

Lamey, Andy. 2007. "Food Fight! Steven Davis vs. Tom Regan on the Ethics of Eating." *Journal of Social Philosophy* 38: 331–348.

LayWel Project. 2006. "Welfare Implications of Changes in Production Systems for Laying Hens" (The LayWel Report). Brussels, Belgium: European Commission.

Leonard, R. B. 1985. "Primary Afferent Receptive Field Properties and Neurotransmitter Candidates in a Vertebrate Lacking Unmyelinated Fibres." *Progress in Clinical and Biological Research* 176: 135–145.

Leopold, Aldo. 1979. "Some Fundamentals of Conservation in the Southwest." *Environmental Ethics* 1: 131–141.

Lewis, David. 1983. "Survival and Identity." In *Philosophical Papers*, vol. 1, pp. 55–72. New York: Oxford University Press.

Lewis, George "Slim," and Byron Fish. 1978. *I Loved Rogues*. Seattle: Superior Publishing.

Linden, Eugene. 1986. *Silent Partners: The Legacy of the Ape Language Experiments*. New York: Times Books.

Locke, John. [1690a] No date. *An Essay Concerning the Human Understanding*. New York: E. P. Dutton.

Locke, John. [1690b] 1952. *The Second Treatise of Government*. Indianapolis: Bobbs-Merrill.

Lockwood, Michael. 1979. "Singer on Killing and the Preference For Life." *Inquiry* 22: 157–170.

MacIntyre, Alasdair. 1981. *After Virtue: A Study in Moral Theory*. Notre Dame, IN: University of Notre Dame Press.

Mackintosh, N. J. 1988. "Approaches to the Study of Animal Intelligence." *British Journal of Psychology* 79: 509–525.

Mackintosh, N. J., and J. Mackintosh. 1964. "Performance of *Octopus* over a Series of Reversals of a Simultaneous Discrimination." *Animal Behaviour* 12: 321–324.

Macphail, Euan. 1982. *Brain and Intelligence in Vertebrates*. New York: Oxford University Press.

Maier, S. F., and M. E. P. Seligman. 1976. "Learned Helplessness: Theory and Evidence." *Journal of Experimental Psychology: General* 105: 3–46.

Matheny, Gaverick. 2003. "Least Harm: A Defense of Vegetarianism from Steven Davis's Omnivorous Proposal." *Journal of Agricultural and Environmental Ethics* 16: 505–511.

McAdams, Dan P. 2003. "Identity and the Life Story." In Robyn Fivush and Catherine A. Haden, eds., *Autobiographical Memory and the Construction of a Narrative Self*, pp. 187–207. Mahwah, NJ: Lawrence Erlbaum Associates.

McComb, Karen, Cynthia Moss, Soila Sayialel, and Lucy Baker. 2000. "Unusually Extensive Networks of Vocal Recognition in African Elephants." *Animal Behaviour* 59: 1103–1109.

McCowan, Brenda, Sean F. Hanser and Laurance R. Doyle. 1999. "Quantitative Tools for Comparing Animal Communication Systems: Information Theory Applied to Bottlenose Dolphin Whistle Repertoires." *Animal Behaviour* 57: 409–419.

McMahan, Jeff. 2002. *The Ethics of Killing: Problems at the Margins of Life*. New York, Oxford University Press.

Melzack, Ronald, and Patrick Wall. 1988. *The Challenge of Pain*, 2nd ed. New York: Penguin Books.

Mench, Joy. 1998. "Thirty Years after Brambell: Whither Animal Welfare Science?" *Journal of Applied Animal Welfare Science* 1: 92–102.

Menzel, E., E. S. Savage-Rumbaugh, and J. Lawson. 1985. "Chimpanzee (*Pan troglodytes*) Spatial Problem Solving with the Use of Mirrors and Televised Equivalents of Mirrors." *Journal of Comparative Psychology* 99: 211–217.

Mercado, E., S. O. Murray, R. K. Uyeyama, A. A. Pack, and L. M. Herman. 1998. "Memory for Recent Actions in the Bottlenosed Dolphin (*Tursiops trunçatus*): Repetition of Arbitrary Behaviors using an Abstract Rule." *Animal Learning & Behavior* 26: 210–218.

Mill, John Stuart. [1859] 1956. *On Liberty*. Indianapolis: Bobbs-Merrill.

Mill, John Stuart. [1861] 1957. *Utilitarianism*. Indianapolis: Bobbs-Merrill.

Miller, Harlan. 1994. "Science, Ethics, and Moral Status." *Between the Species* 10: 10–18.

Moore, G. E. 1903. *Principia Ethica*. New York: Cambridge University Press.

Moss, Cynthia J. 1988. *Elephant Memories: Thirteen Years in the Life of an Elephant Family*. New York: William Morrow.

Mulcahy, Nicholas J., and Joseph Call. 2006. "Apes Save Tools for Future Use." *Science* 312: 1038–1040.

Nagel, Thomas. 1974. "What Is It Like to Be a Bat?" *Philosophical Review* 83: 435–450.

Narveson, Jan. 1977. "Animal Rights." *Canadian Journal of Philosophy* 7: 161–78.

Nass, Roger D., Glenn A. Hood, and Gerald D. Lindsey. 1971. "Fate of Polynesian Rats in Hawaiian Sugarcane Fields During Harvest." *Journal of Wildlife Management* 35: 353–356.

Nietzsche, Friedrich. [1874] 1980. *On the Advantage and Disadvantage of History for Life.* Translated by Peter Preuss. Indianapolis: Hackett Publishing.

Nissani, Moti. 2004. "Theory of Mind and Insight in Chimpanzees, Elephants, and Other Animals?" In Lesley J. Rogers and Gisela Kaplan, eds., *Comparative Vertebrate Cognition: Are Primates Superior to Non-Primates?*, pp. 227–261. New York: Kluwer Academic/Plenum Publishers.

Nissani, Moti, and Donna Hoefler-Nissani. 2007. "Absence of Mirror Self-Referential Behavior in Two Asian Elephants." *Journal of Veterinary Science 1.* Online version: http://www.scientificjournals.org/journals2007/articles/1043.htm accessed January 26, 2011.

Norton, Bryan. 2005. *Sustainability: A Philosophy of Adaptive Ecosystem Management.* Chicago: University of Chicago Press.

Parfit, Derrick. 1976. "Rights, Interests, and Possible People." In Samuel Gorovitz et al., eds., *Moral Problems in Medicine*, 2nd ed., pp. 369–375. Englewood Cliffs, NJ: Prentice-Hall.

Parfit, Derrick. 1984. *Reasons and Persons.* New York: Oxford University Press.

Parfit, Derrick. 1986. "Overpopulation and the Quality of Life." In Peter Singer, ed., *Applied Ethics*, pp. 145–164. New York: Oxford University Press.

Paton, H. J. 1948. "Analysis of the Argument." In Immanuel Kant, *The Moral Law: Kant's Groundwork of the Metaphysic of Morals*, translated by H. J. Paton, pp. 13–52. New York: Hutchinson's University Library.

Patterson, F. G. P., and R. H. Cohn. 1994. "Self-recognition and Self-awareness in Lowland Gorillas." In Sue Taylor Parker, Robert W. Mitchell, and Maria L. Boccia, eds., *Self-Awareness in Animals and Humans: Developmental Perspectives*, pp. 273–290. New York: Cambridge University Press.

Patterson, Francine, and Eugene Linden. 1981. *The Education of Koko.* New York: Holt, Rinehart and Winston.

Payne, Katharine B., William R. Langbauer, Jr., and Elizabeth M. Thomas. "Infrasonic Calls of the Asian Elephant (*Elephas maximus*)." *Behavioral Ecology and Sociobiology* 18: 297–301.

Pepperberg, Irene. 1994. "Numerical Competence in an African Grey Parrot (*Psittacus erithacus*)." *Journal of Comparative Psychology* 108: 36–44.

Pepperberg, Irene. 2002. "Cognitive and Communicative Abilities of Grey Parrots." In Marc Bekoff, Colin Allen, and Gordon M. Burghardt, eds., *The Cognitive Animal: Empirical and Theoretical Perspectives on Animal Cognition*: pp. 247–253. Cambridge, MA: Bradford Books.

Pepperberg, I. M., S. E. Garcia, E. C. Jackson, and S. Marconi. 1995. "Mirror Use by African Grey Parrots (*Psittacus erithacus*)." *Journal of Comparative Psychology* 109(2): 182–195.

Perner, Josef, and Birgit Lang. 1999. "Development of Theory of Mind and Executive Control." *Trends in Cognitive Science* 3: 337–344.

Perry, Ralph Barton. 1926. *General Theory of Value.* New York: Longman, Green.

Pew Commission on Industrial Farm Animal Production. 2008. "Putting Meat on the Table: Industrial Farm Animal Production in America." Washington, DC: The Pew Charitable Trusts. Available online: http://www.pewtrusts.org/uploadedFiles/wwwpewtrustsorg/Reports/Industrial_Agriculture/PCIFAP_FINAL.pdf.

Plotnik, Rodney J., and Robert B. Tallarico. 1966. "Object-quality Learning-set Formation in the Young Chicken." *Psychonomic Science* 5: 195–196.

Plotnik, Joshua M., Frans B.M. de Waal, and Diana Reiss. 2006. "Self-recognition in an Asian Elephant." *Proceedings of the National Academy of Sciences* 103: 17053–17057.

Pluhar, Evelyn. 1995. *Beyond Prejudice: The Moral Significance of Human and Nonhuman Animals*. Durham, NC: Duke University Press.

Pohl, Frederick, and C. M. Kornbluth. 1952. *The Space Merchants*. New York: Ballantine Books.

Poole, Joyce H. 1998. "An Exploration of a Commonality Between Ourselves and Elephants." *Ethica & Animali* 9: 85–110.

Poole, Joyce H., and Cynthia J. Moss. 2008. "Elephant Sociality and Complexity: The Scientific Evidence." In Christen Wemmer and Catherine Christen, eds., *Elephants and Ethics: Toward a Morality of Coexistence*, pp. 69–98. Baltimore, MD: The Johns Hopkins University Press.

Poole, Joyce H., Katherine Payne, William R. Langbauer Jr., and Cynthia J. Moss. 1988. "The Social Context of Some Very Low Frequency Calls of African Elephants." *Behavioral Ecology and Sociobiology* 22: 385–392.

Povinelli, Daniel J. 1989. "Failure to Find Self-Recognition in Asian Elephants (*Elephas maximus*) in Contrast to Their Use of Mirror Clues to Discover Hidden Food." *Journal of Comparative Psychology* 103: 122–131.

Povinelli, Daniel J., and Richard Davis. 1994. "Differences Between Chimpanzees (*Pan troglodytes*) and Humans (*Homo sapiens*) in the Resting State of the Index Finger: Implications for Pointing." *Journal of Comparative Psychology* 108: 134–139.

Povinelli, Daniel J., and T. J. Eddy. 1996. "What Young Chimpanzees Know about Seeing." *Monographs of the Society for Research in Child Development* 61: 1–152. (No. 2, Serial No. 247)

Price, D. 2002. "Central Neural Mechanisms That Interrelate Sensory and Affective Dimensions of Pain." *Molecular Interventions* 2: 392–402.

Raby, C. R., D. M. Alexis, A. Dickinson, and N. S. Clayton. 2007. "Planning for the Future by Western Scrub-jays." *Nature* 445: 919–921.

Rachels, James. 1999. *The Elements of Moral Philosophy*, 3rd ed. Boston: McGraw-Hill.

Rachuonyo, Harold. 2001. "Productivity, Behavior, and Environmental Impact of Outdoor Gestating Sows." Ph.D. diss., Texas Tech University.

Radzik, Linda. 2004. "Making Amends." *American Philosophical Quarterly* 41: 141–154.

Rasmussen, Knud. 1927. *Across Arctic America*. New York: Greenwood Press.

Rawls, John. 1955. "Two Concepts of Rules." *The Philosophical Review* 64: 3–32.

Rawls, John. 1971. *A Theory of Justice*. Cambridge, MA: Harvard University Press.

Rawls, John. 1999. *The Law of Peoples*. Cambridge, MA: Harvard University Press.

Regan, Tom. 1983. *The Case for Animal Rights*. Berkeley: University of California Press.

Reiss, Diana, and Lori Marino. 2001. "Mirror Self-recognition in the Bottlenose Dolphin: A Case of Cognitive Convergence." *Proceedings of the National Academy of Science* 98: 5937–5942.

Rensch, B. 1957. "The Intelligence of Elephants." *Scientific American* 196: 44–49.

Robin, Raizel. 2005. "In Vitro Meat." *New York Times*, December 11 (http.www.nytimes.com/2005/12/11/magazine/11ideas_section2-9.html).

Rollin, Bernard. 1995a. *Farm Animal Welfare: Social, Bioethical, and Research Issues*. Ames: Iowa State University Press.

Rollin, Bernard. 1995b. *The Frankenstein Syndrome: Ethical and Social Issues in the Genetic Engineering of Animals.* New York: Cambridge University Press.

Rolston, Holmes, III. 1979. "Can and Ought We to Follow Nature?" *Environmental Ethics* 1: 7–30.

Rose, Margaret, and David Adams. 1989. "Evidence for Pain and Suffering in Other Animals." In Gill Langley, ed., *Animal Experimentation: The Consensus Changes*, pp. 42–71. New York: Chapman and Hall.

Ross, William David. 1930. *The Right and the Good.* New York: Oxford University Press.

Rubin, David C. 1986. *Autobiographical Memory.* New York: Cambridge University Press.

Sabaté, Joan, Rosemary A. Ratzin-Turner, and Jack E. Brown. 2001. "Vegetarian Diets: Descriptions and Trends." In Joan Sabaté, ed., *Vegetarian Nutrition*, pp. 3–17. Boca Raton, FL: CRC Press.

Salt, Henry. [1892] 1980. *Animals' Rights: Considered in Relation to Social Progress.* Clarks Summit, PA: Society for Animal Rights.

Salt, Henry. [1914] 1976. "The Logic of the Larder." In Tom Regan and Peter Singer, eds., *Animal Rights and Human Obligations*, pp. 185–189. Englewood Cliffs, NJ: Prentice-Hall, Inc.

Sandøe, P., B. L. Nielsen, L. G. Christensen, and P. Sørensen. 1999. "Staying Good While Playing God—The Ethics of Breeding Farm Animals." *Animal Welfare* 8: 313–328.

Santos, Laurie R., Jonathan I. Flombaum, and Webb Phillips. 2006. "The Evolution of Human Mindreading: How Non-Human Primates Can Inform Social Cognitive Neuroscience." In S. Platek, eds., *Evolutionary Cognitive Neuroscience*, pp. 433–456. Cambridge, MA: MIT Press.

Savage-Rumbaugh, E. Sue, Jeannine Murphy, Rose A. Sevcik, Karen E. Brakke, Shelly L. Williams, and Duane M. Rumbaugh. 1993. "Language Comprehension in Ape and Child." *Monographs of the Society for Research in Child Development* 58(3–4): 1–220.

Savage-Rumbaugh, E. Sue, Stuart G. Shanker, and Talbot J. Taylor. 1998. *Apes, Language, and the Human Mind.* New York: Oxford University Press.

Savage-Rumbaugh, Sue, and Roger Lewin. 1994. *Kanzi: The Ape at the Brink of the Human Mind.* New York: John Wiley & Sons.

Savory, Allen and Jody Butterfield. 1999. *Holistic Management: A New Framework for Decision Making.* Washington, DC: Island Press.

Scanes, Colin G., George Brant, and M.W. Ensminger. 2004. *Poultry Science.* Upper Saddle River, NJ: Pearson/Prentice Hall.

Schechtman, Marya. 1994. "The Truth about Memory." *Philosophical Psychology* 7(1): 3. Online serial: <http://search.ebscohost.com/login.aspx?direct=true&db=a9h&AN=9410282752&site=ehost-live>, accessed April 12, 2009.

Schechtman, Marya. 1996. *The Constitution of Selves.* Ithaca, NY: Cornell University Press.

Schedler, George. 2005. "Does Ethical Meat Eating Maximize Utility?" *Social Theory and Practice* 31: 499–411.

Schusterman, R. J. 1964. "Successive Discrimination-reversal Training and Multiple Discrimination Training in One-Trial Learning by Chimpanzees." *Journal of Comparative & Physiological Psychology* 58: 153–156.

Schwartz, Bennett L., Megan L. Hoffman, and Siân Evans. 2005. "Episodic-like Memory in a Gorilla: A Review and New Findings." *Learning and Motivation* 36: 226–244.

Segerdahl, Pär, William Fields, and Sue Savage-Rumbaugh. 2005. *Kanzi's Primal Language: The Cultural Initiation of Primates into Language.* New York: Palgrave/Macmillan.

Seligman, M. E. P., and Maier, S. F. 1967. "Failure to Escape Traumatic Shock." *Journal of Experimental Psychology* 74: 1–9.

Sharman, G. A. M. 1983. "Slaughter of Deer." *Veterinary Record* 112: 500–502.

Shermer, Michael. 2004. *The Science of Good and Evil.* New York: Henry Holt.

Shillito, D. J., G. G. Gallup, Jr., and B. B. Beck. 1999. "Factors Affecting Mirror Behavior in Western Lowland Gorillas, *Gorilla gorilla*." *Animal Behavior* 57: 999–1004.

Shriver, Adam. 2006. "Minding Mammals." *Philosophical Psychology* 19: 433–442.

Shriver, Adam. 2009. "Knocking Out Pain in Livestock: Can Technology Succeed Where Morality Has Stalled?" *Neuroethics* 2: 115–124.

Sidgwick, Henry [1874] 1907. *The Methods of Ethics*, 7th ed. London: Macmillan.

Siegel, J. M. 2001. "The REM Sleep-memory Consolidation Hypothesis." *Science* 294: 1058–1063.

Simonet, Patricia. 2000. "Self-recognition in Asian Elephants, Preliminary Findings." *Elephants: The Publication of the Elephant Research Foundation* 2(4): 103.

Sinclair, Upton. [1906] 1981. *The Jungle*. New York: Bantam Books.

Singer, Marcus G. 1961. *Generalization in Ethics: An Essay in the Logic of Ethics, with the Rudiments of a System of Moral Philosophy*. New York: Atheneum.

Singer, Peter. 1972. "Famine, Affluence and Morality." *Philosophy and Public Affairs* 1: 229–243.

Singer, Peter. 1975. *Animal Liberation*, 1st ed. New York: Avon Books.

Singer, Peter. 1979. *Practical Ethics*, 1st ed. New York: Cambridge University Press.

Singer, Peter. 1980. "Right to Life?" *The New York Review of Books, letters to the editor*, August 14, pp. 53–54.

Singer, Peter. 1987a. "Animal Liberation or Animal Rights?" *Monist* 70: 3–14.

Singer, Peter. 1987b. "Life's Uncertain Voyage." In Philip Pettit, Richard Sylvan, and Jean Norman, eds., *Metaphysics and Morality: Essays in Honor of J.J.C. Smart*, pp. 154–172. New York: Oxford University Press.

Singer, Peter. 1990. *Animal Liberation*, 2nd ed. New York: Avon Books.

Singer, Peter. 1993. *Practical Ethics*, 2nd ed. New York: Cambridge University Press.

Singer, Peter. 1999. "A Response." In Dale Jamieson, ed., *Singer and His Critics*, pp. 269–335. Malden, MA: Blackwell Publishers.

Smart, J. J. C. 1973. "An Outline of a System of Utilitarian Ethics." In J. J. C. Smart and Bernard Williams, *Utilitarianism: For and Against*, pp. 1–74. New York: Cambridge University Press.

Smith, Jane A., and Kenneth M. Boyd, eds. 1991. *Lives in the Balance: The Ethics of Using Animals in Biomedical Research*. New York: Oxford University Press.

Sneddon, Lynne U., Victoria A. Braithwaite, and Michael J. Gentle. 2003a. "Do Fish Have Nociceptors: Evidence for the Evolution of a Vertebrate Sensory System." *Proceedings of the Royal Society of London B* 270: 1115–1121.

Sneddon, Lynne U., Victoria A. Braithwaite, and Michael J. Gentle. 2003b. "Novel Object Test: Examining Nociception and Fear in the Rainbow Trout," *Journal of Pain* 4: 431–440.

Snow, Peter J., Mark B. Plenderleith, Layne L. Wright. 1993. "Quantitative Study of Primary Sensory Neuron Populations of Three Species of Elasmobranch Fish." *Journal of Comparative Neurology* 334: 97–103.

Steadman, Lyle. 1993. "Natural Selection and the Evolutionary Status of Culture." *Presentation to the Human Behavior and Evolution Society Meeting*, Binghamton, NY. August 5.

Steadman, Lyle. 1994. "Social Behavior and Sacrifice." *Presentation to the Human Behavior and Evolution Society Meeting*, Ann Arbor, MI. June 6.

Stettner, Laurence Jay, and Kenneth A. Matyniak 1980. "The Brain of Birds." *Scientific American* (June 1968), reprinted in *Birds*, pp. 192–199. San Francisco: W.H. Freeman.

Stevenson, Charles L. 1945. *Ethics and Language*. New Haven: Yale University Press.

Stoerig, Petra, and Alan Cowey. 1995. "Visual Perception and Phenomenal Consciousness." *Behavioural Brain Research* 71: 147–156.

Stone, Valerie E., Simon Baron-Cohen, and Robert T. Knight. 1998. "Frontal Lobe Contributions to Theory of Mind." *Journal of Cognitive Neuroscience* 10: 640–656.

Suarez, S. and G. G. Gallup, Jr. 1981. "Self-recognition in Chimpanzees and Orangutans, but Not Gorillas." *Journal of Human Evolution* 10: 157–188.

Suddendorf, Thomas, and Janie Busby. 2005. "Making Decisions with the Future in Mind: Developmental and Comparative Identification of Mental Time Travel." *Learning and Motivation* 36: 110–125.

Taylor, Charles. 1976. "Responsibility for the Self." In Amelie Oksenberg Rorty, ed., *The Identities of Persons*, pp. 281–299. Berkeley: University of California Press.

Terrace, Herbert S. 1979a. *Nim*. New York: Alfred A.Knopf.

Terrace, Herbert S. 1979b. "Is Problem-Solving Language?" *Journal of the Experimental Analysis of Behavior* 31: 161–175.

Tew, T. E., and D. W. Macdonald. 1993. "The Effects of Harvest on Arable Wood Mice *Apodemus sylvaticus*." *Biological Conservation* 65: 279–283.

Thompson, Frank H. 1970. *The Jungle: Notes*. Lincoln, NE: Cliffs Notes.

Thompson, Lucien T., James R. Moyer, and John F. Disterhoft. 1996. "Trace Eyeblink Conditioning in Rabbits Demonstrates Heterogeneity of Learning Ability Both Between and Within Age Groups." *Neurobiology of Aging* 17: 619–629.

Thompson, Paul B. 1995. *The Spirit of the Soil: Agriculture and Environmental Ethics*. London: Routledge.

Tooley, Michael. 1983. *Abortion and Infanticide*. New York: Oxford University Press.

Tschudin, Alain, Josep Call, R. I. M. Dunbar, Gabrielle Harris, and Charmaine van der Elst. 2001. "Comprehension of Signs by Dolphins (*Tursiops truncatus*)." *Journal of Comparative Psychology* 115: 100–105.

Tulving, Endel. 1972. "Episodic and Semantic Memory." In Tulving and W. Donaldson (eds.), *Organisation of Memory*. London: Academic Press.

Tulving, Endel. 1985. "Memory and Consciousness." *Canadian Psychology* 26: 1–12.

Varner, Gary. 1994a. "The Prospects for Consensus and Convergence in the Animal Rights Debate." *The Hastings Center Report* January/February: 23–27.

Varner, Gary. 1994b. "What's Wrong with Animal *Byproducts*?" *Journal of Agricultural and Environmental Ethics* 7: 7–17.

Varner, Gary. 1994c. "In Defense of the Vegan Ideal: Rhetoric and Bias in the Nutrition Literature." *Journal of Agricultural and Environmental Ethics* 7: 29–40.

Varner, Gary. 1994d. "Rejoinder to Kathryn Paxton George." *Journal of Agricultural and Environmental Ethics* 7: 83–86.

Varner, Gary. 1998. *In Nature's Interests? Interests, Animal Rights, and Environmental Ethics*. New York: Oxford University Press.

Varner, Gary. 2002a. "Animals." In Gary Comstock, ed., *Life Science Ethics*, pp. 141–168. Ames: Iowa State University Press.

Varner, Gary. 2002b. "Pets, Companion Animals, and Domesticated Partners." In David Benatar, ed., *Ethics for Everyday*, pp. 450–475. Boston: McGraw-Hill.

Varner, Gary. 2003. Review of *Nicholas Agar, Life's Intrinsic Value: Science, Ethics, and Nature* (Columbia University Press, 2001). *Environmental Ethics* 25: 413–416.

Varner, Gary. Forthcoming. *Sustaining Animals: Envisioning Humane, Sustainable Communities* (Oxford University Press).

Vega, Garcilaso de la. [1609] 1961. *The Incas: The Royal Commentaries of the Inca*. Translated by Maria Jolas. New York: The Orion Press.

Walraven, V., L. van Elsacker, and R. Verheyen. 1995. "Reactions of a Group of Pygmy Chimpanzees (*Pan paniscus*) to Their Mirror Images: Evidence of Self-recognition." *Primates* 36: 145–150.

Walters, E. T. 1996. "Comparative and Evolutionary Aspects of Nociceptor Function." In C. Belmonte and F. Cervero, eds., *Neurobiology of Nociceptors*, pp. 92–114. New York: Oxford University Press.

Warren, J. M. 1965. "Primate Learning in Comparative Perspective." In Allan M. Schrier, Harry F. Harlow, and Fred Stollnitz, *Behavior of Nonhuman Primates: Modern Research Trends*, pp. 249–281. New York: Academic Press.

Wattles, Jeffrey. 1996. *The Golden Rule*. New York: Oxford University Press.

Wheeler, Mark A., Donald T. Stuss, and Endel Tulving. 1997. "Toward a Theory of Episodic Memory: The Frontal Lobes and Autonoetic Consciousness." *Psychological Bulletin* 121: 331–354.

White, Thomas I. 2007. *In Defense of Dolphins*. Malden, MA: Blackwell Publishing.

Whiten, Andrew. 1997. "The Machiavellian Mindreader." In Andrew Whiten and Richard W. Byrne, eds., *Machiavellian Intelligence II: Extensions and Evaluations*, pp. 144–173. New York: Cambridge University Press.

Whiten, Andrew, and Richard W. Byrne. 1988. "The Manipulation of Attention in Primate Tactical Deception." In Byrne and Whitten, eds., *Machiavellian Intelligence: Social Expertise and the Evolution of Intellect in Monkeys, Apes, and Humans*, pp. 211–223. New York: Oxford University Press.

Whiten, Andrew, and Richard W. Byrne. 1997. *Machiavellian Intelligence II: Extensions and Evaluations*. New York: Cambridge University Press.

Williams, Bernard. 1973. "A Critique of Utilitarianism," in J. J. C. Smart and Bernard Williams, eds., *Utilitarianism: For and Against*, pp. 75–150. New York: Cambridge University Press.

Williams, Bernard. 1981. *Moral Luck*. New York: Cambridge University Press.

Wilson, B. J. 1978. "Complex Learning in Birds." Ph.D. diss., University of Sussex.

Wilson, Frank. 1998. *The Hand: How Its Use Shapes the Brain, Language and Human Culture*. New York: Pantheon Books.

Wittgenstein, Ludwig. [1953] 1958. *Philosophical Investigations*. Translated by G. E. M. Anscombe.New York: Basil Blackwell & Mott.

Wittgenstein, Ludwig. [1921] 1961. *Tractatus Logico-Philosophicus*. Translated by D. F. Pears and B. F. McGuinness. New York: Routledge & Kegan Paul.

Wooley, J. B., Jr., L. B. Best, and W. R. Clark. 1984. "Impacts of No-Till Row Cropping on Upland Wildlife." *Transactions of the North American Wildlife and Natural Resources Conference* 50: 157–168.

Zeldin, Robert K., and David S. Olton. 1986. "Rats Acquire Spatial Learning Sets." *Journal of Experimental Psychology* 12: 412–419.

Zental, T. R., T. S. Clement, R. S. Bhatt, and J. Allen. 2001. "Episodic-like Memory in Pigeons." *Psychonomic Bulletin & Review* 8: 685–690.

{INDEX}